Airway Smooth Muscle

Airway Smooth Muscle in Asthma and COPD:

Biology and Pharmacology

Edited by

Kian Fan Chung
Airway Disease Section,
National Heart and Lung Institute
Imperial College London, UK

John Wiley & Sons, Ltd

Other Wiley Editorial Offices

John Wiley & Sons Inc., 111 River Street, Hoboken, NJ 07030, USA

Jossey-Bass, 989 Market Street, San Francisco, CA 94103-1741, USA

Wiley-VCH Verlag GmbH, Boschstr. 12, D-69469 Weinheim, Germany

John Wiley & Sons Australia Ltd, 33 Park Road, Milton, Queensland 4064, Australia

John Wiley & Sons (Asia) Pte Ltd, 2 Clementi Loop #02-01, Jin Xing Distripark, Singapore 129809

John Wiley & Sons Canada Ltd, 6045 Freemont Blvd, Mississauga, Ontario, L5R 4J3

Wiley also publishes its books in a variety of electronic formats. Some content that appears in print may
not be available in electronic books.

British Library Cataloguing in Publication Data

A catalogue record for this book is available from the British Library

ISBN 978-0-470-06066-7 (H/B)

Typeset in 10.5/13pt Times by Aptara Inc., New Delhi, India
Printed and bound in Great Britain by Antony Rowe Ltd., Chippenham, Wiltshire
This book is printed on acid-free paper

Contents

7   Airway smooth muscle interaction with mast cells   127

Roger Marthan, Patrick Berger, Pierre-Olivier Girodet and J. Manuel
Tunon-de-Lara

8   Airway smooth muscle synthesis of inflammatory mediators   141

Alison E. John, Deborah L. Clarke, Alan J. Knox and Karl Deacon

9   Airway smooth muscle in experimental models   159

Anne-Marie Lauzon and James G. Martin

10   Altered properties of airway smooth muscle in asthma   181

Judith Black, Janette Burgess, Brian Oliver and Lyn Moir

List of Contributors

Ian M. Adcock Cell and Molecular Biology, Airways Disease Section, National Heart & Lung Institute, Imperial College London, Dovehouse Street, London SW3 6LY, UK

Steven S. An Johns Hopkins Bloomberg School of Public Health, 615 N. Wolfe Street, Baltimore, MD 21205, USA

Patrick Berger Université Bordeaux 2, Laboratoire de Physiologie Cellulaire Respiratoire, F-33076 Bordeaux, France; INSERM, U 885, F-33076 Bordeaux, France; and CHU de Bordeaux, Hôpital du Haut-Lévêque, Pôle Cardio-Thoracique, F-33604 Pessac, France

Pank Bhavsar Cell and Molecular Biology, Airways Disease Section, National Heart & Lung Institute, Imperial College London, Dovehouse Street, London SW3 6LY, UK

Judith L. Black NH&MRC Senior Principal Research Fellow, and Head, Cells Group, Discipline of Pharmacology, Faculty of Medicine, Room 211 Blackburn Building D06, University of Sydney, Woolcock Institute of Medical Research, NSW 2006, Australia

Melanie Brown Section of Critical Care, Department of Pediatrics, University of Chicago, 5841 S. Maryland Avenue, Chicago, IL 60637, USA

Janette Burgess Wright Fellow, Discipline of Pharmacology, Faculty of Medicine, Room 211 Blackburn Building D06, University of Sydney, Woolcock Institute of Medical Research, NSW 2006, Australia

Kian Fan Chung Airway Disease Section, National Heart & Lung Institute, Imperial College London, Dovehouse Street, London SW3 6LY, UK

Deborah L. Clarke Division of Respiratory Medicine, Clinical Sciences Building, Nottingham City Hospital, Hucknall Road, Nottingham NG5 1PB, UK

David A. Cooper Pulmonary, Allergy & Critical Care Division, University of Pennsylvania, Airways Biology Initiative, 125 South 31st Street, TRL Suite 1200, Philadelphia, PA 19104-3403, USA

Karl Deacon Division of Respiratory Medicine, Clinical Sciences Building, Nottingham City Hospital, Hucknall Road, Nottingham NG5 1PB, UK

Jeffrey Fredberg Professor of Bioengineering and Physiology, Program in Molecular, Integrative & Physiological Sciences (MIPS), Department of Environmental Health, Harvard School of Public Health, 665 Huntington Avenue, Boston, MA 02115, USA

Pierre-Olivier Girodet Université Bordeaux 2, Laboratoire de Physiologie Cellulaire Respiratoire, F-33076 Bordeaux, France; INSERM, U 885, F-33076 Bordeaux, France; and CHU de Bordeaux, Hôpital du Haut-Lévêque, Pôle Cardio-Thoracique, F-33604 Pessac, France

Reinoud Gosens Departments of Physiology and Internal Medicine, Section of Respiratory Diseases, and National Training Program in Allergy and Asthma, University of Manitoba, Winnipeg, MB, Canada; Biology of Breathing Theme, Manitoba Institute of Child Health, Winnipeg, MB, Canada; Current address: Department of Molecular Pharmacology, University of Groningen, The Netherlands

Susan J. Gunst Department of Cellular & Integrative Physiology, Indiana University, School of Medicine, 635 Barnhill Drive – MS 313, Indianapolis, IN 46202-5120, USA

Andrew J. Halayko Associate Professor, Canada Research Chair in Airway Cell and Molecular Biology, Section of Respiratory Disease, Health Sciences Centre, RS321 – 810 Sherbrook Street, Winnipeg, MB, Canada R3A 1R8; and Biology of Breathing Theme, Manitoba Institute of Child Health, Winnipeg, MB, Canada

Stuart J. Hirst Reader in Respiratory Cell Pharmacology, MRC & Asthma UK Centre in Allergic Mechanisms of Asthma, Division of Asthma, Allergy & Lung Biology, The Guy's, King's College & St. Thomas' School of Medicine, King's College London, 5th Floor, Thomas Guy House, Guy's Hospital Campus, London SE1 9RT, UK

Alison E. John Division of Respiratory Medicine, Clinical Sciences Building, Nottingham City Hospital, Hucknall Road, Nottingham NG5 1PB, UK

Malcolm W. Johnson GlaxoSmithKline Research & Development, Respiratory Medicines, Development Centre, Greenford Road, Greenford UB6 0HE, UK

Alan J. Knox Professor in Respiratory Medicine and Head of Division, Division of Respiratory Medicine, Clinical Sciences Building, Nottingham City Hospital, Hucknall Road, Nottingham NG5 1PB, UK

Anne-Marie Lauzon Meakins-Christie Laboratories, Department of Medicine, McGill University, 3626 St Urbain Street, Montréal, Québec, Canada, H2X 2P2

Roger Marthan Professor of Physiology – Staff specialist, Laboratoire de Physiologie Cellulaire Respiratoire, INSERM, U 885, Université Bordeaux 2, 146 rue Léo Saignat, 33076 Bordeaux Cedex, France; and CHU de Bordeaux, Hôpital du Haut-Lévêque, Pôle Cardio-Thoracique, F-33604 Pessac, France. Email: roger.marthan@lpcr.u-bordeaux2.fr; roger.marthan@u-bordeaux2.fr

James G. Martin Meakins-Christie Laboratories, Department of Medicine, McGill University, 3626 St Urbain Street, Montréal, Québec, Canada, H2X 2P2

Robert L. Mayock Pulmonary, Allergy & Critical Care Division, University of Pennsylvania, Airways Biology Initiative, 125 South 31st Street, TRL Suite 1200, Philadelphia, PA 19104-3403, USA

Lyn Moir Research Fellow, Discipline of Pharmacology, Faculty of Medicine, Room 211 Blackburn Building D06, University of Sydney, Woolcock Institute of Medical Research, NSW 2006, Australia

Brian Oliver Research Fellow, Discipline of Pharmacology, Faculty of Medicine, Room 211 Blackburn Building D06, University of Sydney, Woolcock Institute of Medical Research, NSW 2006, Australia

Reynold A. Pannetieri, Jr. Pulmonary, Allergy & Critical Care Division, University of Pennsylvania, Airways Biology Initiative, 125 South 31st Street, TRL Suite 1200, Philadelphia, PA 19104-3403, USA

Julian Solway Professor of Medicine and Pediatrics, University of Chicago, 5841 S. Maryland Avenue, Chicago, IL 60637, USA

Maria B. Sukkar Airway Disease Section, National Heart & Lung Institute, Faculty of Medicine, Imperial College London, London SW3 6LY, UK

Thai Tran Departments of Physiology and Internal Medicine, Section of Respiratory Diseases, and National Training Program in Allergy and Asthma, University of Manitoba, Winnipeg, MB, Canada; Biology of Breathing Theme, Manitoba Institute of Child Health, Winnipeg, MB, Canada; Current address: Department of Physiology, National University of Singapore, Singapore

Loukia Tsaprouni Cell and Molecular Biology, Airways Disease Section, National Heart & Lung Institute, Imperial College London, Dovehouse Street, London SW3 6LY, UK

J. Manuel Tunon-de-Lara Université Bordeaux 2, Laboratoire de Physiologie Cellulaire Respiratoire, F-33076 Bordeaux, France; INSERM, U 885, F-33076 Bordeaux, France; and CHU de Bordeaux, Hôpital du Haut-Lévêque, Pôle Cardio-Thoracique, F-33604 Pessac, France

Jane E. Ward Lecturer, Department of Pharmacology, Faculty of Medicine, Dentistry and Health Sciences, Level 8, Medical Building, Corner of Grattan Street and Royal Parade, University of Melbourne, Victoria 3010, Australia

Wenwu Zhang Department of Cellular & Integrative Physiology, Indiana University, School of Medicine, 635 Barnhill Drive – MS 313, Indianapolis, IN 46202-5120, USA

1

Biophysical basis of airway smooth muscle contraction and hyperresponsiveness in asthma

Steven S. An[1] **and Jeffrey J. Fredberg**[2]

[1]*Johns Hopkins Bloomberg School of Public Health, Baltimore, MD, USA*
[2]*Harvard School of Public Health, Boston, MA, USA*

1.1 Introduction

It is self-evident that acute narrowing of the asthmatic airways and shortening of the airway smooth muscle are inextricably linked. Nonetheless, it was many years ago that research on the asthmatic airways and research on the biophysics of airway smooth muscle had a parting of the ways (Seow and Fredberg, 2001). The study of smooth muscle biophysics took on a life of its own and pursued a deeply reductionist agenda, one that became focused to a large extent on myosin II and regulation of the actomyosin cycling rate. The study of airway biology pursued a reductionist agenda as well, but one that became focused less and less on contractile functions of muscle and instead emphasized immune responses, inflammatory cells and mediators, and, to the extent that smooth muscle remained

Airway Smooth Muscle   Edited by Kian Fan Chung

of interest, that interest centred mainly on its synthetic, proliferative and migratory functions (Amrani and Panettieri, 2003; Black and Johnson, 1996; 2000; Black *et al.*, 2001; Holgate *et al.*, 2003; Kelleher *et al.*, 1995; Zhu *et al.*, 2001). Inflammatory remodelling of the airway wall was also recognized as being a key event in the asthmatic diathesis (Dulin *et al.*, 2003; Homer and Elias, 2000; James *et al.*, 1989; McParland *et al.*, 2003; Moreno *et al.*, 1986; Paré *et al.*, 1991; Wang *et al.*, 2003).

To better understand the impact of inflammatory remodelling processes upon smooth muscle shortening and acute airway narrowing, computational models of ever increasing sophistication were formulated, but, remarkably, the muscle compartment of these models remained at a relatively primitive level, being represented by nothing more than the classical relationship of active isometric force versus muscle length (Lambert and Paré, 1997; Lambert *et al.*, 1993; Macklem, 1987; 1989; 1990; 1996; Wiggs *et al.*, 1992). As discussed below, this description is now considered to be problematic because the very existence of a well-defined static force–length relationship has of late been called into question, as has the classical notion that the muscle possesses a well-defined optimal length. Rather, other factors intrinsic to the airway smooth muscle cell, especially muscle dynamics and mechanical plasticity, as well as unanticipated interactions between the muscle and its load, are now understood to be major factors affecting the ability of smooth muscle to narrow the airways (An *et al.*, 2007; Fredberg, 2000a; Fredberg *et al.*, 1999; Pratusevich *et al.*, 1995; Seow and Fredberg, 2001; Seow and Stephens, 1988; Seow *et al.*, 2000).

The topics addressed in this chapter are intended to highlight recent discoveries that bring airway biology and smooth muscle biophysics into the same arena once again. Here we do not provide an exhaustive review of the literature, but rather emphasize key biophysical properties of airway smooth muscle as they relate to excessive airway narrowing in asthma. This is appropriate because, in the end, if airway inflammation did not cause airway narrowing, asthma might be a tolerable disease. But asthma is not a tolerable disease. In order to understand the multifaceted problem of airway hyperresponsiveness in asthma, therefore, an integrative understanding that brings together a diversity of factors will be essential.

1.2 Airway hyperresponsiveness

It was recognized quite early that the lung is an irritable organ and that stimulation of its contractile machinery in an animal with an open chest can cause an increase in lung recoil, an expelling of air, a rise in intratracheal pressure, and an increase in airways resistance (Colebatch *et al.*, 1966; Dixon and Brodie, 1903; Mead, 1973; Otis, 1983). The fraction of the tissue volume that is attributable to contractile machinery is comparable for airways, alveolated ducts and blood vessels in the lung parenchyma (Oldmixon *et al.*, 2001); the lung parenchyma, like the airways, is a contractile tissue (Colebatch and Mitchell, 1971; Dolhnikoff *et al.*, 1998; Fredberg *et al.*, 1993; Ludwig *et al.*, 1987; 1988). Although airway smooth muscle was first described in 1804 by Franz Daniel Reisseisen (as related by Otis (1983)) and its functional properties first considered by Einthoven (1892) and Dixon and Brodie (1903), until the second half of the last century this muscle embedded in the airways was not regarded as being a tissue of any particular significance in respiratory mechanics (Otis, 1983). A notable exception in that regard was Henry Hyde Salter, who, in 1859, was well aware of the 'spastic' nature of airway smooth muscle and its potential role in asthma (Salter, 1868). The airway smooth muscle is now recognized as being the major end-effector of acute airway narrowing in asthma (Lambert and Paré, 1997; Macklem, 1996). There is also widespread agreement that shortening of the airway smooth muscle cell is the proximal cause of excessive airway narrowing during an asthmatic attack (Dulin *et al.*, 2003), and swelling of airway wall compartments and plugging by airway liquid or mucus are important amplifying factors (Lambert and Paré, 1997; Yager *et al.*, 1989). It remains unclear, however, why in asthma the muscle can shorten excessively.

'Airway hyperresponsiveness' is the term used to describe airways that narrow too easily and too much in response to challenge with nonspecific contractile agonists (Woolcock and Peat, 1989). Typically, a graph of airways resistance versus dose is sigmoid in shape (Figure 1.1); the response shows a plateau at high levels of contractile stimulus. The existence of the plateau, in general, is interpreted to mean that the airway smooth muscle is activated maximally and, thereby, has shortened as much as it can against a given elastic load. Once on the plateau, therefore, any further increase in stimulus can produce no additional active force, muscle shortening, or airway resistance.

To say that airways narrow too easily, on the one hand, means that the graph of airways resistance versus dose of a nonspecific contractile stimulus is shifted to the left along the dose axis, and that airways respond appreciably to levels of stimulus at which the healthy individual would be unresponsive; this phenomenon is called hypersensitivity. To say that airways narrow too much, on the other hand, means that the level of the plateau response is elevated, or that the plateau is abolished altogether, regardless of the position of the curve along the dose axis; this phenomenon is called hyperreactivity. As distinct from hypersensitivity, it is

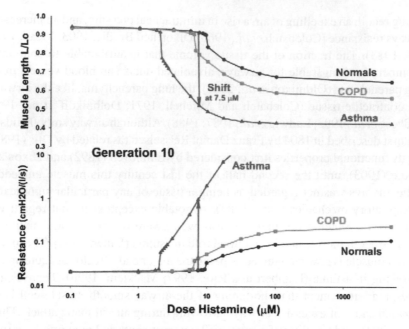

Figure 1.1 Computation of airway hyperresponsiveness in asthma. A computational result showing airway length (top) and airway resistance (bottom) as a function of agonist concentration for a 10th-generation airway (Mijailovich, 2003). The cases shown depict airways from a normal, an asthmatic and a COPD (Chronic obstructive pulmonary disease) lung. In this computation, the effects of tidal breathing and deep inspirations (6/min) upon myosin binding dynamics are taken into account explicitly (Mijailovich, 2003). As explained in the text, such an airway exhibits both hyperreactivity and hypersensitivity. (Reproduced courtesy of the American Journal of Respiratory and Critical Care Medicine 167, A183.)

this ability of the airways to narrow excessively, with an elevated or abolished plateau, that accounts for the morbidity and mortality associated with asthma (Sterk and Bel, 1989).

It has long been thought that the factors that cause hypersensitivity versus hyperreactivity are distinct, the former being associated with receptor complement and downstream signalling events but the latter being associated with purely mechanical factors, including the contractile apparatus, the cytoskeleton (CSK), and the mechanical load against which the muscle shortens (Armour *et al.*, 1984; Lambert and Paré, 1997; Macklem, 1996; Wiggs *et al.*, 1992). Macklem has pointed out that, once the muscle has become maximally activated, it is the active force and the load that become all-important, and the plateau response becomes essentially uncoupled from underlying biochemistry, signalling and cell biology

(Macklem, 1987; 1990; 1996). However, as described below, there is reason to think that these distinctions may not be as clear as once believed.

Although asthma is usually defined as an inflammatory disease, the link between the immunological phenotype and the resulting mechanical phenotype associated with disease presentation, including airway hyperresponsiveness, remains unclear; indeed, it is now established that airway hyperresponsiveness can be uncoupled from airway inflammation (Bryan et al., 2000; Crimi et al., 1998; Holloway et al., 1999; Leckie et al., 2000). It remains equally unclear whether airway hyperresponsiveness is due to fundamental changes within the smooth muscle itself, as might be caused by inflammatory mediators, chemokines and cytokines (Fernandes et al., 2003), or due to changes external to the muscle, such as a reduced mechanical load against which the smooth muscle contracts. Still another possibility supported by recent evidence is that there is an interaction of the two wherein the contractile machinery within the smooth muscle cell adapts in response to a change in its mechanical microenvironment (Dulin et al., 2003; Fredberg et al., 1999; Lakser et al., 2002; Pratusevich et al., 1995; Seow and Fredberg, 2001; Seow et al., 2000; Wang et al., 2001). Moreover, Tschumperlin et al. (2002; 2003) have provided evidence that bronchospasm can lead to mechanically induced pro-inflammatory signalling events in the airway epithelium, in which case airway inflammation may cause bronchospasm, but bronchospasm in turn may amplify or even activate specific inflammatory pathways.

In the balance of this chapter, we address the classical picture of smooth muscle behaviour and then go on to describe what we know about nonclassical behaviour in a dynamic setting and, in particular, the ability of the muscle cell to adapt rapidly to changes in its mechanical microenvironment. We do not address the increasing evidence that now suggests that cytokines such as interleukin (IL)-1β and tumor necrosis factor (TNF)-α augment responses to bronchoconstrictor agonists while attenuating the bronchodilation that can be effected by hormones and paracrine agents such as epinephrine and PGE$_2$ (Shore et al., 1997). Such cytokines, along with growth factors and other inflammatory mediators, also result in smooth muscle hyperplasia, at least in culture systems (Kelleher et al., 1995). In culture, extracellular matrix proteins have been shown not only to regulate synthetic (Chan et al., 2006; Peng et al., 2005), proliferative (Freyer et al., 2001; Hirst et al., 2000; Nguyen et al., 2005) and migratory (Parameswaran et al., 2004) functions of the airway smooth muscle cell, but also to modulate the protein expressions and biochemical pathways that are implicated in muscle maturation and contraction (Freyer et al., 2004; Halayko and Solway, 2001; Halayko et al., 1999; Hirst et al., 2000; Tran et al., 2006). Whether airway inflammation and matrix remodelling in the asthmatic airways can result in a hypercontractile phenotype of the airway smooth muscle cell remains to be established.

1.3 Classical behaviour of airway smooth muscle and the balance of static forces

The microstructure of striated muscle is highly ordered, whereas there is abundant evidence in the literature demonstrating that the cytoskeletal microstructure of smooth muscle is quite disordered (Small, 1995; Small and Gimona, 1998); it is, after all, its amorphous structure that gives 'smooth' muscle its name. Moreover, the airway smooth muscle CSK is in a continuous state of remodelling, a point to which we return below. Despite these differences, it has been widely presumed that to a first approximation Huxley's sliding-filament model of muscle contraction (Huxley, 1957) describes the function of both smooth and striated muscle (Murphy, 1988; 1994; Mijailovich *et al.*, 2000). For many of the biophysical phenomena observed in airway smooth muscle, such as active force generation and shortening velocity, Huxley's model represents a useful tool for thought (Huxley, 1957; Mijailovich *et al.*, 2000), while for others, such as mechanical plasticity, it does not.

As in the case of striated muscle contraction, the principal biophysical parameters that characterize smooth muscle contraction include the maximum active isometric force (or stress, which is simply the force carried per unit area), the length at which the muscle can attain that maximal force (i.e., the optimum length (Lo)), and the shortening capacity of the muscle. The sliding-filament model of Huxley is the starting point for understanding each of these phenomena. As described by Huxley (1957), isometric force, as well as muscle stiffness, is proportional to the number of actomyosin cross links per unit volume. This is true because, assuming rigid filaments, all bridges within a given contractile unit must act mechanically in parallel, with their displacements being identical and their forces being additive. The maximum active stress supported by smooth versus striated muscle is approximately the same and is of the order 10^5 Pa. In striated muscle, Lo is attributed to the extent of overlap between the myosin filament and the actin filament, Lo corresponding to a maximum number of myosin heads finding themselves within the striking distance of an available binding site on the actin filament, and the maximum capacity of the muscle to shorten being limited by the collision of the myosin filament end with the z-disc. Smooth muscle possesses no structure comparable to the z-disc, however, although actin filaments terminate in dense bodies, which might come into play in limiting muscle shortening. Whereas unloaded striated muscle can shorten perhaps 20 per cent from its optimum length, unloaded smooth muscle can shorten as much as 70 per cent (Stephens, 1987; Stephens and Seow, 1993; Uvelius, 1976). Several physical factors may come into play to limit the capacity for unloaded shortening of smooth muscle. Small (1995) has shown that actin filaments of the contractile apparatus connect to the CSK at cytoplasmic dense bodies and with the longitudinal rib-like arrays of dense plaques of the membrane skeleton that couple to the extracellular matrix. Moreover, the side-polar configuration of the myosin filament (Tonino et al., 2002; Xu et al., 1996) is likely to be involved. Still other factors coming into play include length-dependent activation (An and Hai, 1999; 2000; Mehta et al., 1996; Youn et al., 1998), length-dependent rearrangements of the CSK and contractile machinery (Gunst et al., 1995; Pratusevich et al., 1995), and length-dependent internal loads (Stephens and Kromer, 1971; Stephens and Seow, 1993; Warshaw et al., 1988).

What are the extramuscular factors that act to limit airway smooth muscle shortening? The basic notion, of course, is that muscle shortening stops when the total force generated by the muscle comes into a static balance with the load against which the muscle has shortened, both of which vary with muscle length. The factors setting the load include the elasticity of the airway wall, elastic tethering forces conferred by the surrounding lung parenchyma, active tethering forces conferred by contractile cells in the lung parenchyma (Nagase et al., 1994; Romero and Ludwig, 1991), mechanical coupling of the airway to the parenchyma by the peribronchial adventitia, and buckling of the airway epithelium and submucosa (Ding et al., 1987; Robatto et al., 1992; Wiggs et al., 1992). In addition, the airway smooth muscle itself is a syncytium comprised mostly of smooth muscle cells, aligned roughly along the axis of muscle shortening, and held together by an intercellular connective tissue network. In order to conserve volume, as the muscle shortens, it must also thicken. And as the muscle shortens and thickens, the intercellular connective tissue network must distort accordingly. Meiss (1999) has shown that at the extremes of muscle shortening it may be the loads associated with radial expansion (relative to the axis of muscle shortening) of the intercellular connective tissue network that limit the ability of the muscle to shorten further.

In the healthy, intact dog, airway smooth muscle possesses sufficient force-generating capacity to close all airways (Brown and Mitzner, 1998; Warner and Gunst, 1992). This fact may at first seem to be unremarkable, but it is not easily reconciled with the observation that when healthy animals or people are challenged with inhaled contractile agonists in concentrations thought to be sufficient to activate the muscle maximally, the resulting airway narrowing is limited in extent, and that limit falls far short of airway closure (Moore *et al.*, 1997; 1998). Breathing remains unaccountably easy. Indeed, it is this lightness of breathing in the healthy challenged lung, rather than the labored breathing that is characteristic of the asthmatic lung, that in many ways presents the greater challenge to our understanding of the determinants of acute airway narrowing (Fredberg and Shore, 1999). Brown and Mitzner (1998) have suggested that the plateau of the dose-response curve reflects uneven or limited aerosol delivery to the airways. Another possibility, however, is that some mechanisms act to limit the extent of muscle shortening in the healthy, breathing lung, whereas these mechanisms become compromised in the asthmatic lung. It has been suspected that the impairment of that salutary mechanism, if it could only be understood, might help to unlock some of the secrets surrounding excessive airway narrowing in asthma, as well as the morbidity and mortality associated with that disease (Fish *et al.*, 1981; Lim *et al.*, 1987; Nadel and Tierney, 1961; Skloot *et al.*, 1995). This brings us to muscle dynamics and the factors that could account for airway hyperresponsiveness in asthma.

1.4 Shortening velocity and other manifestations of muscle dynamics

The oldest and certainly the simplest explanation of airway hyperresponsiveness would be that muscle from the asthmatic airways is stronger than muscle from the healthy airways, but evidence in support of that hypothesis remains equivocal (Black and Johnson, 1996; 2000; De Jongste *et al.*, 1987; Solway and Fredberg, 1997). Indeed, a number of earlier studies, in which tissues were obtained post-mortem or surgically, have reported normal contractility (Bai, 1990; Bjorck *et al.*, 1992) and even hypocontractility (Goldie *et al.*, 1986; Whicker *et al.*, 1988) of muscle from the asthmatic airways. Accordingly, studies from the laboratory of Stephens and colleagues (Antonissen *et al.*, 1979; Fan *et al.*, 1997; Jiang *et al.*, 1992; Ma *et al.*, 2002; Seow and Stephens, 1988) have emphasized that the force-generation capacity of allergen-sensitized airway smooth muscle of the dog, or of human asthmatic muscle, is no different from that of control muscle. As a result, the search for an explanation turned to other factors, and several alternative hypotheses have been advanced. These fall into three broad classes, each of which is consistent with remodelling events induced by the inflammatory microenvironment, and they include an increase of muscle mass (Johnson *et al.*, 2001;

Lambert *et al.*, 1993; Thomson *et al.*, 1996; Wiggs *et al.*, 1992), a decrease of the static load against which the muscle shortens (Ding *et al.*, 1987; Macklem, 1996; Wiggs *et al.*, 1992; 1997), and a decrease of the fluctuating load that perturbs myosin binding during breathing (Fredberg, 1998; 2000a; 2000b; Fredberg *et al.*, 1999; Mijailovich *et al.*, 2000). Taken together, these hypotheses are attractive because they suggest a variety of mechanisms by which airway smooth muscle can shorten excessively even while the isometric force-generating capacity of the muscle remains essentially unchanged.

Aside from changes in the static load and/or the dynamic load, however, a consistent association has been noted between airway hyperresponsiveness and unloaded shortening velocity of the muscle (Antonissen *et al.*, 1979; Duguet *et al.*, 2000; Fan *et al.*, 1997; Ma *et al.*, 2002; Wang *et al.*, 1997). This association suggests that the problem with airway smooth muscle in asthma may be that it is too fast rather than too strong. But how shortening velocity – a dynamic property of the muscle – might cause excessive airway narrowing – a parameter that was thought to be determined by a balance of static forces – remains unclear. To account for increased shortening capacity of unloaded cells, Stephens and colleagues have reasoned that upon activation virtually all muscle shortening is completed within the first few seconds (Ma *et al.*, 2002). As such, the faster the muscle can shorten within this limited time window, the more it will shorten. However, in isotonic loading conditions at physiological levels of load, muscle shortening is indeed most rapid at the very beginning of the contraction, but appreciable shortening continues for at least 10 min after the onset of the contractile stimulus (Fredberg *et al.*, 1999). An alternative hypothesis to explain why intrinsically faster muscle might shorten more comes from consideration of the temporal fluctuations of the muscle load that are attributable to the action of spontaneous breathing (Fredberg *et al.*, 1997; 1999; Solway and Fredberg, 1997). Load fluctuations that are attendant on spontaneous breathing are the most potent of all known bronchodilating agencies (Gump *et al.*, 2001; Shen *et al.*, 1997). Among many possible effects, these load fluctuations perturb the binding of myosin to actin, causing the myosin head to detach from actin much sooner than it would have during an isometric contraction. But the faster the myosin cycling (i.e., the faster the muscle), the more difficult it is for imposed load fluctuations to perturb the acto-myosin interaction. This is because the faster the intrinsic rate of cycling, the faster will a bridge, once detached, reattach and contribute once again to active force and stiffness.

Why is muscle from the allergen-sensitized animal or asthmatic subject faster? For technical reasons, in their study on the single airway smooth muscle cell freshly isolated from bronchial biopsies obtained from an asthmatic subject,

Ma *et al.* 2002) did not measure protein expression levels of myosin light-chain kinase (MLCK), but their finding of increased content of message strongly implicates MLCK. Although regulation of myosin phosphorylation is a complex process with multiple kinases and phosphatases, this finding substantially narrows the search for the culprit that may account for the mechanical changes observed in these cells. Moreover, these studies seem to rule out changes in the distribution of myosin heavy-chain isoforms; content and isoform distributions of message from asthmatic cells showed the presence of smooth muscle myosin heavy-chain A (SM-A), but not SM-B, the latter of which contains a seven-amino-acid insert that is typical of phasic rather than tonic smooth muscle, and is by far the faster of the two isoforms (Lauzon *et al.*, 1998; Murphy *et al.*, 1997).

Using laser capture microdissection of airway smooth muscle from bronchial biopsies obtained from normal versus mild-to-moderate asthmatics, Woodruff *et al.*, 2004) also found no differences in the expressions profile of a panel of genes that are often considered markers of hypercontractile phenotype (including MLCK, however) but did detect a nearly twofold increase in the number of airway smooth muscle cells in the asthmatics. Although the source of the increased cell number (increased proliferation, decreased apoptosis, and/or increased migration) remains unclear (Hirst *et al.*, 2004; Johnson *et al.*, 2001; Lazaar and Panettieri, 2005; Madison, 2003; Woodruff *et al.*, 2004; Zacour and Martin, 1996), increased muscle mass alone is sufficient to predispose to airway hyperresponsiveness in asthma (James *et al.*, 1989; Lambert *et al.*, 1993; Moreno *et al.*, 1986). The question of whether muscle mass (quantity) and muscle contractility (quality) might covary remains to be elucidated, however. For example, it is likely that the airway smooth muscle cell in the proliferative/synthetic/maturational state might be less contractile than similar cells differentiated into a fully contractile state – an effect that would be compensatory – but no mechanical data are available to support that possibility.

1.5 Biophysical characterization of airway smooth muscle: bronchospasm in culture?

With recent technological advances, such as atomic force microscopy (Alcaraz *et al.*, 2003; Smith *et al.*, 2005), two-point and laser-tracking microrheology (Van Citters *et al.*, 2006; Yamada *et al.*, 2000), magnetic tweezers (Bausch *et al.*, 1998; 1999), and traction microscopy (Butler *et al.*, 2002; Tolic-Norrelykke *et al.*, 2002), a single living cell in culture can now be characterized biophysically. While the use of cultured cells has certain limitations, they do offer the advantage that, when passaged in culture, airway smooth muscle cells retain functional responses to a wide panel of agonists and signalling pathways that are implicated in asthma (Halayko *et al.*, 1999; Hubmayr *et al.*, 1996; Panettieri *et al.*, 1989; Shore *et al.*, 1997; Tao *et al.*, 1999; 2003; Tolloczko *et al.*, 1995). In our laboratories, to probe deeper into the mechanical properties of the airway smooth muscle cell, we use a technology that has its roots in an early contribution of Francis H. C. Crick.

Before his well-known work on the double helical structure of deoxyribonucleic acid (DNA) (Watson and Crick, 1953a; 1953b), Crick measured the viscosity and elasticity of the medium inside cells by observing internalized magnetic particles and how they rotate in reaction to an applied magnetic field (Crick and Hughes, 1950). Extending this approach, Valberg and his colleagues studied populations of particles internalized into populations of cells, and measured induced bead rotations by remote sensing, namely, by means of changes in the horizontal projection of the remanent magnetic field produced by the magnetized particles as they rotate (Valberg, 1984; Valberg and Feldman, 1987). In a major step forward, we subsequently adapted this technique still further (Fabry *et al.*, 2001; Wang *et al.*, 1993) by using ligand-coated, ferrimagnetic microbeads – not internalized as before – but rather bound to the CSK via membrane-spanning integrin receptors. And more recently still, we showed that changes in cell stiffness measured in this way correlate well with stiffness changes in the same cells measured by atomic force microscopy (Alcaraz *et al.*, 2003) and with force changes measured with traction microscopy (Wang *et al.*, 2002). This method is now known as magnetic twisting cytometry (MTC), and it has evolved into a useful tool to probe the mechanical properties of a variety of cell types, both cultured and freshly isolated, through different receptor systems, and with a variety of experimental interventions (Deng *et al.*, 2006; Fabry *et al.*, 2001; Laudadio *et al.*, 2005; Puig-de-Morales *et al.*, 2004).

The principle of MTC is straightforward (Figure 1.2). A ferrimagnetic microbead (4.5 μm in diameter) is coated with a synthetic peptide containing the sequence Arg–Gly–Asp (RGD) and is then allowed to bind to the cell. Such an RGD-coated bead binds avidly to cell-surface integrin receptors (Wang *et al.*, 1993), forms focal adhesions (Matthews *et al.*, 2004), and becomes well integrated into the cytoskeletal scaffold (Maksym *et al.*, 2000): it displays tight functional coupling to stress-bearing cytoskeletal structures and the contractile apparatus (An *et al.*, 2002; Hu *et al.*, 2003). By imposition of a uniform magnetic field upon the magnetized bead, a small torque is applied and resulting bead motions deform structures deep in the cell interior (Hu *et al.*, 2003). Such *forced* bead motions are impeded by mechanical stresses developed within the cell body, and the ratio of

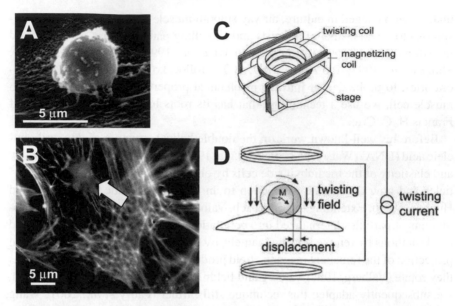

Figure 1.2 Optical magnetic twisting cytometry (OMTC). (A) An RGD-coated bead (4.5 μm in diameter) binds to the surface of the adherent cell. (B) Such bead (white arrow) becomes well-integrated into underlying actin lattice (phalloidin staining). (C) The bead is magnetized horizontally (parallel to the surface on which cells are plated) and then twisted in a vertically aligned homogenous magnetic field that is varying sinusoidally in time. (D) This sinusoidal twisting field causes both a rotation and a pivoting displacement of the bead. As the bead moves, the cell develops internal stresses which in turn resist bead motions. Here the ratio of specific torque to lateral bead displacement is computed and is expressed as cell stiffness in Pa/nm. (Reproduced with permission of J. Appl. Physiol., Vol. 91, p. 988, © 2001 The American Physiological Society and with permission of Phys. Rev. Lett., Vol. 87, p. 148102-1, © 2001 The American Physical Society.)

specific torque to lateral bead displacements is taken as a measure of cell stiffness (Fabry *et al.*, 2001).

By this technique, it has been previously demonstrated that airway smooth muscle cells in culture exhibit pharmacomechanical coupling to a wide panel of contractile and relaxing agonists (An *et al.*, 2002; Hubmayr *et al.*, 1996). For example, cell stiffness increases in response to agonists reported to increase intracellular Ca^{2+} concentration ($[Ca^{2+}]_i$) or inositol 1,4,5-trisphosphate (IP_3) formation and decreases in response to agonists that are known to increase intracellular cAMP or cGMP levels (An *et al.*, 2002; Hubmayr *et al.*, 1996; Shore *et al.*, 1997). Although stiffness is an indirect measure of contractility (Fredberg *et al.*, 1997), changes in cell stiffness range appreciably from maximally relaxed to maximally activated states (Fabry *et al.*, 2001), and such stiffening responses require, as in intact tissues, actin polymerization as well as myosin activation (An *et al.*, 2002; Mehta and Gunst, 1999). Indeed, active stresses within individual airway smooth muscle cells, as measured by traction microscopy, span a similarly wide range (Figure 1.3) and closely track changes in cell stiffness as measured by MTC (Wang *et al.*, 2002). Altogether, the mechanical responsiveness of airway smooth muscle cells measured in culture is consistent with physiological

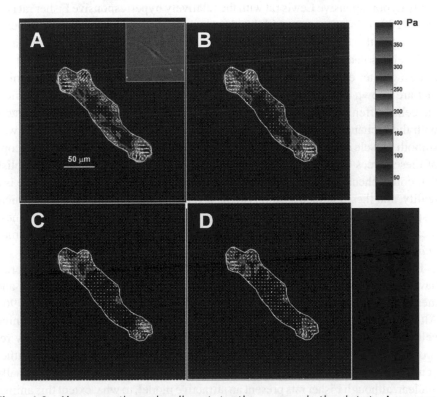

Figure 1.3 Airway smooth muscle cell exerts traction upon an elastic substrate. A representative changes in traction field of a single human airway smooth muscle cell in response to isoproterenol at (A) 0 μM, (B) 0.1 μM, (C) 1 μM and (D) 10 μM. The traction field was computed from the displacement field using Fourier transform traction cytometry (FTTC) (Butler et al., 2002; Tolic-Norrelykke et al., 2002; Wang et al., 2002). The cell boundary is shown by the white line. Colors show the magnitude of the tractions in Pascal (Pa) (see color scale). Arrows show the direction and relative magnitude of the tractions. In general, the greatest tractions are at the cell periphery and directed centripetally. *Inset.* A phase-contrast image of the respective airway smooth muscle cell. Scale bar: 50 μm. (Reproduced with permission of Am. J. Respir. Cell Mol. Biol., Vol. 35, p. 59, ⓒ 2006 The American Thoracic Society.) (For a colour reproduction of this figure, please see the colour section, located towards the centre of the book).

responses measured at tissue and organ levels (Fredberg *et al.*, 1996; Mehta and Gunst, 1999). As such, these biophysical methods are unparalleled in their ability to characterize mechanical properties of airway smooth muscle at the level of the single cell *in vitro*.

Does mechanical responsiveness of the airway smooth muscle cell predict airway hyperresponsiveness? To address this question, we have recently contrasted the biophysical properties of the airway smooth muscle cell isolated from the relatively hyporesponsive Lewis rat with the relatively hyperresponsive Fisher rat (An *et al.*, 2006). In agreement with biochemical changes that have been previously reported in these cells (Tao *et al.*, 1999; 2003; Tolloczko *et al.*, 1995), compared with cells isolated from Lewis rat, those isolated from the Fisher rat demonstrate in turn greater extent of the stiffening response to a panel of contractile agonists that are known to increase $[Ca^{2+}]_i$ or IP_3 formation: Fisher airway smooth muscle cells stiffen fast and also stiffen more (Figure 1.4). Furthermore, consistent with these changes in cell stiffness, the relatively hyperresponsive Fisher airway smooth muscle cells also exert bigger contractile forces and exhibit greater scope of these forces (An *et al.*, 2006). Taken together, these findings firmly establish that comprehensive biophysical characterization of bronchospasm in culture is a reality, and these characterizations at the level of the single cell show mechanical responses that are consistent with phenotypic differences in airway responsiveness measured at tissue and organ levels (Dandurand *et al.*, 1993a; 1993b; Eidelman *et al.*, 1991; Jia *et al.*, 1995; Tao *et al.*, 1999).

Like human asthmatics (Johnson *et al.*, 2001; Woodruff *et al.*, 2004), Fisher rats have abundant smooth muscle cells in their airways (Eidelman *et al.*, 1991), and these cells show great capacity to proliferate in culture (Zacour and Martin, 1996). Although these features, together with increased muscle dynamics (shortening velocity as well as contractile force), may account for the enhanced airway responsiveness of Fisher rats, the precise role of airway smooth muscle in the pathogenesis of airway hyperresponsiveness in asthma is ill-defined. It remains equally unclear, although Fisher rats present an attractive model, to what extent this animal model recapitulates the pathophysiology associated with human asthma.

1.6 Mechanical plasticity: a nonclassical feature of airway smooth muscle

When activated muscle in the muscle bath is subjected to progressively increasing load fluctuations approaching the magnitude and frequency expected during normal breathing, the muscle lengthens appreciably in response (Fredberg

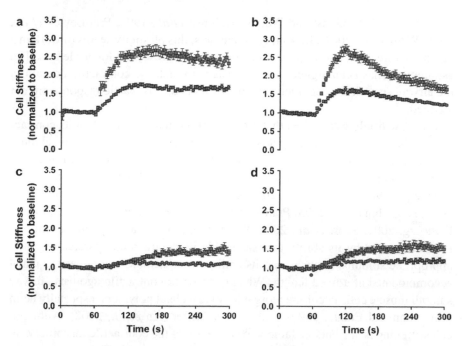

Figure 1.4 Fisher airway smooth muscle cells stiffen fast and also stiffen more. Airway smooth muscle cells isolated from the relatively hyporesponsive Lewis rat (blue closed circles) and the relatively hyperresponsive Fisher rat (red closed squares) were maximally stimulated with a panel of contractile agonists: (A) 5-HT (1 μM), (B) bradykinin (1 μM), (C) acetylcholine (1 μM) and (D) carbachol (100 μM). For each agonist, changes in cell stiffness were normalized to the baseline stiffness of each individual cell before stimulation. (Reproduced with permission of Am. J. Respir. Cell Mol. Biol., Vol. 35, p. 57, © 2006 The American Thoracic Society.)

et al., 1999). But when load fluctuations are progressively reduced, the muscle reshortens somewhat but fails to return to its original length. This incomplete to reshortening is not accounted for by muscle injury; the original operating length can be recovered simply by removing the contractile agonist and allowing the muscle a short interval before contracting again. Nor can incomplete reshortening be accounted for by myosin dynamics; myosin dynamics alone predicts complete reshortening when the load fluctuations are removed (Fredberg et al., 1999). Thus, the failure of activated muscle to reshorten completely is evidence of the plasticity of the contractile response. During a sustained contraction, the operational length of the muscle for a given loading, or the force at a given length, can be reset by loading and the history of that loading (Ford et al., 1994; Fredberg *et al.*, 1997; 1999; Gunst and Wu, 2001; Gunst *et al.*, 1993; Pratusevich *et al.*, 1995; Wang *et al.*, 2001). In healthy individuals, this plasticity seems to work in a favorable direction, allowing activated muscle to be reset to a longer length. The asthmatic, it has been argued, never manages to melt the contractile domain in the airway smooth muscle; therefore, the benefits of this plastic response are not attained.

It is now firmly established that airway smooth muscle can adapt its contractile machinery, as well as the cytoskeletal scaffolding on which that machinery operates, in such a way that the muscle can maintain the same high force over an extraordinary range of muscle length (An *et al.*, 2007; Ford *et al.*, 1994; Fredberg, 1998; Gunst and Wu, 2001; Gunst *et al.*, 1993; 1995; Kuo *et al.*, 2001; 2003; Naghshin *et al.*, 2003; Pratusevich *et al.*, 1995; Qi *et al.*, 2002; Seow and Fredberg, 2001; Seow *et al.*, 2000; Wang *et al.*, 2001); airway smooth muscle is characterized by its ability to disassemble its contractile apparatus when an appropriate stimulus is given, and its ability to reassemble that apparatus when accommodated at a fixed length. When exposed to contractile agonists, airway smooth muscle cells in culture reorganize cytoskeletal polymers, especially actin (Hirshman and Emala, 1999), and become stiffer (An *et al.*, 2002). Although cell stiffening is attributable largely to activation of the contractile machinery, an intact actin lattice has been shown to be necessary, but not sufficient, to account for the stiffening response (An *et al.*, 2002).

The malleability of the cell and its mechanical consequences have been called by various authors mechanical plasticity, remodelling, accommodation or adaptation. Even though the force-generating capacity varies little with length in the fully adapted muscle, the unloaded shortening velocity and the muscle compliance vary with muscle length in such a way as to suggest that the muscle cell adapts by adding or subtracting contractile units that are mechanically in series (Figure 1.5). The mechanisms by which these changes come about and the factors that control the rate of plastic adaptation are unknown, however.

Several hypotheses have been advanced to explain smooth muscle plasticity. Ford and colleagues have suggested that the architecture of the myosin fibres themselves may change (Ford *et al.*, 1994; Kuo *et al.*, 2001; 2003; Pratusevich *et al.*, 1995; Seow *et al.*, 2000), while Gunst and colleagues (Gunst and Wu, 2001; Gunst *et al.*, 1993; 1995) have argued that it is the connection of the actin filament to the focal adhesion plaque at the cell boundary that is influenced by loading history. An alternative notion is that secondary but important molecules stabilize the CSK, and as the contractile domain melts under the influence of imposed load fluctuations, those loads must be borne increasingly by the scaffolding itself, thus reflecting the malleability of the cytoskeletal domain (Fredberg, 2000a; Gunst

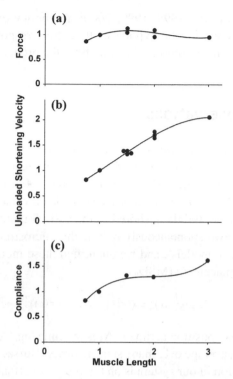

Figure 1.5 Mechanical plasticity of the airway smooth muscle. (A) Isometric force (F), (B) unloaded shortening velocity (V), and (C) compliance (C) of canine tracheal smooth muscle activated over a range of muscle lengths. Filled circles represent data modified from Pratusevich et al. (1995) and Kuo et al. (2003), as compiled by Lambert et al. (2004); solid lines are third-order polynominal functions adjusted to the original data (Silveira and Fredberg, 2005). (Reproduced with permission of Can. J. Physiol. Pharmacol., Vol. 83, p. 924, © 2005 NRC Canada.)

et al., 1995; Halayko and Solway, 2001; Wang and Bitar, 1998). In that connection, a role for the Rho-A pathway has been suggested (Halayko and Solway, 2001; Mehta *et al.*, 2000), and some evidence now suggests that the p38 MAP kinase pathway may be involved (Lakser *et al.*, 2002). For example, airway smooth muscle incubated with an inhibitor of the p38 MAP kinase pathway demonstrates a greater degree of fluctuation-driven muscle lengthening than does control muscle, and upon removal of the force fluctuations it remains at a greater length. Moreover, force fluctuations themselves activate the p38 MAP kinase pathway. It is noteworthy in that connection that heat-shock protein 27 (HSP27), a downstream target of Rho and p38, has been implicated as an essential element in cytoskeletal remodelling of the airway smooth muscle cell (An *et al.*, 2004; Gerthoffer and

Pohl, 1994; Hedges *et al.*, 1998; 1999; 2000; Yamboliev *et al.*, 2000). These findings are consistent with the hypothesis that stress-response pathways may stabilize the airway smooth muscle CSK and limit the bronchodilating effects of deep inspirations.

1.7 Recent observations

Recently, we have made a series of observations in a number of different cell types and reported a functional assay that probes the discrete molecular level remodelling dynamics of the CSK (An *et al.*, 2004; 2005; Bursac *et al.*, 2005; 2007). This assay is based on *spontaneous* nano-scale movements of an individual RGD-coated microbead tightly anchored to the CSK (Figure 1.6): we reasoned that the bead can move spontaneously only if the microstructure to which it is attached rearranges (remodels), and we quantified these motions by calculating its mean square displacement (MSD_b),

$$MSD_b(\Delta t) = \langle (r(t + \Delta t) - r(t))^2 \rangle \qquad (1)$$

where $r(t)$ is the bead position at time t, Δt is the time lag ($\Delta t = 1/12$s), and the brackets indicate an average over many starting times (Bursac *et al.*, 2005; 2007). The limit of resolution in our system is on the order of $\sim$10 nm, but for $\Delta t \sim 4$ s most beads had displaced a much greater distance. Accordingly, we analysed data for time lags greater than 4 s and up to t_{max}. As shown below, MSD of most beads increases with time according to a power-law relationship.

$$MSD(\Delta t) = D^*(\Delta t / \Delta t_o)^\alpha \qquad (2)$$

The coefficient D^* and the exponent α of an individual bead are estimated from a least-square fit of a power-law to the MSD data for Δt between 4 s and $t_{max}/4$. Here we take Δt_o to be 1 s and express D^* in units of nm^2. As shown in Figure 1.6, the ensembled average of all MSD_b (MSD) increased faster than linearly with time ($\sim t^{1.6}$), exhibiting superdiffusive motions. Such anomalous motions were also observed on cells seeded on a micropatterned substrate on which a cell could adhere but not crawl (Bursac *et al.*, 2007; Parker *et al.*, 2002). Taken together, unlike simple, diffusive, thermal Brownian motion that increases its MSD linearly with time (Kubo, 1986), *spontaneous* motions of an individual RGD-coated bead are nonthermal in nature and, instead, consistent with the notion that these anomalous motions report molecular-level reorganization (remodelling) of the underlying CSK (An *et al.*, 2004; 2005; Bursac *et al.*, 2005; 2007).

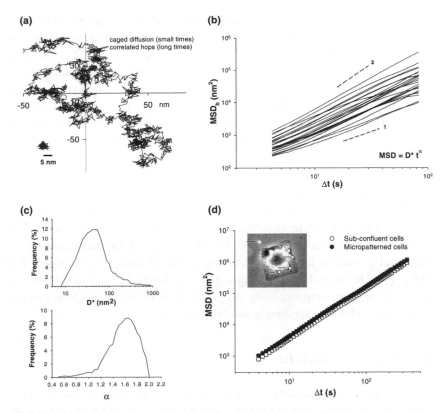

Figure 1.6 Cytoskeleton remodeling of the airway smooth muscle cell. (A) Spontaneous motions of a representative bead show intermittent dynamics, with periods of confinement alternating with hopping; a bead glued to the coverslip is taken to represent the upper limit of measurement noise (bottom left). (B) MSD_b calculated from Equation 1 is shown for representative beads. (C) The histograms of diffusion coefficient D^* and exponent α estimated from a least-square fits of a power-law (Equation 2) to the MSD_b data. (D) Ensemble average of all MSD_b (MSD) increased faster than linearly with time ($\sim t^{1.6}$); beads attached to a cell seeded on a micropatterned substrate (50 μm x 50 μm), on which it could adhere but not crawl, exhibited the same anomalous motions. (Reproduced with permission of Biochem. Biophys. Res. Comm., Vol. 355, p. 326, © 2007 Elsevier Inc., and with permission of Nature Mater., Vol. 4, p. 559, © 2005 Nature Publishing Group.)

By this method, we have demonstrated that the rate of cytoskeletal remodelling is appreciably different between airway smooth muscle cells isolated from the relatively hyporesponsive Lewis rat and those from the relatively hyperresponsive Fisher rat: Fisher cells exhibit faster remodelling dynamics (An *et al.*, 2006). Furthermore, such remodelling is dependent on the levels of intracellular ATP content (An *et al.*, 2006; Bursac *et al.*, 2005) and also becomes progressively slow with phosphorylation of HSP27 (An *et al.*, 2004). Indeed, evidence supporting the notion of a highly malleable cell is accumulating rapidly, but a molecular basis to explain this malleability is only beginning to emerge. Most recently, we observed that, in response to a transient stretch-unstretch maneuver with zero residual macroscale strain, the airway smooth muscle cell promptly fluidizes and then slowly re-solidifies (Trepat *et al.*, 2007). At the same time, the rate of spontaneous nano-scale structural rearrangements promptly accelerates and then slowly decays in a scale-free manner (Trepat *et al.*, 2007). Taken together, these findings suggest that fluidization provides freedom of the cell to reorganize contractile units, stress fibers and focal adhesions in response to mechanical stress (Trepat *et al.*, 2007). Regardless of the specific molecules and mechanisms invoked to explain the plasticity of the contractile responses, therefore, the melting of the contractile domain would appear to be a necessary (or permissive) event, but one that by itself is not sufficient to explain the effects of the history of tidal loading. How these molecular changes and malleability of the airway smooth muscle cell, in turn, correlate with the progression of asthma pathophysiology are currently under investigation in our laboratories.

1.8 Future directions

To understand the multifaceted problem of airway hyperresponsiveness in asthma, an integrative understanding that brings together a diversity of factors is essential. We have outlined here an emerging picture of smooth muscle biophysics as it relates to excessive airway narrowing in asthma, but we need to keep in mind that asthma is a chronic inflammatory disorder; therefore, understanding the impact of inflammatory remodelling of the airway wall and the airway smooth muscle cell on disease presentation is vital. Fortunately, with recent technological advances, we are now equipped with both biochemical and biophysical tools to address nagging questions that have often separated the fields of airway biology and smooth muscle biophysics.

References

Alcaraz, J., Buscemi, L., Grabulosa, M. *et al.* (2003). Microrheology of human lung epithelial cells measured by atomic force microscopy. *Biophys J* **84**, 2071–2079.

Amrani, Y. and Panettieri, R. A. (2003). Airway smooth muscle: contraction and beyond. *Int J Biochem Cell Biol* **35**, 272–276.

An, S. S., Bai, T. R., Bates, J. H. T. *et al.* (2007). Airway smooth muscle dynamics: a final common pathway of airway obstruction in asthma. *Eur Respir J* (in press).

An, S. S., Fabry, B., Mellema, M. *et al.* (2004). Role of heat shock protein 27 in cytoskeletal remodelling of the airway smooth muscle cell. *J Appl Physiol* **96**, 1707–1713.

An, S. S., Fabry, B., Trepat, X. *et al.* (2006). Do biophysical properties of the airway smooth muscle in culture predict airway hyperresponsiveness? *Am J Respir Cell Mol Biol* **35**, 55–64.

An, S. S. and Hai, C. M. (1999). Mechanical strain modulates maximal phosphatidylinositol turnover in airway smooth muscle. *Am J Physiol* **277**, L968–L974.

An, S. S. and Hai, C. M. (2000). Mechanical signals and mechanosensitive modulation of intracellular [Ca^{2+}] in smooth muscle. *Am J Physiol* **279**, C1375–C1384.

An, S. S., Laudadio, R. E., Lai, J. *et al.* (2002). Stiffness changes in cultured airway smooth muscle cells. *Am J Physiol* **283**, C792–C801.

An, S. S., Pennella, C. M., Gonnabathula, A. *et al.* (2005). Hypoxia alters biophysical properties of endothelial cells via p38 MAPK- and Rho kinase-dependent pathways. *Am J Physiol* **289**, C521–C530.

Antonissen, L. A., Mitchell, R. W., Kroeger, E. A. *et al.* (1979). Mechanical alterations of airway smooth muscle in a canine asthmatic model. *J Appl Physiol* **46**, 681–687.

Armour, C. L., Black, J. L., Berend, N. *et al.* (1984). The relationship between bronchial hyperresponsiveness to methacholine and airway smooth muscle structure and reactivity. *Respir Physiol* **58**, 223–233.

Bai, T. R. (1990). Abnormalities in airway smooth muscle in fatal asthma. *Am Rev Respir Dis* **141**, 552–557.

Bausch, A. R., Ziemann, F., Boulbitch, A. A. *et al.* (1998). Local measurements of viscoelastic parameters of adherent cell surfaces by magnetic bead microrheology. *Biophys J* **75**, 2038–2049.

Bausch, A. R., Moller, W. and Sackmann, E. (1999). Measurement of local viscoelasticity and forces in living cells by magnetic tweezers. *Biophys J* **76**, 573–579.

Bjorck, T., Gustafsson, L. E. and Dahlen, S. E. (1992). Isolated bronchi from asthmatics are hyperresponsive to adenosine, which apparently acts indirectly by liberation of leukotrienes and histamine. *Am Rev Respir Dis* **145**, 1087–1091.

Black, J. L. and Johnson, P. R. (1996). Airway smooth muscle in asthma. *Respirology* **1**, 153–158.

Black, J. L. and Johnson, P. R. (2000). What determines asthma phenotype? Is it the interaction between allergy and the smooth muscle? *Am J Respir Crit Care Med* **161**, S207–S210.

Black, J. L., Johnson, P. R. and Armour, C. L. (2001). Factors controlling transduction signaling and proliferation of airway smooth muscle. *Curr Allergy Asthma Rep* **1**, 116–121.

Brown, R. H. and Mitzner, W. (1998). The myth of maximal airway responsiveness in vivo. *J Appl Physiol* **85**, 2012–2017.

Bryan, S. A., O'Connor, B. J., Matti, S. *et al.* (2000). Effects of recombinant human interleukin-12 on eosinophils, airway hyper-responsiveness, and the late asthmatic response. *Lancet* **356**, 2149–2153.

Bursac, P., Lenormand, G., Fabry, B. *et al.* (2005). Cytoskeletal remodelling and slow dynamics in the living cell. *Nat Mater* **4**, 557–561.

Bursac, P., Fabry, B., Trepat, X. *et al.* (2007). Cytoskeleton dynamics: fluctuations within the network. *Biochem Biophys Res Commun* **355**, 324–330.

Butler, J. P., Tolic-Norrelykke, I. M., Fabry, B. *et al.* (2002). Traction fields, moments, and strain energy that cells exert on their surroundings. *Am J Physiol* **282**, C595–C605.

Chan, V., Burgess, J. K., Ratoff, J. C. *et al.* (2006). Extracellular matrix regulates enhanced eotaxin expression in asthmatic airway smooth muscle cells. *Am J Respir Crit Care Med* **174**, 379–385.

Colebatch, H. J. H. and Mitchell, C. A. (1971). Constriction of isolated living liquid-filled dog and cat lungs with histamine. *J Appl Physiol* **30**, 691–702.

Colebatch, H. J. H., Olsen, C. R. and Nadel, J. A. (1966). Effect of histamine, serotonin, and acetylcholine on the peripheral airways. *J Appl Physiol* **21**, 217–226.

Crick, F. H. C. and Hughes, A. F. W. (1949). The physical properties of cytoplasm: a study by means of the magnetic particle method. *Exp Cell Res* **1**, 37–80.

Crimi, E., Spanevello, A., Neri, M. *et al.* (1998). Dissociation between airway inflammation and airway hyperresponsiveness in allergic asthma. *Am J Respir Crit Care Med* **157**, 4–9.

Dandurand, R. J., Xu, L. J., Martin, J. G. *et al.* (1993a). Airway–parenchymal interdependence and bronchial responsiveness in two highly inbred rat strains. *J Appl Physiol* **74**, 538–544.

Dandurand, R. J., Wang, C. G., Phillips, N. C. *et al.* (1993b). Responsiveness of individual airways to methacholine in adult rat lung explants. *J Appl Physiol* **75**, 364–372.

De Jongste, J. C., Mons, H., Bonta, I. L. *et al.* (1987). In vitro responses of airways from an asthmatic patient. *Eur J Respir Dis* **71**, 23–29.

Deng, L., Trepat, X., Butler, J. P. *et al.* (2006). Fast and slow dynamics of the cytoskeleton. *Nat Mater* **5**, 636–640.

Ding, D. J., Martin, J. G. and Macklem, P. T. (1987). Effects of lung volume on maximal methacholine-induced bronchoconstriction in normal humans. *J Appl Physiol* **62**, 1324–1330.

Dixon, W. E. and Brodie, T. G. (1903). Contributions to the physiology of the lungs. I. The bronchial muscles, their innervation, and the action of drugs upon them. *J Physiol* **29**, 97–173.

Dolhnikoff, M., Morin, J. and Ludwig, M. S. (1998). Human lung parenchyma responds to contractile stimulation. *Am J Respir Crit Care Med* **158**, 1607–1612.

Duguet, A., Biyah, K., Minshall, E. *et al.* (2000). Bronchial responsiveness among inbred mouse strains. Role of airway smooth-muscle shortening velocity. *Am J Respir Crit Care Med* **161**, 839–848.

Dulin, N. O., Fernandes, D. J., Dowell, M. *et al.* (2003). What evidence implicates airway smooth muscle in the cause of BHR? *Clin Rev Allergy Immunol* **24**, 73–84.

Eidelman, D. H., Dimaria, G. U., Bellofiore, S. *et al.* (1991). Strain-related differences in airway smooth muscle and airway responsiveness in the rat. *Am Rev Respir Dis* **144**, 792–796.

Einthoven, W. (1892). Ueber die Wirkung der Bronchialmuskeln, nach einer neuen Methode untersucht, und ueber Asthma nervosum. *Pfluegers Arch* **51**, 367–444.

Fabry, B., Maksym, G. N., Butler, J. P. *et al.* (2001). Scaling the mircrorheology of living cells. *Phys Rev Lett* **87**, 1481021–1481024.

Fan, T., Yang, M., Halayko, A. *et al.* (1997). Airway responsiveness in two inbred strains of mouse disparate in IgE and IL-4 production. *Am J Respir Cell Mol Biol* **17**, 156–163.

Fernandes, D. J., Mitchell, R. W., Lakser, O. *et al.* (2003). Do inflammatory mediators influence the contribution of airway smooth muscle contraction to airway hyperresponsiveness in asthma? *J Appl Physiol* **95**, 844–853.

Fish, J. E., Ankin, M. G., Kelly, J. F. *et al.* (1981). Regulation of bronchomotor tone by lung inflation in asthmatic and nonasthmatic subjects. *J Appl Physiol* **50**, 1079–1086.

Ford, L. E., Seow, C. Y. and Pratusevich, V. R. (1994). Plasticity in smooth muscle, a hypothesis. *Can J Physiol Pharmacol* **72**, 1320–1324.

Fredberg, J. J. (1998). Airway smooth muscle in asthma: flirting with disaster. *Eur Respir J* **12**, 1252–1256.

Fredberg, J. J. (2000a). Airway smooth muscle in asthma. Perturbed equilibria of myosin binding. *Am J Respir Crit Care Med* **161**, S158–S160.

Fredberg, J. J. (2000b). Frozen objects: small airways, big breaths, and asthma. *J Allergy Clin Immunol* **106**, 615–624.

Fredberg, J. J., Bunk, D., Ingenito, E. *et al.* (1993). Tissue resistance and the contractile state of lung parenchyma. *J Appl Physiol* **74**, 1387–1397.

Fredberg, J. J., Inouye, D., Miller, B. *et al.* (1997). Airway smooth muscle, tidal stretches, and dynamically determined contractile states. *Am J Respir Crit Care Med* **156**, 1752–1759.

Fredberg, J. J., Jones, K. A., Nathan, M. *et al.* (1996). Friction in airway smooth muscle: mechanism, latch and implications in asthma. *J Appl Physiol* **81**, 2703–2712.

Fredberg, J. J., Inouye, D. S., Mijailovich, S. M. *et al.* (1999). Perturbed equilibrium of myosin binding in airway smooth muscle and its implications in bronchospasm. *Am J Respir Crit Care Med* **159**, 959–967.

Fredberg, J. J. and Shore, S. A. (1999). The unbearable lightness of breathing. *J Appl Physiol* **86**, 3–4.

Freyer, A. M., Johnson, S. R. and Hall, I. P. (2001). Effects of growth factors and extracellular matrix on survival of human airway smooth muscle cells. *Am J Respir Cell Mol Biol* **25**, 569–576.

Freyer, A. M., Billington, C. K., Penn, R. B. *et al.* (2004). Extracellular matix modulates β_2-adrenergic receptor signaling in human airway smooth muscle cells. *Am J Respir Cell Mol Biol* **31**, 440–445.

Gerthoffer, W. T. and Pohl, J. (1994). Caldesmon and calponin phosphorylation in regulation of smooth muscle contraction. *Can J Physiol Pharmacol* **72**, 1410–1414.

Goldie, R. G., Spina, D., Henry, P. J. *et al.* (1986). In vitro responsiveness of human asthmatic bronchus to carbachol, histamine, beta-adrenoceptor agonists and theophylline. *Br J Clin Pharmacol* **22**, 669–676.

Gump, A., Haughney, L. and Fredberg, J. (2001). Relaxation of activated airway smooth muscle: relative potency of isoproterenol vs. tidal stretch. *J Appl Physiol* **90**, 2306–2310.

Gunst, S. J., Meiss, R. A., Wu, M.-F. *et al.* (1995). Mechanisms for the mechanical placticity of tracheal smooth muscle. *Am J Physiol* **268**, C1267–C1276.

Gunst, S. J. and Wu, M. F. (2001). Plasticity of airway smooth muscle stiffness and extensibility: role of length-adaptive mechanisms. *J Appl Physiol* **90**, 741–749.

Gunst, S. J., Wu, M. F. and Smith, D. D. (1993). Contraction history modulates isotonic shortening velocity in smooth muscle. *Am J Physiol* **265**, C467–C476.

Halayko, A. J., Camoretti-Mercado, B., Forsythe, S. M. *et al.* (1999). Divergent differentiation paths in airway smooth muscle culture: induction of functionally contractile myocytes. *Am J Physiol* **276**, L197–L206.

Halayko, A. J. and Solway, J. (2001). Molecular mechanisms of phenotypic plasticity in smooth muscle cells. *J Appl Physiol* **90**, 358–368.

Hedges, J. C., Yamboliev, I. A., Ngo, M. *et al.* (1998). p38 mitogen-activated protein kinase expression and activation in smooth muscle. *Am J Physiol* **275**, C527–C534.

Hedges, J. C., Dechert, M. A., Yamboliev, I. A. *et al.* (1999). A role for p38(MAPK)/HSP27 pathway in smooth muscle cell migration. *J Biol Chem* **274**, 24211–24219.

Hedges, J. C., Oxhorn, B. C., Carty, M. *et al.* (2000). Phosphorylation of caldesmon by ERK MAP kinases in smooth muscle. *Am J Physiol* **278**, C718–C726.

Hirshman, C. A. and Emala, C. W. (1999). Actin reorganization in airway smooth muscle cells involves Gq and Gi-2 activation of Rho. *Am J Physiol* **277**, L653–L661.

Hirst, S. J., Twort, C. H. C. and Lee, T. H. (2000). Differential effects of extracellular matrix proteins on human airway smooth muscle cell proliferation and phenotype. *Am J Respir Cell Mol Biol* **23**, 335–344.

Hirst, S. J., Martin, J. G., Bonacci, J. V. *et al.* (2004). Proliferative aspects of airway smooth muscle. *J Allergy Clin Immunol* **114**, S2–S17.

Holgate, S. T., Peters-Golden, M., Panettieri, R. A. *et al.* (2003). Roles of cysteinyl leukotrienes in airway inflammation, smooth muscle function, and remodeling. *J Allergy Clin Immunol* **111**, S18–S34.

Holloway, J. W., Beghe, B. and Holgate, S. T. (1999). The genetic basis of atopic asthma. *Clin Exp Allergy* **29**, 1023–1032.

Homer, R. J. and Elias, J. A. (2000). Consequences of long-term inflammation. Airway remodeling. *Clin Chest Med* **21**, 331–343.

Hu, S., Chen, J., Fabry, B. *et al.* (2003). Intracellular stress tomography reveals stress focusing and structural anisotropy in the cytoskeleton of living cells. *Am J Physiol* **285**, C1082–C1090.

Hubmayr, R. D., Shore, S. A., Fredberg, J. J. *et al.* (1996). Pharmacological activation changes stiffness of cultured human airway smooth muscle cells. *Am J Physiol* **271**, C1660–C1668.

Huxley, A. F. (1957). Muscle structure and theories of contraction. *Progr Biophys Biophys Chem* **7**, 255–318.

James, A. L, Paré, P. D. and Hogg, J. C. (1989). The mechanics of airway narrowing in asthma. *Am Rev Respir Dis* **139**, 242–246.

Jia, Y., Xu, L., Heisler, S. *et al.* (1995). Airways of a hyperresponsive rat strain show decreased relaxant response to sodium nitroprusside. *Am J Physiol* **269**, L85–L91.

Jiang, H., Rao, K., Halayko, A. J. *et al.* (1992). Ragweed sensitization-induced increase of myosin light chain kinase content in canine airway smooth muscle. *Am J Respir Cell Mol Biol* **7**, 567–573.

Johnson, P. R., Roth, M., Tamm, M. *et al.* (2001). Airway smooth muscle cell proliferation is increased in asthma. *Am J Respir Crit Care Med* **164**, 474–477.

Kelleher, M. D., Abe, M. K., Chao, T. S. *et al.* (1995). Role of MAP kinase activation in bovine tracheal smooth muscle mitogenesis. *Am J Physiol* **268**, L894–L901.

Kubo, R. (1986). Brownian motion and nonequilibrium statistical mechanics. *Science* **233**, 330–334.

Kuo, K. H., Wang, L., Paré, P. D. *et al.* (2001). Myosin thick filament lability induced by mechanical strain in airway smooth muscle. *J Appl Physiol* **90**, 1811–1816.

Kuo, K. H., Herrera, A. M., Wang, L. *et al.* (2003). Structure–function correlation in airway smooth muscle adapted to different lengths. *Am J Physiol* **285**, C384–C390.

Lakser, O. J., Lindeman, R. P. and Fredberg, J. J. (2002). Inhibition of the p38 MAP kinase pathway destabilizes smooth muscle length during physiological loading. *Am J Physiol* **282**, L1117–L1121.

Lambert, R. K. and Paré, P. D. (1997). Lung parenchymal shear modulus, airway wall remodeling, and bronchial hyperresponsiveness. *J Appl Physiol* **83**, 140–147.

Lambert, R. K., Wiggs, B. R., Kuwano, K. *et al.* (1993). Functional significance of increased airway smooth muscle in asthma and COPD. *J Appl Physiol* **74**, 2771–2781.

Laudadio, R. E., Millet, E., Fabry, B. *et al.* (2005). Rat airway smooth muscle cell during actin modulation: rheology and glassy dynamics. *Am J Physiol* **289**, C1388–C1395.

Lauzon, A. M., Tyska, M. J., Rovner, A. S. *et al.* (1998). A 7-amino-acid insert in the heavy chain nucleotide binding loop alters the kinetics of smooth muscle myosin in the laser trap. *J Muscle Res Cell Motil* **19**, 825–837.

Lazaar, A. L. and Panettieri, R. A. (2005). Airway smooth muscle: a modulator of airway remodeling in asthma. *J Allergy Clin Immunol* **116**, 488–495.

Leckie, M. J., ten Brinke, A. and Khan, J, (2000). Effects of an interleukin-5 blocking monoclonal antibody on eosinophils, airway hyper-responsiveness, and the late asthmatic response. *Lancet* **356**, 2144–2148.

Lim, T. K., Pride, N. B. and Ingram, R. H. (1987). Effects of volume history during spontaneous and acutely induced air-flow obstruction in asthma. *Am Rev Respir Dis* **135**, 591–596.

Ludwig, M. S., Dreshaj, I., Solway, J. *et al.* (1987). Partitioning of pulmonary resistance during constriction in the dog: effects of volume history. *J Appl Physiol* **62**, 807–815.

Ludwig, M., Shore, S., Fredberg, J. J. *et al.* (1988). Differential responses of tissue viscance and collateral resistance to histamine and leukotriene C_4. *J Appl Physiol* **65**, 1424–1429.

Ma, X., Cheng, Z., Kong, H. *et al.* (2002). Changes in biophysical and biochemical properties of single bronchial smooth muscle cells from asthmatic subjects. *Am J Physiol* **283**, L1181–L1189.

Madison, J. M. (2003). Migration of airway smooth muscle cells. *Am J Respir Cell Mol Biol* **29**, 8–11.

Macklem, P. T. (1987). Bronchial hyperresponsiveness. *Chest* **91**, 189S–191S.

Macklem, P. T. (1989). Mechanical factors determining maximum bronchoconstriction. *Eur Respir J* **2**, 516S–519S.

Macklem, P. T. (1990). A hypothesis linking bronchial hyperreactivity and airway inflammation: implications for therapy. *Ann Allergy* **64**, 113–116.

Macklem, P. T. (1996). A theoretical analysis of the effect of airway smooth muscle load on airway narrowing. *Am J Respir Crit Care Med* **153**, 83–89.

Maksym, G. N., Fabry, B., Butler, J. P. *et al.* (2000). Mechanical properties of cultured human airway smooth muscle cells from 0.05 to 0.4 Hz. *J Appl Physiol* **89**, 1619–1632.

Matthews, B. D., Overby, D. R., Alenghat, F. J. *et al.* (2004). Mechanical properties of individual focal adhesions probed with a magnetic microneedle. *Biochem Biophys Res Commun* **313**, 758–764.

McParland, B. E., Macklem, P. T. and Paré, P. D. (2003). Airway wall remodeling: friend or foe? *J Appl Physiol* **95**, 426–434.

Mead, J. (1973). Respiration: pulmonary mechanics. *Annu Rev Physiol* **35**, 169–192.

Mehta, D., Wu, M.-F. and Gunst, S. J. (1996). Role of contractile protein activation in the length-dependent modulation of tracheal smooth muscle force. *Am J Physiol* **270**, C243–C252.

Mehta, D. and Gunst, S. J. (1999). Actin polymerization stimulated by contractile activation regulates force development in canine tracheal smooth muscle. *J Physiol* **519**, 829–840.

Mehta, D., Tang, D. D., Wu, M. F. *et al.* (2000). Role of Rho in Ca^{2+}-insensitive contraction and paxillin tyrosine phosphorylation in smooth muscle. *Am J Physiol* **279**, C308–C318.

Meiss, R. A. (1999). Influence of intercellular tissue connections on airway muscle mechanics. *J Appl Physiol* **86**, 3–4.

Mijailovich, S. M. (2003). Dynamics of airway closure: critical smooth muscle activation in normals and asthmatics. *Am J Respir Crit Care Med* **167**, A183.

Mijailovich, S. M., Butler, J. P. and Fredberg, J. J. (2000). Perturbed equilibria of myosin binding in airway smooth muscle: bond-length distributions, mechanics, and ATP metabolism. *Biophys J* **79**, 2667–2681.

Moore, B. J., King, G. G., D'Yachkova, Y. *et al.* (1998). Mechanism of methacholine dose-response plateaus in normal subjects. *Am J Respir Crit Care Med* **158**, 666–669.

Moore, B. J., Verburgt, L. M., King, G. G. *et al.* (1997). Effect of deep inspiration on methacholine dose-response curves in normal subjects. *Am J Respir Crit Care Med* **156**, 1278–1281.

Moreno, R., Hogg, J. C. and Paré, P. D. (1986). Mechanics of airway narrowing. *Am Rev Respir Dis* **133**, 1171–1180.

Murphy, R. A. (1988). Muscle cells of hollow organs. *News Physiol Sci* **3**, 124–128.

Murphy, R. A. (1994). What is special about smooth muscle? The significance of covalent crossbridge regulation. *FASEB J* **8**, 311–318.

Murphy, R. A., Walker, J. S. and Strauss, J. D. (1997). Myosin isoforms and functional diversity in vertebrate smooth muscle. *Comp Biochem Physiol* **117B**, 51–60.

Nadel, J. A. and Tierney, D. F. (1961). Effect of a previous deep inspiration on airway resistance in man. *J Appl Physiol* **16**, 717–719.

Nagase, T., Moretto, A. and Ludwig, M. S. (1994). Airway and tissue behavior during induced constriction in rats: intravenous vs. aerosol administration. *J Appl Physiol* **76**, 830–838.

Naghshin, J., Wang, L., Paré, P. D. *et al.* (2003). Adaptation to chronic length change in explanted airway smooth muscle. *J Appl Physiol* **95**, 448–453.

Nguyen, T. T. B., Nguyen, J., Ward, P. T. *et al.* (2005). β1-integrins mediate enhancement of airway smooth muscle proliferation by collagen and fibronectin. *Am J Respir Crit Care Med* **171**, 217–223.

Oldmixon, E. H., Carlson, K., Kuhn, C. *et al.* (2001). α-Actin: disposition, quantities, and estimated effects on lung recoil and compliance. *J Appl Physiol* **91**, 459–473.

Otis, A. B. (1983). A perspective of respiratory mechanics. *J Appl Physiol* **54**, 1183–1187.

Panettieri, R. A., Murray, R. K., DePalo, L. R. *et al.* (1989). A human airway smooth muscle cell line that retains physiological responsiveness. *Am J Physiol* **256**, C329–C335.

Parameswaran, K., Radford, K., Zuo, J. *et al.* (2004). Extracellular matrix regulates human airway smooth muscle cell migration. *Eur Respir J* **24**, 545–551.

Paré, P. D., Wiggs, B. R., James, A. *et al.* (1991). The comparative mechanics and morphology of airways in asthma and in chronic obstructive pulmonary disease. *Am Rev Respir Dis* **143**, 1189–1193.

Parker, K. K., Brock, A. L., Brangwynne, C. *et al.* (2002). Directional control of lamellipodia extension by constraining cell shape and orienting cell tractional forces. *FASEB J* **16**, 1195–1204.

Peng, Q., Lai, D., Nguyen, T. T. B. *et al.* (2005). Multiple β1 integrins mediate enhancement of human airway smooth muscle cytokine secretion by fibronectin and type I collagen. *J Immunol* **174**, 2258–2264.

Pratusevich, V. R., Seow, C. Y. and Ford, L. E. (1995). Plasticity in canine airway smooth muscle. *J Gen Physiol* **105**, 73–94.

Puig-de-Morales, M., Millet, E., Fabry, B. *et al.* (2004). Cytoskeletal mechanics in adherent human airway smooth muscle cells: probe specificity and scaling of protein–protein dynamics. *Am J Physiol* **287**, C643–C654.

Qi, D., Mitchell, R. W., Burdyga, T. *et al.* (2002). Myosin light chain phosphorylation facilitates in vivo myosin filament reassembly after mechanical perturbation. *Am J Physiol* **282**, C1298–C1305.

Robatto, F. M., Simard, S., Orana, H. *et al.* (1992). Effect of lung volume on plateau response of airways and tissue to methacholine in dogs. *J Appl Physiol* **73**, 1908–1913.

Romero, P. and Ludwig, M. S. (1991). Maximal methacholine-induced constriction in rabbit lungs: interactions between airways and tissues? *J Appl Physiol* **70**, 1044–1050.

Salter, H. H. (1868). *On Asthma: Its Pathology and Treatment*, pp. 24-60. John Churchill & Sons, Science Press Limited: London.

Seow, C. Y. and Fredberg, J. J. (2001). Historical perspective on airway smooth muscle: the saga of a frustrated cell. *J Appl Physiol* **91**, 938–952.

Seow, C. Y. and Stephens, N. L. (1988). Velocity–length–time relations in canine tracheal smooth muscle. *J Appl Physiol* **54**, 2053–2057.

Seow, C. Y., Pratusevich, V. R. and Ford, L. E. (2000). Series-to-parallel transition in the filament lattice of airway smooth muscle. *J Appl Physiol* **89**, 869–876.

Shen, X., Gunst, S. J. and Tepper, R. S. (1997). Effect of tidal volume and frequency on airway responsiveness in mechanically ventilated rabbits. *J Appl Physiol* **83**, 1202–1208.

Shore, S. A., Laporte, J., Hall, I. *et al.* (1997). Effect of IL-1β on responses of cultured human airway smooth muscle cells to bronchodilator agonists. *Am J Respir Cell Mol Biol* **16**, 702–712.

Silveira, P. S. P. and Fredberg, J. J. (2005). Smooth muscle length adaptation and actin filament length: a network model of the cytoskeletal dysregulation. *Can J Physiol Pharmacol* **83**, 923–931.

Skloot, G., Permutt, S. and Togias, A. (1995). Airway hyperresponsiveness in asthma: a problem of limited smooth muscle relaxation with inspiration. *J Clin Invest* **96**, 2393–2403.

Small, J. V. (1995). Structure–function relationships in smooth muscle: the missing links. *Bioessays* **17**, 785–792.

Small, J. V. and Gimona, M. (1998). The cytoskeleton of the vertebrate smooth muscle cell. *Acta Physiol Scand* **164**, 341–348.

Smith, B. A., Tolloczko, B., Martin, J. G. *et al.* (2005). Probing the viscoelastic behavior of cultured airway smooth muscle cells with atomic force microscopy: stiffening induced by contractile agonist. *Biophys J* **88**, 2994–3007.

Solway, J. and Fredberg, J. J. (1997). Perhaps airway smooth muscle dysfunction does contribute to bronchial hyperresponsiveness after all. *Am J Respir Cell Mol Biol* **17**, 144–146.

Stephens, N. L. (1987). Airway smooth muscle. *Am Rev Respir Dis* **135**, 960–975.

Stephens, N. L. and Kromer, U. (1971). Series elastic component of tracheal smooth muscle. *Am J Physiol* **220**, 1890–1895.

Stephens, N. L. and Seow, C. Y. (1993). Airway smooth muscle: physiology, bronchomotor tone, pharmacology, and relation to asthma. In *Bronchial Asthma,* pp. 314–332, E. B. Weiss and M. Stein (eds). Little, Brown: Boston.

Sterk, P. J. and Bel, E. H. (1989). Bronchial hyperresponsiveness: the need to distinguish between hypersensitivity and excessive airway narrowing. *Eur Respir J* **2**, 267–274.

Tao, F. C., Tolloczko, B., Eidelman, D. H. *et al.* (1999). Enhanced Ca^{2+} mobilization in airway smooth muscle contributes to airway hyperresponsiveness in an inbred strain of rat. *Am J Respir Crit Care Med* **160**, 446–453.

Tao, F. C., Shah, S., Pradhan, A. A. *et al.* (2003). Enhanced calcium signaling to bradykinin in airway smooth muscle from hyperresponsive inbred rats. *Am J Physiol* **284**, L90–L99.

Thomson, R. J., Bramley, A. M. and Schellenberg, R. R. (1996). Airway muscle stereology: implications for increased shortening in asthma. *Am J Respir Crit Care Med* **154**, 749–757.

Tolic-Norrelykke, I. M., Butler, J. P., Chen, J. *et al.* (2002). Spatial and temporal traction response in human airway smooth muscle cells. *Am J Physiol* **283**, C1254–C1266.

Tolloczko, B., Jia, Y. L. and Martin, J. G. (1995). Serotonin-evoked calcium transients in airway smooth muscle cells. *Am J Physiol* **269**, L234–L240.

Tonino, P., Simon, M. and Craig, R. (2002). Mass determination of native smooth muscle myosin filaments by scanning transmission electron microscopy. *J Mol Biol* **318**, 999–1007.

Tran, T., McNeill, K. D., Gerthoffer, W. T. *et al.* (2006). Endogenous laminin is required for human airway smooth muscle cell maturation. *Respir Res* **7**, 117.

Trepat, X., Deng, L., An, S.S. *et al.*, (2007). Universal physical responses to stretch in the living cell. *Nature* **447**, 592–595.

Tschumperlin, D. J., Shively, J. D., Swartz, M. A. *et al.* (2002). Bronchial epithelial compression regulates MAP kinase signaling and HB-EGF-like growth factor expression. *Am J Physiol* **282**, L904–L911.

Tschumperlin, D. J., Shively, J. D., Kikuchi, T. *et al.* (2003). Mechanical stress triggers selective release of fibrotic mediators from bronchial epithelium. *Am J Respir Cell Mol Biol* **28**, 142–149.

Uvelius, B. (1976). Isometric and isotonic length–tension relations and variations in cell length in longitudinal smooth muscle from rabbit urinary bladder. *Acta Physiol Scand* **97**, 1–12.

Valberg, P. A. (1984). Magnetometry of ingested particles in pulmonary macrophages. *Science* **224**, 513–516.

Valberg, P. A. and Feldman, H. A. (1987). Magnetic particle motions within living cells. *Biophys J* **52**, 551–561.

Van Citters, K. M., Hoffman, B. D., Massiera, G. *et al.* (2006). The role of F-actin and myosin in epithelial cell rheology. *Biophys J* **91**, 3946–3956.

Wang, C. G., Almirall, J. J., Dolman, C. S. *et al.* (1997). In vitro bronchial responsiveness in two highly inbred rat strains. *J Appl Physiol* **82**, 1445–1452.

Wang, L., McParland, B. E. and Paré, P. D. (2003). The functional consequences of structural changes in the airways: implications for airway hyperresponsiveness in asthma. *Chest* **123**, 356S–362S.

Wang, L., Paré, P. D. and Seow, C. Y. (2001). Effect of chronic passive length change on airway smooth muscle length–tension relationship. *J Appl Physiol* **90**, 734–740.

Wang, N., Butler, J. P. and Ingber, D. E. (1993). Mechanotransduction across the cell surface and through the cytoskeleton. *Science* **260**, 1124–1127.

Wang, N., Tolic-Norrelykke, I. M., Chen, J. *et al.* (2002). Cell prestress. I. Stiffness and prestress are closely associated in adherent contractile cells. *Am J Physiol* **282**, C606–C616.

Wang, P. and Bitar, K. N. (1998). Rho A regulates sustained smooth muscle contraction through cytoskeletal reorganization of HSP27. *Am J Physiol* **275**, G1454–G1462.

Warner, D. O. and Gunst, S. J. (1992). Limitation of maximal bronchoconstriction in living dogs. *Am Rev Respir Dis* **145**, 553–560.

Warshaw, D. M., Rees, D. D. and Fay, F. S. (1988). Characterization of cross-bridge elasticity and kinetics of cross-bridge cycling force development in single smooth muscle cells. *J Gen Physiol* **91**, 761–779.

Watson, J. D. and Crick, F. H. C. (1953a). Molecular structure of nucleic acids. *Nature* **171**, 737–738.

Watson, J. D. and Crick, F. H. C. (1953b). Genetic implications of the structure of deoxyribonucleic acid. *Nature* **171**, 964–967.

Whicker, S. D., Armour, C. L. and Black, J. L. (1988). Responsiveness of bronchial smooth muscle from asthmatic patients to relaxant and contractile agonists. *Pulm Pharmacol* **1**, 25–31.

Wiggs, B. R., Bosken, C., Paré, P. D. *et al.* (1992). A model of airway narrowing in asthma and in chronic obstructive pulmonary disease. *Am Rev Respir Dis* **145**, 1251–1258.

Wiggs, B. R., Hrousis, C. A., Drazen, J. M. *et al.* (1997). On the mechanism of mucosal folding in normal and asthmatic airways. *J Appl Physiol* **83**, 1814–1821.

Woodruff, P. G., Dolganov, G. M., Ferrando, R. E. *et al.* (2004). Hyperplasia of smooth muscle in mild to moderate asthma without changes in cell size or gene expression. *Am J Respir Crit Care Med* **169**, 1001–1006.

Woolcock, A. J. and Peat, J. K. (1989). Epidemiology of bronchial hyperresponsiveness. *Clin Rev Allergy* **7**, 245–256.

Xu, J. Q., Harder, B. A., Uman, P. *et al.* (1996). Myosin filament structure in vertebrate smooth muscle. *J Cell Biol* **134**, 53–66.

Yamada, S., Wirtz, D. and Kuo, S. C. (2000). Mechanics of living cells measured by laser tracking microrheology. *Biophys J* **78**, 1736–1747.

Yager, D., Butler, J. P., Bastacky, J. *et al.* (1989). Amplification of airway constriction due to liquid filling of airway interstices. *J Appl Physiol* **66**, 2873–2884.

Yamboliev, I. A., Hedges, J. C., Mutnick, J. L. *et al.* (2000). Evidence for modulation of smooth muscle force by the p38 MAP kinase/HSP27 pathway. *Am J Physiol* **278**, H1899–H1907.

Youn, T., Kim, S. A. and Hai, C. M. (1998). Length-dependent modulation of smooth muscle activation: effects of agonist, cytochalasin, and temperature. *Am J Physiol* **274**, C1601–C1607.

Zacour, M. E. and Martin, J. G. (1996). Enhanced growth response of airway smooth muscle in inbred rats with airway hyperresponsiveness. *Am J Respir Cell Mol Biol* **15**, 590–599.

Zhu, Z., Lee, C. G., Zheng, T. *et al.* (2001). Airway inflammation and remodeling in asthma. Lessons from interleukin 11 and interleukin 13 transgenic mice. *Am J Respir Crit Care Med* **164**, S67–S70.

2

Dynamics of cytoskeletal and contractile protein organization: an emerging paradigm for airway smooth muscle contraction

Wenwu Zhang and Susan J. Gunst

Department of Cellular and Integrative Physiology, Indiana University School of Medicine, Indianapolis, IN, USA

2.1 Introduction

The airways undergo large changes in shape and volume during breathing, and airway smooth muscle must rapidly adapt its compliance and contractility to accommodate external mechanical forces. The ability of airway smooth muscle to alter its properties in response to mechanical events has been referred to as 'mechanical plasticity' or 'length adaptation' (Bai *et al.*, 2004). Oscillations in the length or load of isolated airway smooth muscle tissues reduces their stiffness and contractility, and the stretch of airway smooth muscle decreases its subsequent responsiveness (Fredberg *et al.*, 1997; Gunst, 1983; 1986; Gunst *et al.*, 1990; 1993;

Airway Smooth Muscle   Edited by Kian Fan Chung
© 2008 John Wiley & Sons, Ltd

Gunst and Fredberg, 2003; Gunst and Wu, 2001; Shen *et al.*, 1997b; Wang *et al.*, 2000). These properties are believed to be particularly important for the normal physiological regulation of airway smooth muscle tone during breathing. Tidal breathing reduces airway responsiveness *in vivo*, and periodic deep inspirations depress airway responsiveness (Gunst *et al.*, 2001a; King *et al.*, 1999; Salerno *et al.*, 1999; Shen *et al.*, 1997a; Skloot *et al.*, 1995; Tepper *et al.*, 1995; Warner and Gunst, 1992). The ability to adapt to mechanical events may enable the airway muscle to undergo large length changes without losing its contractile function, and to retain a high level of compliance over large ranges of length and lung volume (Fredberg *et al.*, 1997; Gunst, 1986; Gunst *et al.*, 1995; 2003; Gunst and Tang, 2000; Gunst and Wu, 2001; Pratusevich *et al.*, 1995). The adaptive property of airway smooth muscle may also be critical for the maintenance of normal low levels of airway reactivity during breathing (Fish *et al.*, 1981; Gunst *et al.*, 2001b; Kapsali *et al.*, 2000; Malmberg *et al.*, 1993; Nadel and Tierney, 1961; Skloot *et al.*, 1995; Tepper *et al.*, 1995; Warner and Gunst, 1992). Aberrations in the adaptive properties of airway smooth muscle have been proposed as a basis for the pathophysiology of asthma (Brusasco *et al.*, 1999; Fish *et al.*, 1981; Gunst and Tang, 2000; Skloot *et al.*, 1995).

The contractile filament system, consisting of the thin filaments (primarily actin) and the thick filaments (primarily myosin), has long been understood to be the engine of shortening and tension generation in both smooth and striated muscles. Decades of research have been directed at characterizing the signalling pathways and processes by which diverse physiological stimuli regulate the activation of actomyosin cross-bridge cycling and the mechanical properties of smooth muscle (Morgan and Gangopadhyay, 2001; Pfitzer *et al.*, 2001; Somlyo and Somlyo, 2003). However, investigation of the mechanisms that underlie the properties of mechanical adaptation and 'plasticity' of airway smooth muscle has led to a significant transformation in our views of the molecular events involved in airway smooth muscle contraction. It has become increasingly clear in recent years that contractile activation of the smooth muscle cell comprises processes far more extensive than thick and thin filament activation and the resulting actomyosin interaction (Gerthoffer and Gunst, 2001; Gunst, 1999; Gunst *et al.*, 2003; Gunst and Fredberg, 2003). Accumulating evidence indicates that smooth muscle contraction involves a broad process of cytoskeletal reorganization that includes the polymerization and remodelling of cytoskeletal filament systems as well as the reorganization and fortification of membrane and cytosolic anchoring structures for cytoskeletal filaments. These cytoskeletal processes have been shown to be essential for force development and tension maintenance in response to a contractile stimulus. They may also enable the muscle cell to adapt its structure and organization to accommodate shape changes initiated by activation of the cross-bridge cycling or by forces imposed on the tissue by external physiological events. Some of these cytoskeletal processes also appear to be critical for activation of the actomyosin system.

This chapter will present an emerging paradigm for the regulation of airway smooth muscle function in which activation of the actomyosin cross-bridge cycling system is integrated with dynamic cytoskeletal processes that act to adapt and organize the structure of the smooth muscle cell to accommodate changes in cell shape imposed by external and internal physiological forces. An overview of the current state of knowledge regarding the structure and organization of the airway smooth muscle contractile/cytoskeletal system and the molecular and cellular transitions that occur within it during the stimulus-induced contractile activation of an airway smooth muscle cell will be presented.

2.2 Molecular structure and organization of contractile and cytoskeletal filaments in the airway smooth muscle cell

Overview of filament organization in the smooth muscle cell

The 'contractile filaments' of smooth muscle cells consist of the 'thick' filaments, which are 12–15 nm in diameter and are composed primarily of myosin, and the 'thin filaments', $\sim$7 nm in diameter, consisting of actin and a number of proteins that bind to actin filaments. In smooth muscle cells, the thin filaments surround the thick filaments to form a 'rosette' structure, and cross-bridge heads can be seen extending from the thick filaments to the thin filaments under high-powered electron microscopy (Craig and Megerman, 1977; Xu et al., 1996). A substantial molar excess of actin relative to myosin exists in all smooth muscle tissues, ranging from as low as 8:1 in chicken gizzard (Nonomura, 1976), to approximately 15:1 in vascular muscle (Somlyo et al., 1973), and to as high as 50:1 in isolated amphibian visceral muscle (Cooke et al., 1987). A subset of filamentous actin exists that does not associate with myosin filaments (Small, 1995). There is evidence from electron micrographic studies that the activation of airway smooth muscle may modulate the ratio of thin to thick filaments (Kuo et al., 2003).

Actin filaments, both within and outside the actomyosin system, anchor at membrane adhesion sites which contain transmembrane integrin proteins that connect them to the extracellular matrix (Bond and Somlyo, 1982; Geiger *et al.*, 1981; Kuo and Seow, 2004). Force generated by the contractile apparatus can be transmitted from the cytoskeletal/contractile system to the extracellular matrix through these junctions, and, conversely, external stresses that are imposed on smooth muscle cells can be transmitted to the cytoskeleton. The molecular organization of these adhesion complexes is believed to be structurally analogous to the focal adhesions of cultured cells, and includes dozens of structural and signalling proteins (Burridge and Chrzanowska-Wodnicka, 1996). Among the proteins that anchor actin filaments to integrins at adhesion complexes are talin and α-actinin, both of which cross-link actin filaments and bind them to integrin proteins (Draeger *et al.*, 1989; Fay *et al.*, 1983). Proteins within this complex also mediate signalling pathways that regulate the organization and activation of cytoskeletal and contractile proteins (Critchley, 2000; DeMali *et al.*, 2003; Gerthoffer and Gunst, 2001; Gunst *et al.*, 2003; Tang *et al.*, 1999; 2003; Tang and Gunst, 2001a; Turner, 2000; Zhang *et al.*, 2005; Zhang and Gunst, 2006).

In the cytoplasm, actin filaments are anchored at α-actinin-rich areas that have been referred to as cytosolic dense bodies (Fay *et al.*, 1983; Fay and Delise, 1973; Kuo and Seow, 2004). Analysis of the movement of dense bodies in contracting isolated smooth muscle cells suggests that they provide mechanical coupling between actin filaments within the cytoplasm (Draeger *et al.*, 1990; Fay *et al.*, 1983; Kargacin *et al.*, 1989). Calponin and desmin have also been found to be associated with cytosolic dense bodies (Mabuchi *et al.*, 1997); however, there is no evidence that these structures provide a locus for signal transduction.

Structure and function of actin

Filamentous actin (F-actin), the backbone of the thin filament, is a polymeric protein composed of asymmetric, bi-lobed, 42-kDa monomers (Holmes *et al.*, 1990; Milligan *et al.*, 1990). Tropomyosin, caldesmon, and calponin bind to filamentous actin with regular repeats, although different subsets of actin filaments differ in their binding to these proteins (Furst *et al.*, 1986; Lehman *et al.*, 1987; Makuch *et al.*, 1991; Rosol *et al.*, 2000; Vibert *et al.*, 1993; Winder and Walsh, 1993; Xu *et al.*, 1999). Caldesmon and calponin have been widely studied in relationship to their possible role in the regulation of actomyosin ATPase activity (Morgan and Gangopadhyay, 2001; Wang, 2001; Winder and Walsh, 1993). Many other actin-binding proteins associate with filamentous actin in smooth muscle and may function to regulate the remodelling or stabilization of actin filaments, their compartmentalization within the cell, or the interaction of actin with other proteins (Carlier and Pantaloni, 1997; Lehman et al., 1996).

Four different isoforms of actin have been identified in smooth muscle tissues: α- and γ- 'contractile' actin, and β- and γ- 'cytoskeleton' actin (Herman, 1993; Kabsch and Vandekerckhove, 1992; Wong et al., 1998). All of these isoforms have also been identified in airway smooth muscle cells and tissues (Wong et al., 1998). The α- and γ- 'contractile' actin isoforms have been considered phenotypic of smooth muscle cells, whereas β- and γ- 'cytoskeletal' actin are expressed ubiquitously among nonmuscle cells; however, subsequent studies have also found α-actin in many nonmuscle cells (Khaitlina, 2001; Wang et al., 2006). It is now recognized that, with the possible exception of smooth muscle myosin heavy chain, no single muscle protein isoform is uniquely expressed in smooth muscle (Owens et al., 2004). The idea of functional differentiation among different actin isoforms was strengthened by studies of smooth muscle cytoskeletal organization by Small and colleagues, who observed distinct subsets of actin filaments localized to different physical domains within the smooth muscle cell (Small, 1995). They reported 'cytoskeletal' β-actin to be selectively localized to the cytoplasmic dense bodies, the membrane adhesion sites, and the longitudinal channels linking consecutive dense bodies, which are also occupied by filamin and desmin. 'Contractile' (α- or γ-) actin was associated with myosin filaments in complementary positions to those occupied by 'cytoskeletal' actin (North et al., 1994). Furthermore, Small and colleagues showed that the actin-binding protein caldesmon was preferentially associated with 'contractile' actin, whereas the actin-binding protein calponin preferentially associated with 'cytoskeletal' actin, suggesting that different actin-binding proteins may compartmentalize various subsets of actin filaments to different subdomains or diverse functions within the cell (North et al., 1994). This idea has been supported by biochemical evidence that subsets of actin filaments are selectively enriched with either calponin and filamin or caldesmon (Lehman, 1991). However, this paradigm has been challenged by studies showing that the distribution of actin isoforms within thin filaments isolated from stomach smooth muscle tissue is random, and that there is no significant clustering among actin isoforms within or between filaments (Drew et al., 1991). The functional distinction between the various isoforms of actin remains unsettled, and there is still no conclusive evidence as to whether the isoforms of actin localize to different regions of the cell, or whether they subserve different cellular functions.

As in most cell types, the polymeric state of actin in smooth muscle is dynamic. Actin polymerizes at the filament ends and subunit exchange occurs even after a steady state is achieved in which net polymerization ceases (Pollard et al., 2000).

Actin filament lengths are precisely determined and maintained by capping proteins such as tropomodulin and CapZ (Pollard and Cooper, 1986; Schafer *et al.*, 1995). Agonists can regulate the polymer state through the activation and inhibition of proteins that nucleate, cap, or sever actin filaments (Zigmond, 1996). The modulation of the actin cytoskeletal network structure has been proposed as a mechanism for modulating the cytoskeletal structure of airway smooth muscle to enable it to adapt to changes in its physical environment (Gerthoffer and Gunst, 2001; Gunst *et al.*, 1995; 2003; Gunst and Wu, 2001; Herrera *et al.*, 2004; Mehta and Gunst, 1999; Tang *et al.*, 1999).

Structure and function of myosin

Smooth muscle myosin, the primary constituent of the thick filaments of the contractile system, is a large, asymmetric protein (molecular mass of $\sim$520 kDa) that is made up of six polypeptide chains: two $\sim$205-kDa heavy chains that form a dimer, and two pairs of light chains, the 20-kDa 'regulatory' light chains and the 17-kDa 'essential' light chains (Warrick and Spudich, 1987). The dimeric myosin heavy chain makes up the main body of the molecule: each heavy chain contains a slightly elongated globular head at the amino terminus that connects to a long α-helical coiled tail of $\sim$120 kDa. Myosin heavy chains polymerize to form the rod-like backbone of the thick filament. Each myosin globular head contains the functional motor domains of the molecule that include the nucleotide- and actin-binding regions. A single essential and regulatory light chain is associated with each myosin head. The light chains are localized along an α-helical segment of the heavy chain at the junction of the globular head and the rod-like backbone referred to as the 'neck' region (Rayment *et al.*, 1993).

Myosin generates force and/or motion by mechanical cycles, during which the myosin head repetitively attaches to actin, undergoes a conformational change that results in a power stroke, and then detaches itself (Guilford and Warshaw, 1998; Hartshorne, 1987; Rayment *et al.*, 1993; 1996). The energy required for mechanical power is generated by the enzymatic hydrolysis of adenosine triphosphate (ATP) by the globular myosin head. In smooth muscle, the phosphorylation of serine-19 and threonine-18 in the N-terminus of the regulatory light chain acts as a switch for turning on actin-activated myosin ATPase activity and is a prerequisite for the initiation of cross-bridge cycling and force development or active shortening (Murphy, 1989; 1994; Warshaw *et al.*, 1990). The degree of myosin light chain phosphorylation correlates with the rate of myosin ATPase activity and the rate of smooth muscle shortening (Murphy, 1994; Warshaw *et al.*, 1990), and thus the stimulus-specific regulation of the degree of myosin light chain phosphorylation provides a mechanism for the regulation of force and shortening velocity in airway smooth muscle.

Phosphorylation of the regulatory light chains of smooth muscle can be regulated by Ca^{2+}-calmodulin-mediated activation of myosin light chain kinase, as well as by several Ca^{2+}-independent signalling pathways, many of which act through the regulation of myosin light chain phosphatase (Somlyo and Somlyo, 2000; 2003). The physiological importance of Ca^{2+}-independent pathways in the regulation of myosin light chain phosphorylation and cross-bridge cycling has been established as an important mechanism for the regulation of contraction in vascular smooth muscle (Somlyo and Somlyo, 2000; 2003). Although there is also evidence for these pathways in airway smooth muscle; the Ca^{2+}-sensitization of myosin light chain phosphorylation may not be the predominant mechanism for the Ca^{2+}-sensitization of contraction in airway muscle (Sanderson et al., 2008; McFawn et al., 2003; Yoshimura et al., 2001).

Four smooth muscle myosin heavy chain isoforms are generated by alternative mRNA splicing of a single gene. These isoforms differ both at the carboxyl terminus (SM1 and SM2 isoforms) and at the amino terminus (SM-A and SM-B isoforms) (Babu et al., 2000). Two of these isoforms differ by the presence (+) or absence (−) of a seven-amino-acid insert in the motor domain (White et al., 1993). The presence of a seven-amino-acid insert in the motor domain of the myosin heavy chain molecule is associated with a faster rate of actin propulsion in the in vitro motility assay (Rovner et al., 1997). There are also two isoforms of the essential light chain of myosin (LC17), and differences in the expression of myosin LC17 isoform expression have also been associated with differences in shortening velocity in some smooth muscle tissues (Malmqvist and Arner, 1991). The smooth muscle myosin heavy chain isoforms are differentially expressed during smooth muscle development in different smooth muscle cell types (Babu et al., 2000; Eddinger and Murphy, 1991; Eddinger and Wolf, 1993). The expression of the head domain insert and LC17 isoforms has been shown to be tissue-specific (Babij, 1993).

Pathophysiological changes in contractile proteins associated with airway hyperresponsiveness

An increase in airway smooth muscle shortening velocity or force generation would be predicted to lead to greater airway smooth muscle shortening and increase airway narrowing. Therefore, possible alterations in the function or regulation of the actomyosin system, long viewed as the primary regulator of force generation and shortening velocity in airway smooth muscle, have been the subject of considerable research as possible contributing causes of the airway hyperresponsiveness associated with asthma. There is evidence that alterations both in myosin isoforms and in the expression and activation of proteins that regulate the myosin activation pathway can lead to airway hyperresponsiveness. An increased expression of the (+) insert myosin heavy chain isoform in Fisher rats is associated with increased airway responsiveness relative to Lewis rats (Gil *et al.*, 2006). Cells taken from asthmatic patients have been shown to exhibit an increase in the expression of myosin light chain kinase, which catalyses myosin light chain phosphorylation (Ma *et al.*, 2002). Increased levels of myosin LC20 phosphorylation and increased expression of myosin light chain kinase have also been shown in airway smooth muscle tissues from a canine model of allergic asthma (Jiang *et al.*, 1995). A sizable body of evidence suggests that alterations in the isoforms and expression levels of contractile proteins and their activating molecules may contribute to increased airway contractility and airway hyperresponsiveness.

2.3 Cytoskeletal dynamics and airway smooth muscle contraction

A new paradigm for smooth muscle contraction

The actomyosin cross-bridge interaction and associated regulatory processes provide a mechanism that can account for tension development and active shortening of the smooth muscle cell, but cross-bridge interactions cannot account for the malleability of airway smooth muscle and its ability to adapt to environmental influences. The ability of airway smooth muscle to alter its stiffness and contractility in response to mechanical oscillation and stretch have been widely observed and described, and this ability is considered to be critically important for the regulation of normal airway responsiveness (Fredberg *et al.*, 1997; Gunst *et al.*, 2003; Skloot *et al.*, 1995). Such adaptive properties have been proposed to result from cytoskeletal processes outside the actomyosin interaction (Gerthoffer and Gunst, 2001; Gunst, 2002; Gunst *et al.*, 2003; Gunst and Fredberg, 2003). Thus, it is important to consider the actomyosin interaction and cross-bridge cycling as a component of a complex and integrated series of cytoskeletal events that occur during the contraction of the airway smooth muscle cell. These concerted events not only result in tension generation and shortening of the smooth muscle cell but also modulate the cell's shape and stiffness, and can thereby affect airway compliance and responsiveness.

A current mechanistic paradigm for cytoskeletal processes that transpire during airway muscle contraction entails a dynamic process in which the response of the smooth muscle cell to changes in shape imposed by external physiological forces is mediated by local mechanotransduction events that serve to catalyse actin polymerization and cytoskeletal remodelling (Gerthoffer and Gunst, 2001; Gunst *et al.*, 1995; 2003; Gunst and Fredberg, 2003; Gunst and Wu, 2001) (see Figure 2.1). Signalling events triggered by the pharmacological activation of smooth muscle cells elicit coordinated responses throughout the cell that modulate the structure and conformation of the actin filament lattice, fix its structure, and increase its rigidity, in addition to activating cross-bridge cycling and the sliding of the thick and thin filaments. The remodelling of the actin cytoskeleton serves to set the shape of the airway smooth muscle cell and may occur concurrently with the remodelling of other cytoskeletal filament systems, including myosin and

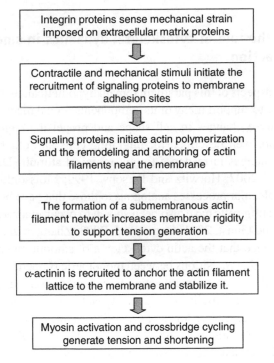

Figure 2.1 Proposed mechanism for the regulation of airway smooth muscle contraction

intermediate filaments (Seow, 2005; Tang *et al.*, 2004; Xu *et al.*, 1997). Additional cytoskeletal processes cell fortify the strength of the connections between membrane adhesion junctions and actin filaments within the contractile apparatus and cytoskeletal network, thus providing a strong and rigid framework for the transmission of force generated by the interaction of myosin and actin filaments to the outside of the cell (Choquet *et al.*, 1997; Galbraith *et al.*, 2002; Opazo *et al.*, 2004; Zhang and Gunst, 2006). According to this paradigm, mechanotransduction events at integrin adhesion sites initiate the cytoskeletal events that enable the cell to adapt and remodel its cytoskeletal structure and organization to conform to forces generated by external and internal physiological processes (Gunst *et al.*, 1995; 2003). Actin polymerization and remodelling may occur preferentially at points of tension or mechanical strain in the cell membrane, and the strengthening of points of tension transmission to the exterior of the cell may occur at discrete points of membrane stress or strain (DeMali *et al.*, 2002; 2003; Galbraith *et al.*, 2002; Zhang *et al.*, 2005; Zhang and Gunst, 2006).

Evidence for the role of cytoskeletal dynamics in smooth muscle contraction

Although many aspects of this paradigm remain hypothetical, there is solid experimental evidence to support many of its components. Local mechanotransduction events that catalyse changes in cell shape and propel the edges of the membrane forward are well described for migrating cells and other cell types that undergo shape changes in response to physiological stimuli (DesMarais *et al.*, 2002; Gerthoffer, 2007; Horwitz and Parsons, 1999; Moissoglu and Schwartz, 2006). Many molecular processes analogous to those that have been documented to occur in migrating cells have also been shown to occur in airway smooth muscle (Gerthoffer and Gunst, 2001; Gunst *et al.*, 2003; Zhang *et al.*, 2005). There is compelling evidence that the actin cytoskeleton of smooth muscle remains in a dynamic state and that transitions between monomeric (G-actin) and filamentous (F) actin are regulated by contractile stimulation and are sensitive to mechanical forces (An *et al.*, 2002; Cipolla *et al.*, 2002; Flavahan *et al.*, 2005; Herrera *et al.*, 2004; Hirshman *et al.*, 2001; Jones *et al.*, 1999; Mehta and Gunst, 1999; Tang and Gunst, 2004c; Zeidan *et al.*, 2003; Zhang *et al.*, 2005). Biochemical assays performed in differentiated airway smooth muscle tissues indicate that in unstimulated muscle tissues, approximately 20 per cent of actin is in soluble monomeric form (G actin), and that the activation of the smooth muscle tissues stimulates the polymerization of about 30 per cent of the G actin into filamentous form (F actin), resulting in an increase in F actin of less than 10 per cent (Zhang *et al.*, 2005). Increases in actin filament number in response to contractile stimulation have also been detected by quantitative morphometric analysis of electron micrographs of intact airway smooth muscle tissues, and by fluorescence analysis of primary cultures of airway smooth muscle cells (Herrera *et al.*, 2004; Hirshman and Emala, 1999). There also evidence that actin polymerization is mechanosensitive. The degree of tension depression caused by the inhibition of actin polymerization depends on the mechanical strain on the muscle, suggesting that actin polymerization may be important in regulating the length sensitivity of tension development (Hai, 2000; Mehta and Gunst, 1999; Silberstein and Hai, 2002).

Much of the experimental evidence suggests that the newly formed actin filaments perform a function that is distinct from that of the actin filaments that interact with myosin and participate in cross-bridge cycling. The inhibition of actin polymerization by latrunculin or cytochalasin inhibits tension development with little or no effect on intracellular Ca^{2+}, on myosin light chain phosphorylation, or on signalling processes involved in the activation of the contractile apparatus (An *et al.*, 2004; Mehta and Gunst, 1999; Zhang *et al.*, 2005). Myosin ATPase activity can also be activated in stimulated airway tissues in which new actin polymerization has been inhibited (Tang *et al.*, 2002). Studies of cultured airway smooth muscle cells also indicate that signalling pathways that regulate actin polymerization are distinct from those that regulate myosin light-chain phosphorylation and cross-bridge cycling and can be independently activated (An *et al.*, 2002). These data are consistent with the idea that the small fraction of actin that undergoes polymerization resides in a cytoskeletal pool that is largely distinct from the actin that interacts with myosin, and that its polymerization does not affect myosin activation or cross-bridge cycling.

The contractile stimulation of airway smooth muscle initiates the recruitment of a number of cytoskeletal regulatory and structural proteins to the peripheral submembranous area of the cell (Opazo *et al.*, 2004; Zhang *et al.*, 2005; Zhang and Gunst, 2006). Actin polymerization requires the activation of the actin-nucleating protein, neuronal Wiskott–Aldrich syndrome protein (N-WASp), and can be completely inhibited by peptide inhibitors of N-WASp activation (Hufner *et al.*, 2001; Pollard *et al.*, 2000; Zhang *et al.*, 2005). Proteins that associate with integrin adhesion complexes, including paxillin and vinculin, regulate both actin polymerization and N-WASp activation (DeMali *et al.*, 2002; 2003; Opazo *et al.*, 2004; Tang *et al.*, 2002; 2003; Tang and Gunst, 2004a; 2004b). The recruitment of the actin-regulatory proteins N-WASp, vinculin, and paxillin to the submembranous region of the airway smooth muscle cell is initiated by contractile stimulation of the cell (Opazo *et al.*, 2004; Zhang *et al.*, 2005; Zhang and Gunst, 2006). The inhibition of the recruitment of paxillin or vinculin to the membrane inhibits both actin polymerization and tension development in airway smooth muscle, but has no effect on myosin light-chain phosphorylation (Opazo *et al.*, 2004). Contractile activation also stimulates the recruitment of the actin-cross-linking and integrin-binding proteins, α-actinin and talin, to the periphery of airway smooth muscle cells and increases their localization in integrin complexes (Opazo *et al.*, 2004; Zhang and Gunst, 2006). The fact that proteins that are involved in initiating and regulating the actin polymerization process and those involved in cross-linking actin filaments and anchoring them to adhesion complexes all localize to the membrane during contractile stimulation suggests that the new actin polymerization that occurs during contractile stimulation may occur predominantly in the submembranous area of the cell. The actin polymerization process that occurs in smooth muscle cells may thus be analogous to the formation of 'cortical actin' that has been observed with the stimulation of other contractile cell types, such as endothelial cells (Flanagan *et al.*, 2001; Garcia *et al.*, 2001; Miki *et al.*, 1996). The formation of a network of submembranous actin has been proposed to function to enhance membrane rigidity (Morone *et al.*, 2006). In airway smooth muscle, the formation of this actin network may function to regulate cell shape and rigidity, and to connect the contractile and cytoskeletal filament lattice to the membrane to transmit force generated by cross-bridge cycling. Actin polymerization might also occur focally at points at which membrane tension is greatest. In airway smooth muscle, the phosphorylation of paxillin and its kinase, focal adhesion kinase (FAK), is also mechanosensitive, suggesting a mechanism whereby mechanical signals may be sensed by integrin proteins at membrane adhesion sites and transduced to regulate the local polymerization of actin (Tang and Gunst, 2001b; Tang *et al.*, 1999; 2002; 2003). There is currently very little direct experimental evidence documenting the function or cellular localization of the actin that is newly polymerized in response to a contractile stimulation. However, it is clear that the cytoskeletal processes of adhesion junction fortification and filament formation that are initiated in response to a contractile stimulus play critical roles in the process of tension development, as the disruption of any of these cytoskeletal events inhibits tension development in response to a contractile stimulus.

Implications for airway disorder

Can this paradigm for the regulation of airway smooth muscle function provide novel perspectives to explain the normal physiological behaviour of the airways and pathophysiological properties of the airways in asthma? The effects of tidal volume oscillations and periodic stretch of the airways in decreasing airway responsiveness in experimental animals and human subjects are well documented, and an absence of stretch or a decrease in tidal volume amplitude can cause airway hyperresponsiveness (Gunst *et al.*, 2001b; Shen *et al.*, 1997a; Skloot *et al.*, 1995). There is unequivocal evidence establishing the importance of mechanical oscillations of airway smooth muscle and bronchial segments in maintaining a low level of airway muscle stiffness and contractility (Gunst *et al.*, 1990; Gunst and Stropp, 1988; Gunst and Wu, 2001; Ramchandani *et al.*, 2002; Shen *et al.*, 1997a). Clearly, the ability of airway smooth muscle to modulate its stiffness and compliance to accommodate changes in lung volume during breathing is of critical importance for normal airway function. In fact, the regulation of airway compliance during breathing may be the most critical physiological function of airway smooth muscle. If dynamic cytoskeletal processes form the basis for this phenomenon, then pathophysiological processes that disturb normal cytoskeletal dynamics may result in abnormalities in the ability of the airways to regulate their compliance normally. Airway hyperresponsiveness could be a byproduct of such disturbances. However, much remains to be determined regarding the mechanistic basis for the properties of mechanical adaptation of airway smooth muscle, and the role of dynamic cytoskeletal processes in the regulation of the compliance and contractility of airway smooth muscle. A better understanding of these processes could provide novel insights into the basis for the normal physiological properties of the airways and in disturbances in their function that occur during disease.

References

An, S. S., Fabry, B., Mellema, M. *et al.* (2004). Role of heat shock protein 27 in cytoskeletal remodeling of the airway smooth muscle cell. *J Appl Physiol* **96**, 1701–1713.

An, S. S., Laudadio, R. E., Lai, J. *et al.* (2002). Stiffness changes in cultured airway smooth muscle cells. *Am J Physiol Cell Physiol* **283**, C792–C801.

Babij, P. (1993). Tissue-specific and developmentally regulated alternative splicing of a visceral isoform of smooth muscle myosin heavy chain. *Nucleic Acids Res* **21**, 1467–1471.

Babu, G. J., Warshaw, D. M. and Periasamy, M. (2000). Smooth muscle myosin heavy chain isoforms and their role in muscle physiology. *Microsc Res Tech* **50**, 532–540.

Bai, T. R., Bates, J. H., Brusasco, V. *et al.* (2004). On the terminology for describing the length–force relationship and its changes in airway smooth muscle. *J Appl Physiol* **97**, 2029–2034.

Bond, M. and Somlyo, A. V. (1982). Dense bodies and actin polarity in vertebrate smooth muscle. *J Cell Biol* **95** (2 pt 1), 403–413.

Brusasco, V., Crimi, E., Barisione, G. *et al.* (1999). Airway responsiveness to methacholine: effects of deep inhalations and airway inflammation. *J Appl Physiol* **87**, 567–573.

Burridge, K. and Chrzanowska-Wodnicka, M. (1996). Focal adhesions, contractility, and signaling. *Annu Rev Cell Dev Biol* **12**, 463–518.

Carlier, M. F. and Pantaloni, D. (1997). Control of actin dynamics in cell motility. *J Mol Biol* **269**, 459–467.

Choquet, D., Felsenfeld, D. P. and Sheetz, M. P. (1997). Extracellular matrix rigidity causes strengthening of integrin–cytoskeleton linkages. *Cell* **88**, 39–48.

Cipolla, M. J., Gokina, N. I. and Osol, G. (2002). Pressure-induced actin polymerization in vascular smooth muscle as a mechanism underlying myogenic behavior. *FASEB J* **16**, 72–76.

Cooke, P. H., Kargacin, G., Craig, R. *et al.* (1987). Molecular structure and organization of filaments in single, skinned smooth muscle cells. *Prog Clin Biol Res* **245**, 1–25.

Craig, R. and Megerman, J. (1977). Assembly of smooth muscle myosin into side-polar filaments. *J Cell Biol* **75**, 990–996.

Critchley, D. R. (2000). Focal adhesions – the cytoskeletal connection. *Curr Opin Cell Biol* **12**, 133–139.

DeMali, K. A., Wennerberg, K. and Burridge, K. (2003). Integrin signaling to the actin cytoskeleton. *Curr Opin Cell Biol* **15**, 572–582.

DeMali, K. A., Barlow, C. A. and Burridge, K. (2002). Recruitment of the Arp2/3 complex to vinculin: coupling membrane protrusion to matrix adhesion. *J Cell Biol* **159**, 881–891.

DesMarais, V., Ichetovkin, I., Condeelis, J. *et al.* (2002). Spatial regulation of actin dynamics: a tropomyosin-free, actin-rich compartment at the leading edge. *J Cell Sci* **115**, 4649–4660.

Draeger, A., Amos, W. B., Ikebe, M. *et al.* (1990). The cytoskeletal and contractile apparatus of smooth muscle: contraction bands and segmentation of the contractile elements. *J Cell Biol* **111** (6 Pt 1), 2463–2473.

Draeger, A., Stelzer, E. H., Herzog, M. *et al.* (1989). Unique geometry of actin-membrane anchorage sites in avian gizzard smooth muscle cells. *J Cell Sci* **94** (pt 4), 703–711.

Drew, J. S., Moos, C. and Murphy, R. A. (1991). Localization of isoactins in isolated smooth muscle thin filaments by double gold immunolabeling. *Am J Physiol* **260** (6 pt 1), C1332–1340.

Eddinger, T. J. and Murphy, R. A. (1991). Developmental changes in actin and myosin heavy chain isoform expression in smooth muscle. *Arch Biochem Biophys* **284**, 232–237.

Eddinger, T. J. and Wolf, J. A. (1993). Expression of four myosin heavy chain isoforms with development in mouse uterus. *Cell Motil Cytoskeleton* **25**, 358–368.

Fay, F. S. and Delise, C. M. (1973). Contraction of isolated smooth-muscle cells – structural changes. *Proc Natl Acad Sci USA* **70**, 641–645.

Fay, F. S., Fujiwara, K., Rees, D. D. *et al.* (1983). Distribution of alpha-actinin in single isolated smooth muscle cells. *J Cell Biol* **96**, 783–795.

Fish, J. E., Ankin, M. G., Kelly, J. F. *et al.* (1981). Regulation of bronchomotor tone by lung inflation in asthmatic and nonasthmatic subjects. *J Appl Physiol* **50**, 1079–1086.

Flanagan, L. A., Chou, J., Falet, H. *et al.* (2001). Filamin A, the Arp2/3 complex, and the morphology and function of cortical actin filaments in human melanoma cells. *J Cell Biol* **155**, 511–518.

Flavahan, N. A., Bailey, S. R., Flavahan, W. A. *et al.* (2005). Imaging remodeling of the actin cytoskeleton in vascular smooth muscle cells after mechanosensitive arteriolar constriction. *Am J Physiol Heart Circ Physiol* **288**, H660–H669.

Fredberg, J. J., Inouye, D., Miller, B. *et al.* (1997). Airway smooth muscle, tidal stretches, and dynamically determined contractile states. *Am J Respir Crit Care Med* **156**, 1752–1759.

Furst, D. O., Cross, R. A., De Mey, J. *et al.* (1986). Caldesmon is an elongated, flexible molecule localized in the actomyosin domains of smooth muscle. *EMBO J* **5**, 251–257.

Galbraith, C. G., Yamada, K. M. and Sheetz, M. P. (2002). The relationship between force and focal complex development. *J Cell Biol* **159**, 695–705.

Garcia, J. G. N., Liu, F., Verin, A. D. *et al.* (2001). Sphingosine 1-phosphate promotes endothelial cell barrier integrity by Edg-dependent cytoskeletal rearrangement. *J Clin Invest* **108**, 689–701.

Geiger, B., Dutton, A. H., Tokuyasu, K. T. *et al.* (1981). Immunoelectron microscope studies of membrane–microfilament interactions: distributions of alpha-actinin, tropomyosin, and vinculin in intestinal epithelial brush border and chicken gizzard smooth muscle cells. *J Cell Biol* **91** (3 pt 1), 614–628.

Gerthoffer, W. T. (2007). Mechanisms of vascular smooth muscle cell migration. *Circ Res* **100**, 607–621.

Gerthoffer, W. T. and Gunst, S. J. (2001). Invited review: focal adhesion and small heat shock proteins in the regulation of actin remodeling and contractility in smooth muscle. *J Appl Physiol* **91**, 963–972.

Gil, F. R., Zitouni, N. B., Azoulay, E. *et al.* (2006). Smooth muscle myosin isoform expression and LC20 phosphorylation in innate rat airway hyperresponsiveness. *Am J Physiol Lung Cell Mol Physiol* **291**, L932–L940.

Guilford, W. H. and Warshaw, D. M. (1998). The molecular mechanics of smooth muscle myosin. *Comp Biochem Physiol B Biochem Mol Biol* **119**, 451–458.

Gunst, S. J. (1983). Contractile force of canine airway smooth muscle during cyclical length changes. *J Appl Physiol* **55**, 759–769.

Gunst, S. J. (1986). Effect of length history on contractile behavior of canine tracheal smooth muscle. *Am J Physiol* **250** (1 pt 1), C146–154.

Gunst, S. J. (1999). Applicability of the sliding filament/crossbridge paradigm to smooth muscle. *Rev Physiol Biochem Pharmacol* **134**, 7–61.

Gunst, S. J. (2002). Role of airway smooth muscle mechanical properties in the regulation of airway caliber. In *Mechanics of Breathing: Pathophysiology, Diagnosis and Treatment*, pp. 34–44, A. Alverti (ed.). Springer-Verlag Italia: Milan.

Gunst, S. J. and Fredberg, J. J. (2003). The first three minutes: smooth muscle contraction, cytoskeletal events, and soft glasses. *J Appl Physiol* **95**, 413–425.

Gunst, S. J., Meiss, R. A., Wu, M. F. *et al.* (1995). Mechanisms for the mechanical plasticity of tracheal smooth muscle. *Am J Physiol* **268** (5 pt 1), C1267–1276.

Gunst, S. J., Shen, X., Ramchandani, R. *et al.* (2001a). Bronchoprotective and bronchodilatory effects of deep inspiration in rabbits subjected to bronchial challenge. *J Appl Physiol* **91**, 2511–2516.

Gunst, S. J., Shen, X. and Tepper, R. S. (2001b). Bronchoprotective and bronchodilatory effects of deep inspiration in rabbits subjected to methacholine challenge. *J Appl Physiol* **91**, 2511–2516.

Gunst, S. J. and Stropp, J. Q. (1988). Pressure–volume and length–stress relationships in canine bronchi in vitro. *J Appl Physiol* **64**, 2522–2531.

Gunst, S. J., Stropp, J. Q. and Service, J. (1990). Mechanical modulation of pressure–volume characteristics of contracted canine airways in vitro. *J Appl Physiol* **68**, 2223–2229.

Gunst, S. J. and Tang, D. D. (2000). The contractile apparatus and mechanical properties of airway smooth muscle. *Eur Respir J* **15**, 600–616.

Gunst, S. J., Tang, D. D. and Opazo, S. A. (2003). Cytoskeletal remodeling of the airway smooth muscle cell: a mechanism for adaptation to mechanical forces in the lung. *Respir Physiol Neurobiol* **137**, 151–168.

Gunst, S. J. and Wu, M. F. (2001). Plasticity of airway smooth muscle stiffness and extensibility: role of length-adaptive mechanisms. *J Appl Physiol* **90**, 741–749.

Gunst, S. J., Wu, M. F. and Smith, D. D. (1993). Contraction history modulates isotonic shortening velocity in smooth muscle. *Am J Physiol* **265** (2 pt 1), C467–476.

Hai, C. M. (2000). Mechanosensitive modulation of receptor-mediated crossbridge activation and cytoskeletal organization in airway smooth muscle. *Arch Pharm Res* **23**, 535–547.

Hartshorne, D. J. (1987). Biochemistry of the contractile process in smooth muscle. In *Physiology of the Gastrointestinal Tract* (2nd edn), vol. 1, pp. 423–482, L. R. Johnson (ed.). Raven Press: New York.

Herman, I. M. (1993). Actin isoforms. *Curr Opin Cell Biol* **5**, 48–55.

Herrera, A. M., Martinez, E. C. and Seow, C. Y. (2004). Electron microscopic study of actin polymerization in airway smooth muscle. *Am J Physiol Lung Cell Mol Physiol* **286**, L1161–L1168.

Hirshman, C. A. and Emala, C. W. (1999). Actin reorganization in airway smooth muscle cells involves Gq and Gi-2 activation of Rho. *Am J Physiol* **277** (3 pt 1), L653–661.

Hirshman, C. A., Zhu, D., Panettieri, R. A. *et al.* (2001). Actin depolymerization via the beta-adrenoceptor in airway smooth muscle cells: a novel PKA-independent pathway. *Am J Physiol Cell Physiol* **281**, C1468–C1476.

Holmes, K. C., Popp, D., Gebhard, W. *et al.* (1990). Atomic model of the actin filament. *Nature* **347**, 44–49.

Horwitz, A. R. and Parsons, J. T. (1999). Cell migration – movin' on. *Science* **286**, 1102–1103.

Hufner, K., Higgs, H. N., Pollard, T. D. *et al.* (2001). The verprolin-like central (vc) region of Wiskott–Aldrich syndrome protein induces Arp2/3 complex-dependent actin nucleation. *J Biol Chem* **276**, 35761–35767.

Jiang, H., Rao, K., Liu, X. *et al.* (1995). Increased Ca^{2+} and myosin phosphorylation, but not calmodulin activity in sensitized airway smooth muscles. *Am J Physiol* **268** (5 pt 1), L739–746.

Jones, K. A., Perkins, W. J., Lorenz, R. R. *et al.* (1999). F-actin stabilization increases tension cost during contraction of permeabilized airway smooth muscle in dogs. *J Physiol* **519** (pt 2), 527–538.

Kabsch, W. and Vandekerckhove, J. (1992). Structure and function of actin. *Annu Rev Biophys Biomol Struct* **21**, 49–76.

Kapsali, T., Permutt, S., Laube, B. *et al.* (2000). Potent bronchoprotective effect of deep inspiration and its absence in asthma. *J Appl Physiol* **89**, 711–720.

Kargacin, G. J., Cooke, P. H., Abramson, S. B. *et al.* (1989). Periodic organization of the contractile apparatus in smooth muscle revealed by the motion of dense bodies in single cells. *J Cell Biol* **108**, 1465–1475.

Khaitlina, S. Y. (2001). Functional specificity of actin isoforms. *Int Rev Cytol* **202**, 35–98.

King, G. G., Moore, B. J., Seow, C. Y. *et al.* (1999). Time course of increased airway narrowing caused by inhibition of deep inspiration during methacholine challenge. *Am J Respir Crit Care Med* **160**, 454–457.

Kuo, K. H., Herrera, A. M. and Seow, C. Y. (2003). Ultrastructure of airway smooth muscle. *Respir Physiol Neurobiol* **137**, 197–208.

Kuo, K. H. and Seow, C. Y. (2004). Contractile filament architecture and force transmission in swine airway smooth muscle. *J Cell Sci* **117** (pt 8), 1503–1511.

Lehman, W. (1991). Calponin and the composition of smooth muscle thin filaments. *J Muscle Res Cell Motil* **12**, 221–224.

Lehman, W., Sheldon, A. and Madonia, W. (1987). Diversity in smooth muscle thin filament composition. *Biochim Biophys Acta* **914**, 35–39.

Lehman, W., Vibert, P., Craig, R. *et al.* (1996). Actin and the structure of smooth muscle thin filaments. In *Biochemistry of Smooth Muscle Contraction*, pp. 47–60, M. Barany (ed.), Academic Press: San Diego, CA.

Ma, X., Cheng, Z., Kong, H. *et al.* (2002). Changes in biophysical and biochemical properties of single bronchial smooth muscle cells from asthmatic subjects. *Am J Physiol Lung Cell Mol Physiol* **283**, L1181–L1189.

Mabuchi, K., Li, B., Ip, W. *et al.* (1997). Association of calponin with desmin intermediate filaments. *J Biol Chem* **272**, 22662–22666.

Makuch, R., Birukov, K., Shirinsky, V. *et al.* (1991). Functional interrelationship between calponin and caldesmon. *Biochem J* **280** (pt 1), 33–38.

Malmberg, P., Larsson, K., Sundblad, B. M. *et al.* (1993). Importance of the time interval between FEV_1 measurements in a methacholine provocation test. *Eur Respir J* **6**, 680–686.

Malmqvist, U. and Arner, A. (1991). Correlation between isoform composition of the 17 kDa myosin light chain and maximal shortening velocity in smooth muscle. *Pflugers Arch – Eur J Physiol* **418**, 523–530.

McFawn, P. K., Shen, L., Vincent, S. G., *et al.* (2003). Calcium-independent contraction and sensitization of airway smooth muscle by p21-activated protein kinase. *Am J Physiol Lung Cell Mol Physiol.* **284**, L863–L870.

Mehta, D. and Gunst, S. J. (1999). Actin polymerization stimulated by contractile activation regulates force development in canine tracheal smooth muscle. *J Physiol* **519**
(pt 3), 829–840.

Miki, H., Miura, K. and Takenawa, T. (1996). N-WASp, a novel actin-depolymerizing
protein, regulates the cortical cytoskeletal rearrangement in a PIP2-dependent manner
downstream of tyrosine kinases. *EMBO J* **15**, 5326–5335.

Milligan, R. A., Whittaker, M. and Safer, D. (1990). Molecular structure of F-actin and
location of surface binding sites. *Nature* **348**, 217–221.

Moissoglu, K. and Schwartz, M. A. (2006). Integrin signalling in directed cell migration.
Biol Cell **98**, 547–555.

Morgan, K. G. and Gangopadhyay, S. S. (2001). Invited review: cross-bridge regulation
by thin filament-associated proteins. *J Appl Physiol* **91**, 953–962.

Morone, N., Fujiwara, T., Murase, K. *et al.* (2006). Three-dimensional reconstruction of
the membrane skeleton at the plasma membrane interface by electron tomography. *J
Cell Biol* **174**, 851–862.

Murphy, R. A. (1994). What is special about smooth muscle? The significance of covalent
crossbridge regulation. *FASEB J* **8**, 311–318.

Murphy, R. A. (1989). Contraction in smooth muscle cells. *Annu Rev Physiol* **51**, 275–283.

Nadel, J. A. and Tierney, D. F. (1961). Effect of a previous deep inspiration on airway
resistance in man. *J Appl Physiol* **16**, 717–719.

Nonomura, J. (1976). Fine structure of myofilaments in chicken gizzard smooth muscle.
In *Recent Progress in Electron Microscopy of Cells and Tissues*, pp. 40–48, E. Yamada
et al. (eds). Thieme: Stuttgart.

North, A. J., Gimona, M., Cross, R. A. *et al.* (1994). Calponin is localised in both the
contractile apparatus and the cytoskeleton of smooth muscle cells. *J Cell Sci* **107**
(pt 3), 437–444.

Opazo, S. A., Zhang, W., Wu, Y. *et al.* (2004). Tension development during contractile
stimulation of smooth muscle requires recruitment of paxillin and vinculin to the
membrane. *Am J Physiol Cell Physiol* **286**, C433–C447.

Owens, G. K., Kumar, M. S. and Wamhoff, B. R. (2004). Molecular regulation of vascular smooth muscle cell differentiation in development and disease. *Physiol Rev* **84**,
767–801.

Pfitzer, G. (2001). Invited review: regulation of myosin phosphorylation in smooth muscle.
J Appl Physiol **91**, 497–503.

Pollard, T. D., Blanchoin, L. and Mullins, R. D. (2000). Molecular mechanisms controlling
actin filament dynamics in nonmuscle cells. *Annu Rev Biophys Biomol Struct* **29**,
545–576.

Pollard, T. D. and Cooper, J. A. (1986). Actin and actin-binding proteins. A critical
evaluation of mechanisms and functions. *Annu Rev Biochem* **55**, 987–1035.

Pratusevich, V. R., Seow, C. Y. and Ford, L. E. (1995). Plasticity in canine airway smooth
muscle. *J Gen Physiol* **105**, 73–94.

Ramchandani, R., Wu, M. F., Zhang, L. *et al.* (2002). Persistent alterations in the properties
of isolated bronchi and trachealis muscle induced by chronic mechanical forces. *Am J
Respir Crit Care Med* **165**, A255.

Rayment, I., Rypniewski, W. R., Schmidt-Base, K. *et al.* (1993). Three-dimensional structure of myosin subfragment-1: a molecular motor. *Science* **261**, 50–58.

Rayment, I., Smith, C. and Yount, R. G. (1996). The active site of myosin. *Annu Rev Physiol* **58**, 671–702.

Rosol, M., Lehman, W., Craig, R. *et al.* (2000). Three-dimensional reconstruction of thin filaments containing mutant tropomyosin. *Biophys J* **78**, 908–917.

Rovner, A. S., Freyzon, Y. and Trybus, K. M. (1997). An insert in the motor domain determines the functional properties of expressed smooth muscle myosin isoforms. *J Muscle Res Cell Motil* **18**, 103–110.

Salerno, F. G., Shinozuka, N., Fredberg, J. J. *et al.* (1999). Tidal volume amplitude affects the degree of induced bronchoconstriction in dogs. *J Appl Physiol* **87**, 1674–1677.

Sanderson, M. J., Delmotte, P., Bai, Y. *et al.* (2008). Regulation of airway smooth muscle cell contractility by Ca^{2+} signaling and sensitivity. *Proc Am Thorac Soc.* **5**, 23–31.

Schafer, D. A., Hug, C. and Cooper, J. A. (1995). Inhibition of CapZ during myofibrillogenesis alters assembly of actin filaments. *J Cell Biol* **128**, 61–70.

Seow, C. Y. (2005). Myosin filament assembly in an ever-changing myofilament lattice of smooth muscle. *Am J Physiol Cell Physiol* **289**, C1363–C1368.

Shen, X., Gunst, S. J. and Tepper, R. S. (1997a). Effect of tidal volume and frequency on airway responsiveness in mechanically ventilated rabbits. *J Appl Physiol* **83**, 1202–1208.

Shen, X., Wu, M. F., Tepper, R. S. *et al.* (1997b). Mechanisms for the mechanical response of airway smooth muscle to length oscillation. *J Appl Physiol* **83**, 731–738.

Silberstein, J. and Hai, C. M. (2002). Dynamics of length–force relations in airway smooth muscle. *Respir Physiol Neurobiol* **132**, 205–221.

Skloot, G., Permutt, S. and Togias, A. (1995). Airway hyperresponsiveness in asthma: a problem of limited smooth muscle relaxation with inspiration. *J Clin Invest* **96**, 2393–2403.

Small, J. V. (1995). Structure–function relationships in smooth muscle: the missing links. *Bioessays* **17**, 785–792.

Somlyo, A. P., Devine, C. E., Somlyo, A. V. *et al.* (1973). Filament organization in vertebrate smooth muscle. *Philos Trans R Soc Lond A* **265**, 223–229.

Somlyo, A. P. and Somlyo, A. V. (2000). Signal transduction by G-proteins, rho-kinase and protein phosphatase to smooth muscle and non-muscle myosin II. *J Physiol* **522** (pt 2), 177–185.

Somlyo, A. P. and Somlyo, A. V. (2003). Ca^{2+} sensitivity of smooth muscle and nonmuscle myosin II: modulated by G proteins, kinases, and myosin phosphatase. *Physiol Rev* **83**, 1325–1358.

Tang, D. D., Bai, Y. and Gunst, S. J. (2004). Silencing of p21-activated kinase attenuates phosphorylation and reorientation of the vimentin network during serotonin stimulation of smooth muscle cells. *J Biol Chem* **388**(pt 3), 773–783.

Tang, D. D. and Gunst, S. J. (2004a). Dominant negative Cdc42 mutant inhibits contractile force development and the association of N-WASp with Arp2/3 complex during stimulation of smooth muscle tissues. *Am J Respir Crit Care Med* **169**, no. 7.

Tang, D. D. and Gunst, S. J. (2004b). The adapter protein CrkII regulates the activation of N-WASp and the Arp2/3 complex and tension development in smooth muscle. *J Biol Chem* **279**, 51772–51778.

Tang, D. D. and Gunst, S. J. (2001a). Depletion of focal adhesion kinase by antisense depresses contractile activation of smooth muscle. *Am J Physiol Cell Physiol* **280**, C874–C883.

Tang, D. D. and Gunst, S. J. (2001b). Roles of focal adhesion kinase and paxillin in the mechanosensitive regulation of myosin phosphorylation in smooth muscle. *J Appl Physiol* **91**, 1452–1459.

Tang, D. D., Mehta, D. and Gunst, S. J. (1999). Mechanosensitive tyrosine phosphorylation of paxillin and focal adhesion kinase in tracheal smooth muscle. *Am J Physiol* **276** (1 pt 1), C250–258.

Tang, D. D., Turner, C. E. and Gunst, S. J. (2003). Expression of non-phosphorylatable paxillin mutants in canine tracheal smooth muscle inhibits tension development. *J Physiol* **553**, 21–35.

Tang, D. D. and Gunst, S. J. (2004c). The small GTPase Cdc42 regulates actin polymerization and tension development during contractile stimulation of smooth muscle. *J Biol Chem* **279**, 51722–51728.

Tang, D. D., Wu, M. F., Opazo Saez, A. M. *et al.* (2002). The focal adhesion protein paxillin regulates contraction in canine tracheal smooth muscle. *J Physiol* **542**, 501–513.

Tepper, R. S., Shen, X., Bakan, E. *et al.* (1995). Maximal airway response in mature and immature rabbits during tidal ventilation. *J Appl Physiol* **79**, 1190–1198.

Turner, C. E. (2000). Paxillin and focal adhesion signalling. *Nature Cell Biol* **2**, E231–E236.

Vibert, P., Craig, R. and Lehman, W. (1993). Three-dimensional reconstruction of caldesmon-containing smooth muscle thin filaments. *J Cell Biol* **123**, 313–321.

Wang, C. L. (2001). Caldesmon and smooth-muscle regulation. *Cell Biochem Biophys* **35**, 275–288.

Wang, J., Zohar, R. and McCulloch, C. A. (2006). Multiple roles of alpha-smooth muscle actin in mechanotransduction. *Exp Cell Res* **312**, 205–214.

Wang, L., Pare, P. D. and Seow, C. Y. (2000). Effects of length oscillation on the subsequent force development in swine tracheal smooth muscle. *J Appl Physiol* **88**, 2246–2250.

Warner, D. O. and Gunst, S. J. (1992). Limitation of maximal bronchoconstriction in living dogs. *Am Rev Respir Dis* **145**, 553–560.

Warrick, H. M. and Spudich, J. A. (1987). Myosin structure and function in cell motility. *Annu Rev Cell Biol* **3**, 379–421.

Warshaw, D. M., Desrosiers, J. M., Work, S. S. *et al.* (1990). Smooth muscle myosin cross-bridge interactions modulate actin filament sliding velocity in vitro. *J Cell Biol* **111**, 453–463.

White, S., Martin, A. F. and Periasamy, M. (1993). Identification of a novel smooth muscle myosin heavy chain cDNA: isoform diversity in the S1 head region. *Am J Physiol* **264** (5 pt 1), C1252–1258.

Winder, S. J. and Walsh, M. P. (1993). Calponin: thin filament-linked regulation of smooth muscle contraction. *Cell Signal* **5**, 677–686.

Wong, J. Z., Woodcock-Mitchell, J., Mitchell, J. *et al.* (1998). Smooth muscle actin and myosin expression in cultured airway smooth muscle cells. *Am J Physiol* **274** (5 pt 1), L786–792.

Xu, C., Craig, R., Tobacman, L. *et al.* (1999). Tropomyosin positions in regulated thin filaments revealed by cryoelectron microscopy. *Biophys J* **77**, 985–992.

Xu, J. Q., Gillis, J. M. and Craig, R. (1997). Polymerization of myosin on activation of rat anococcygeus smooth muscle. *J Muscle Res Cell Motil* **18**, 381–393.

Xu, J. Q., Harder, B. A., Uman, P. *et al.* (1996). Myosin filament structure in vertebrate smooth muscle. *J Cell Biol* **134**, 53–66.

Yoshimura, H., Jones, K. A., Perkins, W. J., *et al.* (2001). Calcium sensitization produced by G protein activation in airway smooth muscle. *Am J Physiol Lung Cell Mol Physiol.* **281**, L631–L638.

Zeidan, A., Nordstrom, I., Albinsson, S. *et al.* (2003). Stretch-induced contractile differentiation of vascular smooth muscle: sensitivity to actin polymerization inhibitors. *Am J Physiol Cell Physiol* **284**, C1387–C1396.

Zhang, W., Wu, Y., Du, L. *et al.* (2005). Activation of the Arp2/3 complex by N-WASp is required for actin polymerization and contraction in canine tracheal smooth muscle. *Am J Physiol Cell Physiol* **288**, C1145–C1160.

Zhang, W. W. and Gunst, S. J. (2006). Dynamic association between alpha-actinin and beta-integrin regulates contraction of canine tracheal smooth muscle. *J Physiol* **572**, 659–676.

Zigmond, S. H. (1996). Signal transduction and actin filament organization. *Curr Opin Cell Biol* **8**, 66–73.

3

Airway smooth muscle: role in airway constrictor hyperresponsiveness

Melanie Brown[1] and Julian Solway[1,2]

Departments of Pediatrics[1] and Medicine[2], University of Chicago, Chicago, IL 60637, USA

3.1 What is airway constrictor hyperresponsiveness (AHR)?

Airway constrictor hyperresponsiveness (AHR) is defined as an abnormally large bronchoconstrictor response to external stimuli; AHR can be reflected in the development of significant bronchoconstriction in response to stimuli that usually elicit little or no airflow obstruction, or in the development of larger bronchoconstrictor responses than ever develop in normal individuals. AHR is such a stereotypical feature of asthma that its documentation in the pulmonary function laboratory is frequently sought as a diagnostic measure in patients in whom the presence of asthma is otherwise clinically uncertain. To test whether AHR is present, one can assess the degree of airflow obstruction that occurs in response to increasingly intense bronchoconstrictor stimuli. These stimuli can include natural provocations, like exercise or breathing of cold air, as well as artificial exposures, such as to

Airway Smooth Muscle   Edited by Kian Fan Chung
© 2008 John Wiley & Sons, Ltd

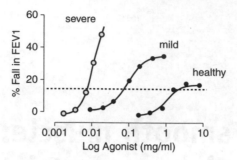

Figure 3.1 Typical bronchoconstrictor dose-response curves to inhaled constrictor agonists found in healthy individuals, or in asthmatics with mild or severe airway constrictor hyperresponsiveness (AHR). The severity of airflow obstruction, reflected in fall of FEV_1 from baseline (vertical axis), increases with the intensity of bronchoconstrictor stimulation (log agonist concentration) in each type of individual. However, normal individuals require substantial bronchoconstrictor stimulation, and have limited maximal constrictor response, whereas individuals with AHR develop bronchoconstriction of more severe degree than normal individuals do, and they do so in response to much smaller stimuli than those that elicit any substantial response in normal people. (Reproduced courtesy of The American Review of Respiratory Disease, 1984 July; 130(1):71–5)

methacholine or histamine aerosols. In the methacholine challenge test, the subject is exposed to increasing doses of the acetylcholine analogue methacholine, and the volume of air that can be forcibly exhaled in the first second (FEV_1) is taken as a measure of airway calibre. Excessively large falls in FEV_1 (substantially greater than 20 per cent from the pre-challenge value) that occur after modest bronchoconstrictor challenges (e.g., inhalation of methacholine at concentrations lower than 8 mg/ml) demonstrate the presence of AHR (Figure 3.1).

Airflow obstruction is one of the hallmark features of asthma (Hargreave *et al.*, 1982), but it can be found in other disorders such as cystic fibrosis (CF), bronchopulmonary dysplasia, chronic obstructive pulmonary disease, and other obstructive lung diseases. There are generally two components of airflow obstruction: a fixed component (likely due to chronic structural changes) and a variable component (due to acute bronchoconstriction or mucus secretion). Presently, only the variable component can be modified by medical therapies. Airway remodelling refers to the structural changes that occur in chronic airway diseases such as asthma or CF. Airway narrowing and bronchial hyperresponsiveness are seen in CF, and these occur in conjunction with excessive accumulation of smooth muscle cells (Hays *et al.*, 2005). Smooth muscle also accumulates to excess in asthma, but whether the excessive smooth muscle mass causes AHR in these diseases remains unknown (Hargreave *et al.*, 1986;

King *et al.*, 1999). Besides its excessive accumulation, airway smooth muscle (ASM) might also contribute to airway inflammation and airway remodelling by elaborating cytokines, chemokines, and growth factors. Thus, ASM is a likely contributor to AHR through many mechanisms.

3.2 Is AHR ever good?

Embryology and AHR in the developing lung

The lung begins to form from the primitive human foregut at about 21 days' gestation. The airways continue to increase in number until about 16 weeks' gestation, by which time the bronchial tree has formed. By post-natal age 2 months, lung development has occurred, although alveolar growth and differentiation generally continue until about age 3 years. Unlike the tonic constrictor activity seen when ASM contracts in adults, fetal ASM contracts in peristaltic waves much like those that occur in the gastrointestinal tract. These waves travel from proximal airways to distal airways, and may function to promote lung growth by directing positive pressure towards the distensible distal lung buds (McCray, 1993). In the developing lung, smooth muscle contractions can be blocked with calcium channel antagonists, and such blockade results in lung hypoplasia (Schittny *et al.*, 2000). Alveolar development is at least partially dependent on the distal distension of the developing lung (Moessinger *et al.*, 1990). It thus stands to reason that ASM contraction may play an important role in the growth and development of the lung. Smooth muscle might also function to clear the airways in both the antenatal and post-natal periods. Whether ASM or AHR play any useful role in the adult human is uncertain.

3.3 Potential mechanisms leading to airflow obstruction

Excessive uniform and non-uniform airway narrowing

The early-phase and late-phase airflow obstruction that results from inhaled allergen exposure is reversed by albuterol and likely is a result of alterations in smooth muscle tone. Airway inflammation in asthma is associated with non-uniform airway narrowing (Henderson *et al.*, 2003), which in turn results in heterogeneous ventilation and so contributes to the inability to maximally dilate the airways. Mathematical models have provided insight into the heterogeneous nature of airflow obstruction in lungs. For example, severe but almost uniform smooth muscle activation throughout the airway tree results in patchy ventilation. However, patchy constriction of terminal bronchioles alone, even if severe, would not result in the frequency dependence of impedance found in obstructive lung diseases; rather, heterogeneous constriction of the rest of the airway tree is necessary to account for oscillatory mechanic changes (Tgavalekos et al., 2007). In 2005, Venegas used a computational model to show that even with uniform activation of ASM in a symmetric bronchial tree, minimal heterogeneity can lead to large contiguous areas of poorly ventilated lung. The interactions of various feedback mechanisms generate these areas of poor ventilation and can cause shifts of airflow that may be a part of the sudden and severe attacks that occur in asthma. These changes have been likened to the catastrophic shifts that are seen as a result of self-organized patchiness in nature (Rietkerk et al., 2004; Venegas et al., 2005). Interestingly, Venegas' model also predicts that the severity of small airway constriction can be reduced by reducing central airway constriction, because increasing overall flow to the region in that way enhances distal tidal volume fluctuations that in turn antagonize small airway narrowing. Venegas' analysis also suggests that delivering a bronchodilator through the respiratory tree by inhalation may serve only to dilate the already dilated areas of ventilation and thus worsen the heterogeneity and overall bronchoconstriction (Venegas et al., 2005). Severe narrowing and closure of small airways appear to be among the causes of heterogeneous ventilation defects in asthmatics, as shown by a computational modelling approach (Tgavalekos et al., 2005). However, it is likely that the patchy ventilation defects seen in asthma are a result not of the independent behaviour of the airways, but of the complex interactions of the entire system (Tgavalekos et al., 2007; Winkler and Venegas, 2007). Differences in ASM contraction, volume-fluctuation-related antagonism of local airway narrowing, local inflammation, and bronchoconstrictor stimulus deposition/distribution all likely contribute to initiation of the heterogeneity observed. Such heterogeneity may cause increased airway reactivity (Downie et al., 2007) and difference in response to bronchodilators among asthmatics (Venegas, 2007).

Airway closure

Computational model analysis of the airway structure of mice has been used to shed light on the factors responsible for bronchoconstriction in experimental models of AHR. Two mouse models that are commonly used are A/J mice, which are a strain known to have innate hyperresponsiveness relative to other strains even in the absence of experimentally induced lung inflammation, and BALB/c mice with allergen-induced airway inflammation. The physiology of the hyperresponsiveness in these mouse models depends on both the route of bronchoconstrictor agonist administration and the strain used (Wagers *et al.*, 2007). In BALB/c mice with allergen-induced airway inflammation, enhanced airway closure of the peripheral airways is a major factor responsible for bronchial hyperresponsiveness (Wagers *et al.*, 2004). In these mice, such airway closure is associated with atelectasis of the subtended portions of lung; although this phenomenon occurred in response to bronchoconstrictor administration to mice without airway inflammation, atelectasis was much more severe in mice with allergen-induced inflammation. Wagers *et al.* (2004) suggest that hyperresponsiveness in allergen-inflamed BALB/c mice can be explained by airway wall/mucosal thickening and the propensity of the airways in the periphery to close without any change in degree of muscle shortening. In such studies, it is important to consider the breathing pattern used, since this can substantially influence bronchoconstrictor responsiveness (Chen *et al.*, 2006). Fredberg *et al.* (1997) suggest that the tidal stretch decreases the number of actin–myosin interactions; perhaps the relative importance of smooth muscle bronchoconstriction versus other mechanisms of airflow obstruction varies with breathing pattern during bronchoconstriction. Other factors might also modulate airway closure. For example, non-asthmatic lungs exhibit more long-lasting diminution of bronchoconstrictor responsiveness after a deep breath than do asthmatic lungs (Black *et al.*, 2004).

Epithelial barrier disruption

Disruption of the epithelial barrier of the lung with a cationic protein such as poly-L-lysine, an artificial analogue of eosinophil cationic protein, leads to increased constrictor response to inhaled, but not IV methacholine, at low doses, but not at high doses (Bates *et al.*, 2006). This is likely due to a decreased epithelial barrier function in the treated mice.

Loss of lung recoil

Elastic recoil describes the natural tendency of the lung to collapse inward. Determinants of the inward recoil of the lung include the active surface forces of the alveolus and the lung parenchyma, where the elastic recoil pressure of the lungs is the alveolar pressure minus the pleural pressure. 'Near fatal asthma' is an episode of asthma requiring intubation and mechanical ventilation secondary to deteriorating respiratory status; it has been associated with a worse prognosis. The loss of lung elastic recoil and hyperinflation at total lung capacity are risk factors for near fatal asthma (Gelb *et al.*, 2004). In those with severe refractory and chronic persistent asthma, expiratory flow limitation is caused at least partially by the loss of lung elastic recoil (Gelb and Zamel, 2002). The mechanisms of this loss of recoil are not known (Gelb *et al.*, 2002).

Fixed airway obstruction due to airway wall fibrosis

The airway wall comprises many structures, and changes in any of these could potentially affect AHR. As early as the 1920s, it was reported that patients with fatal asthma have thickening of the airway subepithelial and basement membranes (Huber and Koessler, 1922). Early investigations focused largely on the structural changes in asthma, many of which are thought to result from chronic airway inflammation. Epithelial damage can stimulate myofibroblast differentiation, collagen gene expression and proliferation in co-culture of bronchial epithelial cells and airway fibroblasts. Not only are basic fibroblast growth factor, platelet derived growth factor, IGF-1, transforming growth factor (TGF)-β2 and endothelin-1 levels increased in the supernatant from damaged epithelial cells, but also, when these factors are blocked, myofibroblast proliferation is inhibited (Zhang *et al.*, 1999). Therefore, epithelial injury might lead to subepithelial thickening due to myofibroblast proliferation. The lamina reticularis is the region associated with subepithelial fibrosis in asthma. It is made up of collagen types I, III, V, VI and VII fibrils, and is distinct from the basal lamina (considered the true basement membrane) (Evans *et al.*, 2000). In 2007, Boulet *et al.* (1997) studied the bronchial biopsies of subjects with varying degrees of asthma along with those of normal subjects. They found a significant correlation between the degree of subepithelial fibrosis and airway responsiveness as measured by the methacholine challenge.

Matrix

TGF-β is a profibrotic cytokine that is over-expressed by eosinophils and other resident cells in asthmatic airways, and that stimulates myofibroblast transformation of airway fibroblasts (Minshall *et al.*, 1997; Ohno *et al.*, 1996). Myofibroblast activation increases extracellular matrix protein secretion from these cells, including collagen and laminin; specific isoforms of the latter are associated with injury and may promote inflammation (Altraja *et al.*, 1996). Since the degree of subepithelial fibrosis in asthma correlates with the expression of TGF-β (Minshall *et al.*, 1997; Ohno *et al.*, 1996), it seems likely that this profibrotic cytokine plays an important role in regulating extracellular matrix and subsequent fibrosis.

Surfactant deficiency

Surfactant is largely secreted by type II alveolar cells. In the alveoli, it modulates surface tension during inspiration and exhalation, helping to prevent alveolar closure. However, surfactant is also drawn into the peripheral airways by virtue of the higher surface tension of their surface-lining fluids. Airway surfactant influences airway function. For example, the deteriorating lung function otherwise seen in sensitized guinea pigs was ameliorated by intratracheal administration of surfactant, and an increase in airway resistance was alleviated by treating immunized guinea pigs with aerosolized surfactant (Hohlfeld, 2002). Surfactant dysfunction after allergen challenge has also been documented in human asthmatics (Hohlfeld *et al.*, 1999), and surfactant administration can abolish the early allergen-induced response in asthma (Babu *et al.*, 2003). However, whether exogenous surfactant administration represents a useful tool in the treatment of severe asthma is unknown.

Mucus

Mucus is part of the first line of defence of the airways, but when excessive in either amount or tenacity, it can greatly contribute to the pathology of airways disease. In a post-mortem study of 93 patients with fatal asthma, Kuyper *et al.* (2003) found that airway luminal obstruction is a frequent occurrence in fatal asthma. Some investigators have reported increased epithelial mucin stores and goblet cell hyperplasia in mild and moderate asthma (Ordonez *et al.*, 2001), while others have shown no difference in goblet cells in endobronchial biopsies of mild asthmatics versus normal control subjects (Lozewicz *et al.*, 1990). In severe asthma, bronchial biopsies reveal larger mucus glands (Benayoun *et al.*, 2003). Whether these are caused by the disease itself or are an initial inciting factor is uncertain. Excess mucus in the airway can undoubtedly contribute to the airway closure that is so important in asthma.

Blood flow

Increases in both bronchial and airway blood flow are seen in allergic sheep after inhaled antigen exposure. This increase is due at least in part to inflammatory mediators (Csete *et al.*, 1991). Most airway blood flow is distributed to the subepithelium. In sheep, the subepithelial tissue accommodates over 85 per cent of airway blood flow in the trachea and bronchi. In humans, stable bronchial asthma is associated with increased blood flow when compared to normal individuals (Wanner *et al.*, 1990). It is conceivable, though not proved, that engorgement of the airway wall contributes to bronchial luminal narrowing in asthma.

Fibrin

Fibrin is known to inhibit surfactant function and is deposited in the airway lumen during airway inflammation. Tissue plasminogen activator (tPA) is fibrinolytic. Wagers *et al.* (2003; 2004) have shown that administration of tPA diminishes airway closure, as evidenced by the decreased widening of the pressure volume (PV) curve that is seen in asthma. In addition, fibrin administration augments the PV curve widening (i.e., increased airway closure). Taken together, these findings led these authors to conclude that fibrin is a significant contributor to the enhanced airway closure in inflamed lungs.

3.4 Potential abnormalities of airway smooth muscle (ASM)

None of the above mechanisms of airflow obstruction and constrictor hyperresponsiveness rely on dysfunction of ASM itself. If ASM were contributory, then what might be the potential abnormalities of ASM, and what would be their effects on airway responses?

Too much there: ASM hypertrophy and/or hyperplasia

Smooth muscle wraps in submucosal bands that are essentially circumferentially oriented around the airways. As mentioned earlier, smooth muscle abnormalities in general, and increased ASM in particular, have been implicated in a number of disease states such as CF, sudden infant death syndrome (SIDS), and asthma. SIDS is an unexpected and unexplained death in a seemingly healthy infant under

1 year of age. There is increased ASM in SIDS (Elliot *et al.*, 1999), and it has been suggested that SIDS may be caused by exaggerated ASM closure (Martinez, 1991), though whether the excessive ASM accumulation or airway closure is causal in SIDS deaths remains unknown. The ASM layer is considerably thicker in asthmatics than in non-asthmatics (Carroll *et al.*, 1993), and excessive ASM accumulation is particularly marked in asthmatics who die from asthma. In culture, bronchial myocytes from asthmatics proliferate at an increased rate compared to those of non-asthmatics (Johnson *et al.*, 2001), and quantitative morphological studies have variably revealed ASM cell hypertrophy and/or hyperplasia in asthmatic airways (Ebina *et al.*, 1993; Woodruff *et al.*, 2004). For example, when comparing airway biopsies from mild-to-moderate asthmatics with those obtained from normal volunteers, Woodruff *et al.* (2004) found increased myocyte numbers but no difference in airway myocyte size. However, Benayoun *et al.* (2003) found larger smooth muscle cell diameter in the airways of severe asthmatics than in mild asthmatics or those with COPD. Whether the increased smooth muscle mass stems from smooth muscle cell hypertrophy, hyperplasia, or both, the contractile capacity of the ASM in asthma would appear to be increased in view of its excessive mass. Curiously, no study has yet proved this. However, mathematical modelling has suggested that amplified maximum smooth muscle tension, as opposed to thickening of the airway wall, is the single most important cause of airway hyperreactivity (Affonce and Lutchen, 2006).

Too much force: asthmatic ASM force generation

Given the paucity of available smooth muscle tissue from asthmatic donors, it is not surprising that the body of literature on human asthmatic muscle strips is relatively limited. Some investigators report that the asthmatic ASM layer has a greater isometric force-generating capacity, even after accounting for in-creased ASM mass (Bai, 1990). However, a greater number have found that, after controlling for cross-sectional area, muscle strips from both asthmatic and non-asthmatic human lung show similar force-generation responses to histamine or methacholine (Black, 1996; Thomson, 1987). IgE sensitization impairs relaxation, stimulates cytokine production, and increases the force generation of ASM (Grunstein *et al.*, 2002).

Accelerated velocity of shortening

The antagonistic activities of myosin light-chain kinase (MLCK) and myosin phosphatase largely determine the phosphorylation state of the regulatory 20-kDa myosin light chain, which affects the rate of actomyosin cross-bridge activity and, for a given load, the velocity of shortening. Severe persistent asthmatics have augmented MLCK expression (Benayoun *et al.*, 2003). Furthermore, individual bronchial smooth muscle cells from asthmatic subjects have significantly increased maximum shortening capacity and velocity compared to those of normal subjects (Ma *et al.*, 2002). It has been proposed that the changes in velocity and shortening contribute to the constrictor hyperresponsiveness seen in asthmatic individuals (Solway and Fredberg, 1997).

Forced fluctuation-induced relengthening

During tidal breathing, the inflation and deflation of lung parenchyma superimposes load fluctuations on ASM, through parenchymal attachments to the airway adventitia. Thus, *in vivo*, ASM stimulated to contract does so against a fluctuating – not a steady – load. Fredberg's group demonstrated unequivocally that such load fluctuations promote the relengthening of contracted smooth muscle that had been allowed to shorten against a steady load (Lakser *et al.*, 2002; Latourelle *et al.*, 2002), and our laboratory showed that even though initial shortening is unaffected by such force fluctuations, force fluctuation-induced relengthening (FFIR) still follows that initial shortening (Dowell *et al.*, 2005). Furthermore, the magnitude of FFIR can be physiologically regulated, as inhibition of actin polymerization, inhibition of p38 MAKP signalling (Lakser *et al.*, 2002), or inhibition of ERK1/2 activation (our unpublished results) all increase the extent of FFIR. If ASM shortening were the principal mechanism of airflow obstruction, it follows that breathing-induced load fluctuations should relengthen the shortened ASM and ameliorate that obstruction. In this regard, it is certain that tidal breathing (especially deeper tidal breathing) mitigates airflow obstruction in normal individuals. However, while deep breaths also relieve experimentally induced airflow obstruction in normal people, they are less effective at doing so in asthmatics. Taken together, these observations suggest that the FFIR of contracting ASM that is normally present is dysfunctional in asthma, though this hypothesis has yet to be tested directly.

Impaired smooth muscle relaxation

Impaired relaxation has been reported in animal models of AHR (Farmer *et al.*, 1987; Jiang and Stephens, 1992). In sensitized airways, the time required for relaxation is prolonged, and in both *in vivo* and *in vitro* studies, asthmatic smooth muscle has an impaired response to beta-agonists (Bai, 1990; Bai *et al.*, 1992; Barnes and Pride, 1983). Potential contributory mechanisms include alterations in bronchodilator delivery, changes in cell-surface receptors or interactions, or differences in cytokine exposure.

3.5 If ASM is dysfunctional, how did it get that way?

Cytokine effects

Even though the weight of evidence suggests that excised ASM from asthmatics does not develop increased isometric force, a variety of cytokines that appear in asthmatic airways or serum potentiate isometric constrictor responses (Grunstein *et al.*, 2002; Hakonarson *et al.*, 1997; 2001). Calcium is the major second messenger that modulates ASM contraction, the degree of intracellular calcium mobilization determines contractility (Tao *et al.*, 1999), and many cytokines appear to exert their effects by enhancing calcium responses. For example, TNF-α, IL-1β, and IL-13 all increase expression of the cell surface protein CD38 (Deshpande *et al.*, 2003; 2004), an enzyme responsible for the synthesis of cyclic ADP-ribose. The latter promotes calcium mobilization upon G protein-coupled receptor agonists in ASM cells. Indeed, CD38-deficient mice were found to have reduced AHR following IL-13 challenge when compared to wild-type (Guedes *et al.*, 2006).

Autocrine secretion

Although immunocytes and various resident airway wall cells can secrete cytokines that modulate ASM function, airway myocytes themselves can also express a range of immunomodulatory molecules, some of which can act on ASM in an autocrine fashion (Panettieri, 2004). Chemokines that can be elaborated by airway myocytes include eotaxin (Ghaffar *et al.*, 1999), RANTES (Berkman *et al.*, 1996), TARC, IL-8, monocyte chemoattractant proteins (MCPs), fractalkine (Sukkar *et al.*, 2004), and stem cell factor (Brightling *et al.*, 2002). Cultured myocytes also synthesize such cytokines as IL-5, IL-1β (Hakonarson *et al.*, 1999), IL-10 (Grunstein *et al.*, 2001), GM-CSF (Saunders *et al.*, 1997), IL-11, IL-6, and leukaemia inhibitory factor (Elias *et al.*, 1997). IgE sensitization stimulates ASM to express IL-13 (Grunstein *et al.*, 2002) and IL-5; the latter in turn acts on the myocytes themselves to induce IL-1β, which can dysregulate constrictor and relaxant responses. However, the degree to which autocrine action of ASM-secreted cytokines alters ASM function or contributes in a quantitatively important fashion to airway inflammation more generally in authentic asthma remains uncertain.

RhoA

Rho-activated kinase potentiates smooth muscle contraction through its influence on the phosphorylation of the myosin regulatory light chain, MLC20. MLC20 is a 20-kDa protein that activates contraction when phosphorylated, and which can be dephosphorylated when the catalytic subunit of myosin phosphatase is directed to myosin by its myosin binding subunit. Rho kinase phosphorylates the myosin binding subunit and so inhibits such binding, thereby preventing MLC20 dephosphorylation and potentiating contraction. This pathway appears to play a role in allergen-induced bronchoconstrictor hyperresponsiveness in experimental animal models (Chiba *et al.*, 2001; Chiba and Misawa, 2004; Hashimoto *et al.*, 2002).

3.6 Summary

AHR is complex and multifactorial. ASM is certainly a significant contributor, not only through intrinsic dysfunctions, such as altered relaxation and perhaps abnormally reduced forced fluctuation induced relengthening, but probably also by responding normally to abnormal dynamic loads and by altering the airway itself through autocrine regulation and changes in the matrix and mass.

Acknowledgements

This work was supported by NIH Grants AI56352 and HL79398, and by the Amos Medical Faculty Development Program of the Robert Wood Johnson Foundation.

References

Affonce, D. A. and Lutchen, K. R. (2006). New perspectives on the mechanical basis for airway hyperreactivity and airway hypersensitivity in asthma. *J Appl Physiol* **101**, 1710–1719.

Altraja, A., Laitinen, A., Virtanen, I. *et al.* (1996). Expression of laminins in the airways in various types of asthmatic patients: a morphometric study. *Am J Respir Cell Mol Biol* **15**, 482–488.

Babu, K. S., Woodcock, D. A., Smith, S. E. *et al.* (2003). Inhaled synthetic surfactant abolishes the early allergen-induced response in asthma. *Eur Respir J* **21**, 1046–1049.

Bai, T. R. (1990). Abnormalities in airway smooth muscle in fatal asthma. *Am Rev Respir Dis* **141**, 552–557.

Bai, T. R., Mak, J. C. and Barnes, P. J. (1992). A comparison of beta-adrenergic receptors and in vitro relaxant responses to isoproterenol in asthmatic airway smooth muscle. *Am J Respir Cell Mol Biol* **6**, 647–651.

Barnes, P. J. and Pride, N. B. (1983). Dose-response curves to inhaled beta-adrenoceptor agonists in normal and asthmatic subjects. *Br J Clin Pharmacol* **15**, 677–682.

Bates, J. H., Wagers, S. S., Norton, R. J. *et al.* (2006). Exaggerated airway narrowing in mice treated with intratracheal cationic protein. *J Appl Physiol* **100**, 500–506.

Benayoun, L., Druilhe, A. Dombret, M. C. *et al.* (2003). Airway structural alterations selectively associated with severe asthma. *Am J Respir Crit Care Med* **167**, 1360–1368.

Berkman, N., Krishnan, V. L., Gilbey, T. *et al.* (1996). Expression of RANTES mRNA and protein in airways of patients with mild asthma. *Am J Respir Crit Care Med* **154**(6 pt 1), 1804–1811.

Black, J. L. (1996). Role of airway smooth muscle. *Am J Respir Crit Care Med* **153** (6 pt 2), S2–4.

Black, L. D., Henderson, A. C., Atileh, H. *et al.* (2004). Relating maximum airway dilation and subsequent reconstriction to reactivity in human lungs. *J Appl Physiol* **96**, 1808–1814.

Brightling, C. E., Bradding, P., Symon, F. A. *et al.* (2002). Mast-cell infiltration of airway smooth muscle in asthma. *N Engl J Med* **346**, 1699–1705.

Boulet, L. P., Laviolette, M., Turcotte, H. *et al.* (1997). Bronchial subepithelial fibrosis correlates with airway responsiveness to methacholine. *Chest* **112**, 45–52.

Carroll, N., Carello, S., Cooke, C. *et al.* (1993). The structure of large and small airways in nonfatal and fatal asthma. *Am Rev Respir Dis* **147**, 405–410.

Chen, B., Liu, G., Shardonofsky, F. *et al.* (2006). Tidal breathing pattern differentially antagonizes bronchoconstriction in C57BL/6J vs. A/J mice. *J Appl Physiol* **101**, 249–255.

Chiba, Y., Sakai, H. and Misawa, M. (2001). Augmented acetylcholine-induced translocation of RhoA in bronchial smooth muscle from antigen-induced airway hyperresponsive rats. *Br J Pharmacol* **133**, 886–890.

Chiba, Y. and Misawa, M. (2004). The role of RhoA-mediated Ca(2+) sensitization of bronchial smooth muscle contraction in airway hyperresponsiveness. *J Smooth Muscle Res* **40**(4–5), 155–167.

Chiba, Y., Sakai, H. and Misawa, M. (2001). Augmented acetylcholine-induced translocation of RhoA in bronchial smooth muscle from antigen-induced airway hyperresponsive rats. *Br J Pharmacol* **133**, 886–890.

Csete, M. E., Chediak, A. D., Abraham, W. M. *et al.* (1991). Airway blood flow modifies allergic airway smooth muscle contraction. *Am Rev Respir Dis* **144**, 59–63.

Deshpande, D. A., Dogan, S., Walseth, T. F. *et al.* (2004). Modulation of calcium signaling by interleukin-13 in human airway smooth muscle: role of CD38/cyclic adenosine diphosphate ribose pathway. *Am J Respir Cell Mol Biol* **31**, 36–42.

Deshpande, D. A., Walseth, T. F., Panettieri, R. A. *et al.* (2003). CD38/cyclic ADP-ribose-mediated Ca^{2+} signaling contributes to airway smooth muscle hyper-responsiveness. *FASEB J* **17**, 452–454.

Dowell, M. L., Lakser, O. J., Gerthoffer, W. T. *et al.* (2005). Latrunculin B increases force fluctuation-induced relengthening of ACh-contracted, isotonically shortened canine tracheal smooth muscle. *J Appl Physiol* **98**, 489–497.

Downie, S. R., Salome, C. M., Verbanck, S. *et al.* (2007). Ventilation heterogeneity is a major determinant of airway hyperresponsiveness in asthma, independent of airway inflammation. *Thorax* **62**, 684–689.

Ebina, M., Takahashi, T., Chiba, T. *et al.* (1993). Cellular hypertrophy and hyperplasia of airway smooth muscles underlying bronchial asthma. A 3-D morphometric study. *Am Rev Respir Dis* **148**, 720–726.

Elias, J. A., Wu, Y., Zheng, T. *et al.* (1997). Cytokine- and virus-stimulated airway smooth muscle cells produce IL-11 and other IL-6-type cytokines. *Am J Physiol* **273**(3 pt 1), L648–655.

Elliot, J., Vullermin, P., Carroll, N. *et al.* (1999). Increased airway smooth muscle in sudden infant death syndrome. *Am J Respir Crit Care Med* **160**, 313–316.

Evans, M. J., Van Winkle, L. S., Fanucchi, M. V. *et al.* (2000). Three-dimensional organization of the lamina reticularis in the rat tracheal basement membrane zone. *Am J Respir Cell Mol Biol* **22**, 393–397.

Farmer, S. G., Hay, D. W., Raeburn, D. *et al.* (1987). Relaxation of guinea-pig tracheal smooth muscle to arachidonate is converted to contraction following epithelium removal. *Br J Pharmacol* **92**, 231–236.

Fredberg, J. J., Inouye, D., Miller, B. *et al.* (1997). Airway smooth muscle, tidal stretches, and dynamically determined contractile states. *Am J Respir Crit Care Med* **156**, 1752–1759.

Gelb, A. F. and Zamel, N. (2002). Lung elastic recoil in acute and chronic asthma. *Curr Opin Pulm Med* **8**, 50–53.

Gelb, A. F., Licuanan, J., Shinar, C. M. *et al.* (2002). Unsuspected loss of lung elastic recoil in chronic persistent asthma. *Chest* **121**, 715–721.

Gelb, A. F., Schein, A., Nussbaum, E. *et al.* (2004). Risk factors for near-fatal asthma. *Chest* **126**, 1138–1146.

Ghaffar, O., Hamid, Q., Renzi, P. M. *et al.* (1999). Constitutive and cytokine-stimulated expression of eotaxin by human airway smooth muscle cells. *Am J Respir Crit Care Med* **159**, 1933–1942.

Grunstein, M. M., Hakonarson, H., Leiter, J. *et al.* (2001). Autocrine signaling by IL-10 mediates altered responsiveness of atopic sensitized airway smooth muscle. *Am J Physiol Lung Cell Mol Physiol* **281**, L1130–1137.

Grunstein, M. M., Hakonarson, H., Leiter, J. *et al.* (2002). IL-13-dependent autocrine signaling mediates altered responsiveness of IgE-sensitized airway smooth muscle. *Am J Physiol Lung Cell Mol Physiol* **282**, L520–528.

Grunstein, M. M., Whelan, R., Grunstein, J. S. *et al.* (2002). IL-13-dependent autocrine signaling mediates altered responsiveness of IgE-sensitized airway smooth muscle. *Am J Physiol Lung Cell Mol Physiol* **282**, L520–528.

Guedes, A. G., Paulin, J., Rivero-Nava, L. *et al.* (2006). CD38-deficient mice have reduced airway hyperresponsiveness following IL-13 challenge. *Am J Physiol Lung Cell Mol Physiol* **291**, L1286–1293.

Hakonarson, H., Halapi, E., Whelan, R. *et al.* (2001). Association between IL-1beta/TNF-alpha-induced glucocorticoid-sensitive changes in multiple gene expression and altered responsiveness in airway smooth muscle. *Am J Respir Cell Mol Biol* **25**, 761–771.

Hakonarson, H., Herrick, D. J., Serrano, P. G. *et al.* (1997). Autocrine role of interleukin 1beta in altered responsiveness of atopic asthmatic sensitized airway smooth muscle. *J Clin Invest* **99**, 117–124.

Hakonarson, H., Maskeri, N., Carter, C. *et al.* (1999). Autocrine interaction between IL-5 and IL-1beta mediates altered responsiveness of atopic asthmatic sensitized airway smooth muscle. *J Clin Invest* **104**, 657–667.

Hargreave, F. E., Dolovich, J., O'Byrne, P. M. *et al.* (1986). The origin of airway hyperresponsiveness. *J Allergy Clin Immunol* **78**(5 pt 1), 825–832.

Hargreave, F. E., Ryan, G., Thomson, N. C. *et al.* (1982). Bronchial responsiveness to histamine or methacholine in asthma: measurement and clinical significance. *Eur J Respir Dis Suppl* **121**, 79–88.

Hashimoto, K., Peebles, R. S., Sheller, J. R. *et al.* (2002). Suppression of airway hyperresponsiveness induced by ovalbumin sensitisation and RSV infection with Y-27632, a Rho kinase inhibitor. *Thorax* **57**, 524–527.

Hays, S. R., Peyrol, S., Gindre, D. *et al.* (2005). Structural changes to airway smooth muscle in cystic fibrosis. *Thorax* **60**, 226–228.

Henderson, A. C., Ingenito, E. P., Atileh, H. *et al.* (2003). How does airway inflammation modulate asthmatic airway constriction? An antigen challenge study. *J Appl Physiol* **95**, 873–882.

Hohlfeld, J. M. (2002). The role of surfactant in asthma. *Respir Res* **3**, 4.

Hohlfeld, J. M., Ahlf, K., Enhorning, G. *et al.* (1999). Dysfunction of pulmonary surfactant in asthmatics after segmental allergen challenge. *Am J Respir Crit Care Med* **159**, 1803–1809.

Huber, H. L. and Koessler, K. K. (1922). The pathology of bronchial asthma. *Arch Intern Med* **30**, 689–760.

Jiang, H. and Stephens, N. L. (1992). Isotonic relaxation of sensitized bronchial smooth muscle. *Am J Physiol* **262**(3 pt 1), L344–350.

Johnson, P. R., Roth, M., Tamm, M. *et al.* (2001). Airway smooth muscle cell proliferation is increased in asthma. *Am J Respir Crit Care Med* **164**, 474–477.

King, G. G., Pare, P. D. and Seow, C. Y. (1999). The mechanics of exaggerated airway narrowing in asthma: the role of smooth muscle. *Respir Physiol* **118**, 1–13.

Kuyper, L. M., Pare, P. D., Hogg, J. C. *et al.* (2003). Characterization of airway plugging in fatal asthma. *Am J Med* **115**, 6–11.

Lakser, O. J., Lindeman, R. P. and Fredberg, J. J. (2002). Inhibition of the p38 MAP kinase pathway destabilizes smooth muscle length during physiological loading. *Am J Physiol Lung Cell Mol Physiol* **282**, L1117–1121.

Latourelle, J., Fabry, B. and Fredberg, J. J. (2002). Dynamic equilibration of airway smooth muscle contraction during physiological loading. *J Appl Physiol* **92**, 771–779.

Lozewicz, S., Wells, C., Gomez, E. *et al.* (1990). Morphological integrity of the bronchial epithelium in mild asthma. *Thorax* **45**, 12–15.

Ma, X., Cheng, Z., Kong, H. *et al.* (2002). Changes in biophysical and biochemical properties of single bronchial smooth muscle cells from asthmatic subjects. *Am J Physiol Lung Cell Mol Physiol* **283**, L1181–1189.

Martinez, F. D. (1991). Sudden infant death syndrome and small airway occlusion: facts and a hypothesis. *Pediatrics* **87**, 190–198.

McCray, P. B., Jr. (1993). Spontaneous contractility of human fetal airway smooth muscle. *Am J Respir Cell Mol Biol* **8**, 573–580.

Minshall, E. M., Leung, D. Y., Martin, R. J. *et al.* (1997). Eosinophil-associated TGF-beta1 mRNA expression and airways fibrosis in bronchial asthma. *Am J Respir Cell Mol Biol* **17**, 326–333.

Moessinger, A. C., Harding, R., Adamson, T. M. *et al.* (1990). Role of lung fluid volume in growth and maturation of the fetal sheep lung. *J Clin Invest* **86**, 1270–1277.

Ohno, I., Nitta, Y., Yamauchi, K. *et al.* (1996). Transforming growth factor beta 1 (TGF beta 1) gene expression by eosinophils in asthmatic airway inflammation. *Am J Respir Cell Mol Biol* **15**, 404–409.

Ordonez, C. L., Khashayar, R., Wong, H. H. *et al.* (2001). Mild and moderate asthma is associated with airway goblet cell hyperplasia and abnormalities in mucin gene expression. *Am J Respir Crit Care Med* **163**, 517–523.

Panettieri, R. A., Jr. (2004). Airway smooth muscle: immunomodulatory cells? *Allergy Asthma Proc* **25**, 381–386.

Rietkerk, M., Dekker, S. C., de Ruiter, P. C. *et al.* (2004). Self-organized patchiness and catastrophic shifts in ecosystems. *Science* **305**(5692), 1926–1929.

Saunders, M. A., Mitchell, J. A., Seldon, P. M. *et al.* (1997). Release of granulocyte-macrophage colony stimulating factor by human cultured airway smooth muscle cells: suppression by dexamethasone. *Br J Pharmacol* **120**, 545–546.

Schittny, J. C., Miserocchi, G. and Sparrow, M. P. (2000). Spontaneous peristaltic airway contractions propel lung liquid through the bronchial tree of intact and fetal lung explants. *Am J Respir Cell Mol Biol* **23**, 11–18.

Solway, J. and Fredberg, J. J. (1997). Perhaps airway smooth muscle dysfunction contributes to asthmatic bronchial hyperresponsiveness after all. *Am J Respir Cell Mol Biol* **17**, 144–146.

Sukkar, M. B., Issa, R., Xie, S., *et al.* (2004). Fractalkine/CX3CL1 production by human airway smooth muscle cells: induction by IFN-gamma and TNF-alpha and regulation by TGF-beta and corticosteroids. *Am J Physiol Lung Cell Mol Physiol* **287**, L1230–1240.

Tao, F. C., Tolloczko, B., Eidelman, D. H. *et al.* (1999). Enhanced Ca(2+) mobilization in airway smooth muscle contributes to airway hyperresponsiveness in an inbred strain of rat. *Am J Respir Crit Care Med* **160**, 446–453.

Tgavalekos, N. T., Musch, G., Harris, R. S. *et al.* (2007). Relationship between airway narrowing, patchy ventilation and lung mechanics in asthmatics. *Eur Respir J* **29**, 1174–1781.

Tgavalekos, N. T., Tawhai, M., Harris, R. S. *et al.* (2005). Identifying airways responsible for heterogeneous ventilation and mechanical dysfunction in asthma: an image functional modeling approach. *J Appl Physiol* **99**, 2388–2397.

Thomson, N. C. (1987). In vivo versus in vitro human airway responsiveness to different pharmacologic stimuli. *Am Rev Respir Dis* **136**(4 pt 2), S58–62.

Venegas, J. (2007). Linking ventilation heterogeneity and airway hyperresponsiveness in asthma. *Thorax* **62**, 653–654.

Venegas, J. G., Winkler, T., Musch, G. *et al.* (2005). Self-organized patchiness in asthma as a prelude to catastrophic shifts. *Nature* **434**(7034), 777–782.

Wagers, S., Lundblad, L. K., Ekman, M. *et al.* (2004). The allergic mouse model of asthma: normal smooth muscle in an abnormal lung? *J Appl Physiol* **96**, 2019–2027.

Wagers, S., Norton, R., Bates, J. *et al.* (2003). Role of fibrin in determining airway closure. *Chest* **123**(3 Suppl), 362S–363S.

Wagers, S. S., Haverkamp, H. C., Bates, J. H. *et al.* (2007). Intrinsic and antigen-induced airway hyperresponsiveness are the result of diverse physiological mechanisms. *J Appl Physiol* **102**, 221–230.

Wagers, S. S., Norton, R. J., Rinaldi, L. M. *et al.* (2004). Extravascular fibrin, plasminogen activator, plasminogen activator inhibitors, and airway hyperresponsiveness. *J Clin Invest* **114**, 104–111.

Wanner, A., Chediak, A. D. and Csete, M. E. (1990). Airway mucosal blood flow: response to autonomic and inflammatory stimuli. *Eur Respir J Suppl* **12**, 618s–623s.

Winkler, T. and Venegas, J. G. (2007). Complex airway behavior and paradoxical responses to bronchoprovocation. *J Appl Physiol* **103**, 655–663.

Woodruff, P. G., Dolganov, G. M., Ferrando, R. E. *et al.* (2004). Hyperplasia of smooth muscle in mild to moderate asthma without changes in cell size or gene expression. *Am J Respir Crit Care Med* **169**, 1001–1006.

Zhang, S., Smartt, H., Holgate, S. T. *et al.* (1999). Growth factors secreted by bronchial epithelial cells control myofibroblast proliferation: an in vitro co-culture model of airway remodeling in asthma. *Lab Invest* **79**, 395–405.

4

Airway smooth muscle phenotypic and functional plasticity

Andrew J. Halayko[1,2], Reinoud Gosens[1−3] and Thai Tran[1,2,4]

[1] *Departments of Physiology and Internal Medicine, Section of Respiratory Diseases, and National Training Program in Allergy and Asthma, University of Manitoba, Winnipeg, MB, Canada*

[2] *Biology of Breathing Theme, Manitoba Institute of Child Health, Winnipeg, MB, Canada*

[3] *Current address – Department of Molecular Pharmacology, University of Groningen, The Netherlands*

[4] *Current Address – Department of Physiology, National University of Singapore, Singapore*

4.1 Introduction

Airway smooth muscle has long been known to be a primary determinant of airway physiology in health and disease by virtue of its ability to contract, and thereby control the diameter of the bronchi and bronchioles that it encircles.

Airway Smooth Muscle   Edited by Kian Fan Chung
© 2008 John Wiley & Sons, Ltd

Classical physiological and pharmacological investigation has focused on factors that determine the contractility of airway smooth muscle, including its mechanical properties, response to biological and chemical spasmogens and relaxing factors, and structure–function relationships in the airway wall. Many of these issues are discussed in other chapters in this volume. The scope of contemporary research involving airway smooth muscle has broadened greatly, accommodating new understanding from the past decade that, unlike striated muscle, differentiated airway smooth muscle cells retain a capacity for phenotype plasticity. This equips myocytes with the functional potential to be engaged in multiple biological processes associated with contraction, inflammation, and wound healing. Thus, in obstructive airways disease, myocytes can control airway diameter acutely via reversible bronchospasm, and chronically as a central driver of airway wall remodelling, which underpins irreversible reduction in airway conductance. This chapter provides an overview of existing paradigms for airway smooth muscle phenotype plasticity, the mechanisms that control phenotype and functional plasticity, and its contribution to airway pathobiology.

4.2 Historical perspective: smooth muscle phenotype plasticity

Over 40 years has passed since Robert Wissler asked whether the arterial medial cell is 'smooth muscle or multifunctional mesenchyme' (Wissler, 1967). Indeed, he hypothesized that arterial smooth muscle cells both subserve a contractile function to regulate vessel diameter, and are the primary effectors of fibroproliferative changes associated with normal vessel growth and the pathogenesis of atherosclerosis. His model evolved from a 'unicellular concept' proposed 10 years earlier by Pease and Paule (1960), who recognized that 'smooth muscle is virtually the only cell type in the arterial media'. In the decade after Wissler's proposal, major advances in developing primary cultures of myocytes from the blood vessels of adult animals and man were made, providing an important experimental system to investigate phenotype diversity, and to identify ultrastructural and biochemical markers of vascular smooth muscle cell populations (Chamley-Campbell *et al.*, 1979). These observations established a paradigm for the existence of a phenotypic spectrum of differentiated myocytes in which diversity results from reversible switching of myocytes between contractile and synthetic functional states (Figure 4.1). Moreover, the phenotypic and functional diversity of arterial smooth muscle cells is now accepted as a

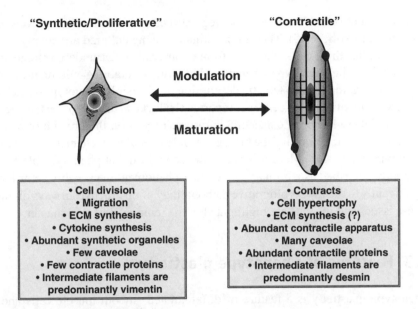

Figure 4.1 Reversible phenotypic plasticity of airway smooth muscle. Modulation of myocytes to a functional 'synthetic/proliferative' state is induced by placing cells in low density primary culture, exposure to mitogens, and adherence to extracellular matrix (ECM) proteins such as fibronectin and collagen-1. Maturation to a functional 'contractile' state occurs in cell culture at high cell density, in response to mitogen withdrawal, exposure to insulin, and adherence to laminin-rich ECM. 'Synthetic/proliferative' myocytes predominate in the developing lung and primary cell culture, and their numbers are thought to be increased in adult airways in association with fibroproliferative disorders. 'Contractile' myocytes predominate in normal adult tissues, and exhibit variable degrees of maturation based on the level of molecular and functional markers. Key functional, ultrastructural and biochemical features of the extreme phenotypic states are shown

central mechanism in atherogenesis and post-angioplasty restenosis (Iyemere *et al.*, 2006).

It was not until after 1980 that Avner *et al.* (1981), and then Brown and colleagues (Tom-Moy *et al.*, 1987) described the ultrastructure, filamentous organization of contractile proteins, and the pharmacological properties of primary cultured canine tracheal smooth muscle cells. In 1989, independent reports from Twort and van Breeman (1989) and Panettieri *et al.* (1989) first characterized primary cultured human airway smooth muscle cells, and investigated their responsiveness to physiological agonists. Over the next few years, the number of studies using cultured airway myocytes increased, and these chiefly aimed to understand the expression and coupling of physiologically relevant receptors and ion channels, and to elucidate factors that regulate cell proliferation (for a review see Hall and Kotlikoff, 1995). Though investigators using cultured airway myocytes recognized that these cells, like those from the arterial media, undergo modulation from a contractile phenotype when placed in serum-containing culture, it was not until 1996 that the first systematic description of reversible phenotype-switching for airway smooth muscle cells was reported (Halayko *et al.*, 1996). In the decade following, there has been considerable research focus on the breadth of airway smooth muscle function, and the biological and molecular mechanisms regulating phenotype expression. Moreover, it is now recognized that phenotype plasticity, originally described as a primary cell culture phenomenon, probably contributes significantly to the fibroproliferative pathobiology of obstructive airways disease. These issues and insights are highlighted in the remainder of this chapter.

4.3 Features of phenotype plasticity

Phenotypic plasticity is a feature of differentiated smooth muscle cells and is manifest as the reversible modulation and maturation of individual myocytes both *in vitro* and *in vivo* (Halayko and Solway, 2001; Owens, 1995). Switching of smooth muscle phenotype requires the differential expression of repertoires of phenotype-specific genes and subsequent accumulation of the proteins that they encode. Primary culture models using smooth muscle cells, including those from the airways, reveal that contractile smooth muscle cells undergo spontaneous phenotype *modulation* when seeded at a subconfluent density in the presence of mitogens (Chamley-Campbell *et al.*, 1979; Halayko *et al.*, 1996; Owens, 1995). Phenotype modulation promotes acquisition of a synthetic phenotype, in which cells are marked by abundant organelles for protein and lipid synthesis, and numerous mitochondria (Figure 4.1). The cells also exhibit a high proliferative index, but lose responsiveness to physiological contractile agonists and exhibit greatly reduced contractile apparatus and associated proteins (Chamley-Campbell *et al.*, 1979; Mitchell *et al.*, 2000). The reversion of differentiated smooth muscle cells to a contractile phenotype is called *maturation*, and in cell culture it occurs as cells grow to confluence and achieve contact inhibition (Figure 4.1). Maturation is marked by accumulation of contractile apparatus and associated proteins, reacquisition of responsiveness to physiological contractile agonists, and decreased abundance of synthetic organelles (Halayko *et al.*, 1999; Mitchell *et al.*, 2000). Some of the best-characterized markers for the contractile phenotype include smooth muscle (sm)-α-actin (Gabbiani *et al.*, 1981), sm-γ-actin (Sawtell and Lessard, 1989), sm-myosin heavy chain (smMHC) (Nagai *et al.*, 1989), calponin

(Gimona *et al.*, 1992), *h*-caldesmon (Frid *et al.*, 1992), SM22 (Gimona *et al.*, 1992; Solway *et al.*, 1995), desmin (Tran *et al.*, 2006), caveolin-1 (Gosens *et al.*, 2006), and smoothelin (van der Loop *et al.*, 1996).

Phenotype maturation of myocytes in confluent cultures is accelerated by the withdrawal of mitogens, and can be promoted by supplementation with insulin (Halayko *et al.*, 1999; Schaafsma *et al.*, 2007). A distinct subset of airway smooth muscle cells mature into a functionally contractile phenotype with elongated morphology, fully reconstituted contractile apparatus, abundant contractile protein content, and cell-surface recoupling of muscarinic M_3 receptors for acetylcholine (Halayko *et al.*, 1999; Ma *et al.*, 1998; Mitchell *et al.*, 2000). Similar reconstitution of contractile phenotype arterial myocytes in culture has also been reported (Li *et al.*, 1999). Maturation appears to require endogenously expressed laminin-2, and is reliant on individual cells expressing a unique repertoire of laminin-binding α-integrin subunits (Tran *et al.*, 2007; Tran and Halayko, 2007). Notably, maturation of the large contractile cells appears to parallel the process of cellular hypertrophy, as it is reliant on signalling pathways that control protein translation (Goldsmith *et al.*, 2006; Halayko *et al.*, 2004; Zhou *et al.*, 2005). Insofar as increased smooth muscle mass is a feature of airways remodelling in asthma, further evaluation of myocyte maturation is likely to contribute better understanding of the remodelling process.

As modulation of contractile smooth muscle cells from adult tissue appears to involve recapitulation of gene-expression profiles that, in part, resemble those in the developing lung, reversible phenotype switching has sometimes been described as 'differentiation' and 'de-differentiation'. It is, however, important to appreciate that these processes are fundamentally distinct, as phenotype plasticity occurs in already differentiated smooth muscle cells in mature organs, a process that is important in disease pathogenesis and physiological modelling concomitant with organ growth. In contrast, differentiation occurs during lung development, during which airway smooth muscle cells are derived from the undifferentiated lung mesenchyme that surrounds growing, fluid-filled epithelial tubes that derive from the laryngotracheal bud. Commitment and differentiation of embryonic lung mesenchymal cells to the smooth muscle lineage are subject to interactions with epithelial cells, which release pro-differentiation factors, such as Sonic hedgehog and transforming growth factor β (TGF-β), and with the lung mesothelium, which releases differentiation suppressor signals, such as fibroblast growth factor-9 (Cardoso and Lu, 2006; Weaver *et al.*, 2003). Behind the leading edge of the growing epithelial tube, committed mesenchymal cells are subsequently triggered to differentiate into smooth muscle by two factors: binding to the newly formed, laminin-rich basement membrane beneath differentiated epithelia, and exposure to mechanical signals imposed by intraluminal fluids (Cardoso and Lu, 2006; Weaver *et al.*, 2003). Thus, factors that affect smooth muscle differentiation are important determinants of lung development and structural modelling, whereas signals that regulate phenotype plasticity are more relevant to disease processes and structural changes associated with normal growth of mature organs.

4.4 Mechanisms for phenotypic plasticity

The airway smooth muscle phenotype is regulated by extracellular stimuli, including growth factors, contractile agonists acting on G protein-coupled receptors, and extracellular matrix proteins. Depending on the type of stimulus, smooth muscle cells can be induced to undergo modulation to a proliferative/synthetic phenotype or to follow a maturation process leading to the expression of a functionally contractile state. Growth factors such as TGF-β and insulin, and extracellular matrix proteins of the laminin family support the induction of a contractile airway smooth muscle phenotype, whereas growth factors, such as platelet-derived growth factor (PDGF) or fetal bovine serum (FBS), and extracellular matrix proteins, such as fibronectin, support proliferative functions of the smooth muscle cell at the expense of the capacity to contract (Dekkers *et al.*, 2007; Gosens *et al.*, 2003; Hirst *et al.*, 2000; Tran *et al.*, 2006). The differential capacity of growth factors and extracellular matrix proteins to modulate or mature the smooth muscle phenotype is probably explained by the induction of unique cassettes of intracellular signalling pathways.

Maturation of airway smooth muscle cells requires the accumulation of contractile and regulatory proteins, such as SM22, sm-α-actin, smMHC, calponin, and desmin. Regulation of the expression of these proteins requires coordinated control of transcriptional and translational processes (Halayko *et al.*, 2006; Halayko and Solway, 2001). A number of intracellular signalling cascades, including the Rho/Rho kinase pathway and Class 1 PI3K (phosphatidyl inositide 3-kinase)-dependent pathways, have been associated with transcription and translation of smooth muscle-specific proteins, respectively (Camoretti-Mercado *et al.*, 2000; 2006; Liu *et al.*, 2003; Halayko *et al.*, 2004; Zhou *et al.*, 2005). These pathways are illustrated in Figure 4.2, and are discussed below.

A common motif shared by 5'-promoter regions of smooth muscle-specific genes is the presence of CArG box elements [CC(A/T)6GG] that bind dimers of serum response factor (SRF); these sites are essential for smooth muscle-specific gene expression, as their mutation renders their promoters inactive (e.g., SM22 and smMHC) (Halayko *et al.*, 2006; Solway *et al.*, 1995; 1998). In the

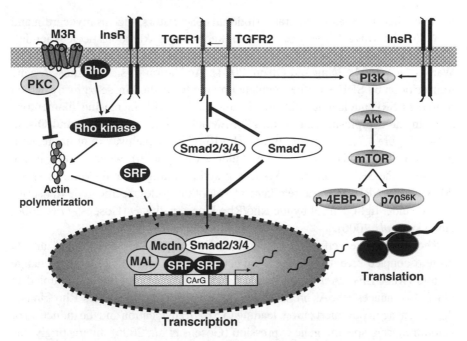

Figure 4.2 Mechanisms of airway smooth muscle phenotype maturation. Maturation of airway smooth muscle cells requires the accumulation of contractile and contraction regulatory proteins; expression is under coordinated control of transcriptional and translational processes. Smooth muscle-specific gene transcription is chiefly regulated by the transcription factor SRF and its coactivators myocardin (Mcdn) and MAL. Nuclear localization of SRF is regulated by Rho/Rho kinase-mediated actin polymerization. Protein kinase C activation can counter effects of Rho/Rho kinase by inducing actin depolymerization. Myocardin binding to SRF can be prevented competitively by phosphorylated Elk-1, which is targeted by p42/p44 MAP kinase (not shown). Smad signalling, activated by TGF-β receptors, can enhance smooth muscle-specific gene transcription by the interaction of SRF with regulatory Smad-2, -3, and -4; inhibitory Smad, Smad-7, suppresses the effects of regulatory Smads on SRF. Translation of smooth muscle-specific gene transcripts into proteins requires intracellular signalling by PI3K-dependent pathways. PI3K signalling also disrupts the interaction of the inhibitory transcription factor, Fox04, with nuclear SRF (not shown). The pathways involved with gene transcription and protein translation are regulated by extracellular stimuli such as acetylcholine, TGF-β, insulin, and growth factors (e.g., PDGF). See text for further details. InsR: insulin receptor; M3R: muscarinic M_3 receptor; mTOR: mammalian target of rapamycin; PI3K: phosphatidyl inositide 3-kinase; SRF: serum response factor; Mcdn: myocardin; PKC: protein kinase C

nucleus, SRF associates with transcriptional coactivators, such as myocardin and MAL/MKL1 (megakaryocytic acute leukaemia/megakaryoblastic leukaemia), which direct SRF to smooth muscle-specific promoters (Miralles *et al.*, 2003; Wang *et al.*, 2004; Wang and Olson, 2004). Activation of smooth muscle gene transcription by SRF is further regulated by the Rho/Rho kinase pathway, which promotes SRF nuclear localization in association with increasing filamentous actin in the cytoplasm (Liu *et al.*, 2003; Mack *et al.*, 2001; Wang and Olson, 2004) (Figure 4.2). The ability of the Rho/Rho kinase pathway to promote actin polymerization leads to a loss of globular actin (g-actin), which results in the release of the SRF coactivator MAL, a g-actin-binding protein (Miralles *et al.*, 2003). Thus, SRF, a central regulator of smooth muscle-specific gene transcription, is under tight control by the Rho/Rho kinase pathway (Gosens *et al.*, 2006; Halayko *et al.*, 2006).

Rho/Rho kinase activation is regulated by receptor tyrosine kinases and G protein-coupled receptors through the action of Rho-specific guanine exchange factors (RhoGEFs). Activation of muscarinic M_3 receptors that are coupled to $G\alpha q$ also induces RhoA, probably via p63RhoGEF, and promotes Rho kinase-dependent actin polymerization, leading to SRF translocation and the induction of smooth muscle-specific gene expression (Gosens *et al.*, 2006). Interestingly, the concomitant induction of conventional protein kinase C isoforms by $G\alpha q$-coupled receptors appears to balance the actin-dependent, pro-transcriptional effects of RhoA, indicating the existence of complex feedback mechanisms (Wang *et al.*, 2004) (Figure 4.2). Insulin-induced expression of contractile phenotype markers and the induction of a functionally hypercontractile phenotype also requires the Rho/Rho kinase pathway, though the GEFs involved have not yet been identified (Gosens *et al.*, 2003; Schaafsma *et al.*, 2007). Collectively, this indicates that stimuli that induce Rho/Rho kinase signalling have the capacity to induce smooth muscle phenotype maturation, probably through action on SRF-mediated gene transcription (Figure 4.2).

TGF-β, a potent inducer of a hypertrophic and hypercontractile airway smooth muscle phenotype, promotes SRF-dependent gene transcription through activation of Smad signalling (Camoretti-Mercado *et al.*, 2006; Goldsmith *et al.*, 2006). TGF-β acts on TGFβ receptor-1, which then dimerizes with the constitutively active TGFβ receptor-2 and induces phosphorylation and nuclear translocation of 'regulatory' Smad-2, -3, and -4. Regulatory Smads can bind with nuclear SRF to promote smooth muscle-specific gene transcription, whereas Smad-7 counteracts this effect (Camoretti-Mercado *et al.*, 2006) (Figure 4.2). TGF-β also regulates smooth muscle maturation by activating TGF-β control elements (TCE) located on the promoters of smooth muscle-specific genes. Collective understanding

indicates that SRF-dependent gene transcription is controlled by multiple signalling pathways, which can be selectively activated by extracellular stimuli.

The transcripts from smooth muscle-specific genes must be translated into proteins to affect myocyte phenotype and function. Protein translation is controlled by
several pathways that converge at the level of the ribosome, and a number of these
are required for smooth muscle maturation. Mature smooth muscle cells express
elevated levels of active, phosphorylated kinases with known effects on protein
translation, including PI3K, Akt1, mTOR (mammalian target of rapamycin), and
p70 ribosomal S6 kinase (Halayko *et al.*, 2004). Pharmacological inhibition of
PI3K and mTOR is sufficient to prevent p70 ribosomal S6 kinase activation and
accumulation of smooth muscle-specific proteins (Halayko *et al.*, 2004). Furthermore, active mTOR can phosphorylate and activate 4E-BP1, a protein that binds to
and activates the eukaryotic initiation factor, eIF4, which initiates protein translation and contractile protein accumulation (Zhou *et al.*, 2005). Activation of these
PI3K-dependent signalling pathways is required for TGF-β and insulin-induced
airway smooth muscle maturation (Goldsmith *et al.*, 2006; Gosens *et al.*, 2003;
Schaafsma *et al.*, 2007), indicating that in addition to SRF-dependent gene transcription, the translation of smooth muscle-specific proteins is tightly regulated
by specific intracellular signalling pathways (Figure 4.2).

Less is known about the specific mechanisms that trigger modulation of smooth
muscle cells to a more immature phenotype. Highly mitogenic growth factors,
such as PDGF or FBS, induce a proliferative airway smooth muscle phenotype
that is accompanied by a loss in contractile function due to loss of contractile
and contraction regulatory proteins (Gosens *et al.*, 2002; Halayko and Solway,
2001). Paradoxically, activation of the early response gene *c-fos*, which regulates
cell proliferation, requires SRF to bind to CArG elements embedded in serum
response elements that are present in the *c-fos* promoter. The apparent duality of
SRF to induce pro-differentiation and pro-proliferative genes can be explained by
the existence of multiple transcriptional coactivators that compete for binding to
SRF, and thereby direct selective induction of gene transcription. When bound to
myocardin, SRF induces smooth muscle-specific genes; in contrast, when bound
to ternary complex factors (TCFs) such as phospho-Elk-1, SRF induces proliferative genes such as *c-fos* (Wang *et al.*, 2004). As Elk-1 is phosphorylated
by p42/p44 MAP (mitogen-activated protein) kinase, the regulation of p42/p44
MAP kinase is a key determinant of the smooth muscle cell phenotype; for example, mitogen-induced phenotype modulation is prevented by inhibitors of the
p42/p44 MAP kinase pathway (Gosens *et al.*, 2002; Roy *et al.*, 2001; Wang *et al.*,
2004; Wang and Olson, 2004). Thus, there appears to be a crucial role for growth
factor-induced p42/p44 MAP kinase signalling in determining SRF-dependent

gene transcription targets, and in controlling pro-contractile to pro-mitogenic functional responses of smooth muscle cells. In contrast to the suppressive effects of p42/p44 MAP kinase, a recent study revealed that PI3K-Akt1 signalling leads to phosphorylation of Fox04 forkhead transcription factor, which in its unphosphorylated state binds to nuclear myocardin and inhibits its interaction with SRF (Liu *et al.*, 2005). Thus, targeting of Fox04 by Akt1 promotes myocardin-SRF binding and transcription of smooth muscle-specific genes, whereas suppression of PI3K-Akt1 (and Foxo4 phosphorylation) probably underpins phenotype modulation. Clearly, transcriptional mechanisms controlling phenotype dynamics of smooth muscle involve multiple pathways and downstream targets and require complex spatial and temporal integration in response to external stimuli.

4.5 Functional plasticity of airway smooth muscle: role in asthma pathogenesis

Mature airway smooth muscle cells retain the ability of phenotype plasticity that confers multifunctional capacity for a range of cellular responses that include contraction, proliferation and hypertrophy, migration, and synthesis of pro-inflammatory and pro-fibrotic cytokines (Figure 4.3). The importance of the broad functional capacity of these cells in the progression of obstructive airways disease is emerging (Halayko *et al.*, 2006). By virtue of their plastic nature myocytes can contribute directly to both reversible, intermittent bronchial spasm and to structural changes in the airway wall that are associated with irreversible loss of lung function in patients with long-standing obstructive airways disease (Figure 4.4). This model is consistent with well-established schema for the role of arterial myocytes in atherogenesis and post-angioplasty restenosis (Campbell *et al.*, 1987; Halayko and Stephens, 1994; Pease and Paule, 1960).

The capacity of airway smooth muscle cells to synthesize and release pro-inflammatory biomolecules in response to a plethora of stimuli is well documented (Howarth *et al.*, 2004). Thus, concomitant with the increased inflammatory microenvironment that develops in asthmatic airways, airway smooth muscle cells that undergo phenotype maturation can produce pro-inflammatory cytokines and support local inflammation. This may also constitute an autocrine mechanism for maintaining airway smooth muscle cells in a synthetic/proliferative phenotype, thereby promoting pathogenic aspects of asthma (Chan *et al.*, 2006; Johnson *et al.*, 2000; 2004; Peng *et al.*, 2005). Details of the repertoire and mechanisms underlying inflammatory mediator expression and release by airway myocytes are

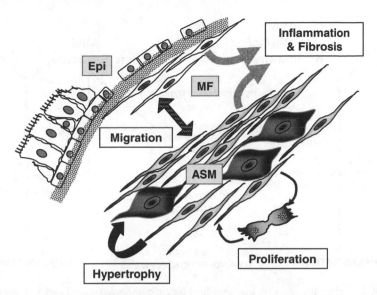

Figure 4.3 Phenotypic and functional plasticity of mesenchymal cells in airway remodelling. ASM cells (dark grey) and myofibroblasts (grey) potentiate inflammation and fibrosis in the airway wall by releasing pro-inflammatory mediators, and increasing expression of cell adhesion molecules and extracellular matrix (ECM) components. Increased ASM mass results from myocyte proliferation and cellular hypertrophy induced by growth factors, pro-inflammatory mediators, and ECM components of the airway wall toward the ASM layer may also contribute to ASM thickening. In this case, upon reaching the ASM compartment, mesenchymal cells undergo maturation to a fully contractile state; the entire process is probably modulated by locally produced cytokines, chemokines, and ECM. Migration of phenotypically modulated ASM to the submucosal compartment leads to increased numbers of subepithelial myofibroblasts. ASM: airway smooth muscle; EPI: epithelium; MF: myofibroblasts

presented in Chapter 8 of this volume, and have been described elsewhere (Halayko and Amrani, 2003).

Cultured airway smooth muscle cells from asthmatics produce increased amounts and an altered composition of extracellular matrix proteins (Johnson, 2001; Johnson *et al.*, 2000). This contributes to fibrotic changes associated with asthma that affect airway and lung function due to thickening of the airway wall, changes in the structural properties of the airway wall, and changes in airway interdependence with the surrounding parenchyma. Notably, the extracellular matrix appears to affect all aspects of the functional repertoire of smooth muscle cells (Figure 4.4). Seeding airway smooth muscle onto fibronectin or collagen type-1 promotes a proliferative phenotype, whereas laminin-rich matrices promote

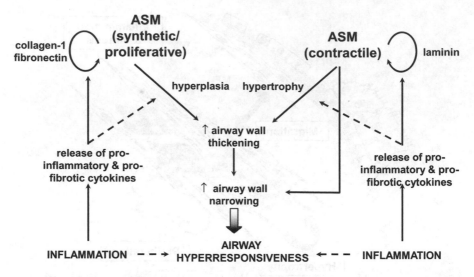

Figure 4.4 Schematic showing the cascade of biologic responses associated with phenotype plasticity of airway smooth muscle cells in the pathogenesis of obstructive airways disease and airways hyperresponsiveness. ASM: airway smooth muscle

retention or maturation of a contractile phenotype (Hirst *et al.,* 2000; Tran *et al.*, 2006; 2007). Details of changes in matrix protein composition and its impact on myocyte function are discussed in Chapter 6; however, it is worth noting that laminin-2, which is required for myocyte maturation (Tran *et al.*, 2006), is increased in the asthmatic airway. Thus, it appears that the expression of this glycoprotein and it receptors may be a central intrinsic mechanism regulating phenotype maturation and hypertrophy in the adult airway. To date, relatively little work has focused on the phenotype dependence of extracellular matrix receptor expression by airway smooth muscle, but this is clearly an area that warrants considerable attention in the future.

Increased airway smooth muscle mass in asthmatic patients results from cellular hyperplasia and hypertrophy, and is a principal factor underlying the excessive airways narrowing that is characteristic of the disease (Dunnill *et al.*, 1969; Ebina *et al.*, 1993). There has been considerable investigation of the molecular signals that regulate airway smooth muscle proliferation *in vitro* to explain increased airway smooth muscle mass in asthmatics (see Chapter 3). Modulation of airway smooth muscle cells toward the proliferative phenotype induced by asthma-associated mitogens promotes an increase in the number of myocytes in the airway wall. Interestingly, airway smooth muscle cells cultured from biopsies taken from asthmatics proliferate at a much faster rate than those cultured from healthy individuals, and this property appears to be directly linked to changes in the profile of extracellular matrix proteins synthesized by 'asthmatic' myocytes (Johnson, 2001; Johnson *et al.*, 2001; 2004). This suggests the existence of an intrinsic, stable abnormality in the phenotype of myocytes in asthmatics that is associated with altered functional responses to asthma-associated biomolecules. Airway smooth muscle cell hypertrophy appears to contribute to an increase in airway smooth muscle mass in asthma (Benayoun *et al.*, 2003; Ebina *et al.*, 1993). Pro-hypertrophic factors identified in cell culture studies include cardiotrophin, TGF-β1 and endothelin-1 (Goldsmith *et al.*, 2006; McWhinnie *et al.*, 2006; Zhou *et al.*, 2003). Increased airway smooth muscle mass, even in mild-to-moderate asthma, is accompanied by increased abundance of proteins associated with contraction, such as myosin light-chain kinase, which underpins increased contractility of airway smooth muscle tissue in asthma and in asthma models (Ammit *et al.*, 2000; Benayoun *et al.*, 2003; Jiang *et al.*, 1992; Woodruff *et al.*, 2004). This is noteworthy, as airway myocyte maturation appears to be a parallel process to myocyte hypertrophy (Halayko *et al.*, 2004); this suggests that increased abundance of contractile phenotype marker proteins may result from myocyte responses that are consistent with the maturation of contractile myocytes that occurs during prolonged, serum-free cell culture (Halayko *et al.*, 1999; 2004).

Additional mechanisms that have been suggested to account for increased in airway smooth muscle mass in asthma also encompass the multifunctional capacity of these cells. Migration of fibroblasts from the submucosa to the muscle layer with concomitant maturation to a contractile myocyte phenotype could underpin accumulation of airway smooth muscle tissue. A similar mechanism, in which circulating mesenchymal stem cells home and migrate into damaged airways, and then undergo differentiation to smooth muscle phenotypes, has been postulated (Stewart, 2004). Similar mechanisms have also been suggested to account for increases in the number of myofibroblasts in the airway submucosa, whereby airway smooth muscle cells are thought to undergo phenotype modulation, with subsequent migration toward the epithelium. Hard evidence for the existence of these cellular responses *in vivo* has not yet been reported; however, there is evidence from *in vitro* systems that support their plausibility. For example, stimuli such as PDGF and leukotrienes, which are upregulated during an acute asthma attack, can induce airway smooth muscle migration (Goncharova *et al.*, 2003; Parameswaran *et al.*, 2004); notably, mitogens promote phenotype modulation, which is consistent with their ability to induce airway smooth muscle cell migration.

4.6 Concluding remarks

The dynamic functional behaviour of airway smooth muscle is a unique feature that may be at the root of biological mechanisms that lead to changes in airway function and structure in obstructive airways disease. Increased contractile responses and chronic changes in airway wall structure that manifest as airways hyperresponsiveness probably results from the plastic and multifunctional behaviour of airway smooth muscles cells. For instance, alterations in myocyte phenotype, in which expression of key regulators of contractility such as myosin light-chain kinase leads to increased contraction and airway narrowing. Similarly, changes in pharmacological responsiveness of airway myocytes occurs in response to asthma-associated mediators, and thus underpins increased contractile sensitivity to inhaled allergic and nonallergic agonists. Notably, airway myocytes are themselves rich sources of pro-inflammatory mediators, and so they are key determinants of the local inflammatory environment. Furthermore, fibroproliferative changes of the airway wall, collectively known as airway remodelling, that are thought to result in the development of irreversible airway dysfunction, include excessive accumulation of extracellular matrix, airway smooth muscle, and submucosal myofibroblasts. Each of these pathological features can be linked to the potential for dynamic changes in airway myocyte phenotype and function. The role of airway smooth muscle cells in all aspects of the pathogenesis of obstructive airways disease is at the forefront of current research in airway biology, and the advances expected in the next decade will undoubtedly yield significant insight that may dictate the direction for development of new and more effective pharmacological interventions.

References

Ammit, A. J., Armour, C. L. and Black, J. L. (2000). Smooth-muscle myosin light-chain kinase content is increased in human sensitized airways. *Am J Respir Crit Care Med* **161**, 257–263.

Avner, B. P., Delongo, J., Wilson, S. *et al.* (1981). A method for culturing canine tracheal smooth muscle cells in vitro: morphologic and pharmacologic observations. *Anat Rec* **200**, 357–370.

Benayoun, L., Druilhe, A., Dombret, M. C. *et al.* (2003). Airway structural alterations selectively associated with severe asthma. *Am J Respir Crit Care Med* **167**, 1360–1368.

Camoretti-Mercado, B., Fernandes, D. J., Dewundara, S. *et al.* (2006). Inhibition of transforming growth factor beta-enhanced serum response factor-dependent transcription by SMAD7. *J Biol Chem* **281**, 20383–20392.

Camoretti-Mercado, B., Liu, H. W., Halayko, A. J. *et al.* (2000). Physiological control of smooth muscle-specific gene expression through regulated nuclear translocation of serum response factor. *J Biol Chem* **275**, 30387–30393.

Campbell, J. H., Campbell, G. R., Kocher, O. *et al.* (1987). Cell biology of smooth muscle in culture: implications for atherogenesis. *Int Angiol* **6**, 73–79.

Cardoso, W. V. and Lu, J. (2006). Regulation of early lung morphogenesis: questions, facts and controversies. *Development* **133**, 1611–1624.

Chamley-Campbell, J., Campbell, G. R. and Ross, R. (1979). The smooth muscle cell in culture. *Physiol Rev* **59**, 1–61.

Chan, V., Burgess, J. K., Ratoff, J. C. *et al.* (2006). Extracellular matrix regulates enhanced eotaxin expression in asthmatic airway smooth muscle cells. *Am J Respir Crit Care Med* **174**, 379–385.

Dekkers, B. G., Schaafsma, D., Nelemans, S. A. *et al.* (2007). Extracellular matrix proteins differentially regulate airway smooth muscle phenotype and function. *Am J Physiol Lung Cell Mol Physiol* **292**, L1405–L1413.

Dunnill, M. S., Massarella, G. R. and Anderson, J. A. (1969). A comparison of the quantitative anatomy of the bronchi in normal subjects, in status asthmaticus, in chronic bronchitis, and in emphysema. *Thorax* **24**, 176–179.

Ebina, M., Takahashi, T., Chiba, T. *et al.* (1993). Cellular hypertrophy and hyperplasia of airway smooth muscles underlying bronchial asthma. A 3-D morphometric study. *Am Rev Respir Dis* **148**, 720–726.

Frid, M. G., Shekhonin, B. V., Koteliansky, V. E. *et al.* (1992). Phenotypic changes of human smooth muscle cells during development: late expression of heavy caldesmon and calponin. *Dev Biol* **153**, 185–193.

Gabbiani, G., Schmid, E., Winter, S. *et al.* (1981). Vascular smooth muscle cells differ from other smooth muscle cells: predominance of vimentin filaments and a specific alpha-type actin. *Proc Natl Acad Sci U S A* **78**, 298–302.

Gimona, M., Sparrow, M. P., Strasser, P. *et al.* (1992). Calponin and SM 22 isoforms in avian and mammalian smooth muscle. Absence of phosphorylation in vivo. *Eur J Biochem* **205**, 1067–1075.

Goldsmith, A. M., Bentley, J. K., Zhou, L. *et al.* (2006). Transforming growth factor-beta induces airway smooth muscle hypertrophy. *Am J Respir Cell Mol Biol* **34**, 247–254.

Goncharova, E. A., Billington, C. K., Irani, C. *et al.* (2003). Cyclic AMP-mobilizing agents and glucocorticoids modulate human smooth muscle cell migration. *Am J Respir Cell Mol Biol* **29**, 19–27.

Gosens, R., Meurs, H., Bromhaar, M. M. *et al.* (2002). Functional characterization of serum- and growth factor-induced phenotypic changes in intact bovine tracheal smooth muscle. *Br J Pharmacol* **137**, 459–466.

Gosens, R., Nelemans, S. A., Hiemstra, M. *et al.* (2003). Insulin induces a hypercontractile airway smooth muscle phenotype. *Eur J Pharmacol* **481**, 125–131.

Gosens, R., Schaafsma, D., Nelemans, S. A. *et al.* (2006). Rho kinase as a drug target for the treatment of airway hyperrespon-siveness in asthma. *Mini Rev Med Chem* **6**, 339–348.

Gosens, R., Stelmack, G. L., Dueck, G. *et al.* (2006). Role of caveolin-1 in p42/p44 MAP kinase activation and proliferation of human airway smooth muscle. *Am J Physiol Lung Cell Mol Physiol* **291**, L523–L534.

Halayko, A. J. and Amrani, Y. (2003). Mechanisms of inflammation-mediated airway smooth muscle plasticity and airways remodelling in asthma. *Respir Physiol Neurobiol* **137**, 209–222.

Halayko, A. J., Camoretti-Mercado, B., Forsythe, S. M. *et al.* (1999). Divergent differentiation paths in airway smooth muscle culture: induction of functionally contractile myocytes. *Am J Physiol* **276**, L197–206.

Halayko, A. J., Kartha, S., Stelmack, G. L. *et al.* (2004). Phosphatidylinositol-3 kinase/mammalian target of rapamycin/p70S6K regulates contractile protein accumulation in airway myocyte differentiation. *Am J Respir Cell Mol Biol* **31**, 266–275.

Halayko, A. J., Salari, H., Ma, X. *et al.* (1996). Markers of airway smooth muscle cell phenotype. *Am J Physiol* **270**, L1040–1051.

Halayko, A. J. and Solway, J. (2001). Molecular mechanisms of phenotypic plasticity in smooth muscle cells. *J Appl Physiol* **90**, 358–368.

Halayko, A. J. and Stephens, N. L. (1994). Potential role for phenotypic modulation of bronchial smooth muscle cells in chronic asthma. *Can J Physiol Pharmacol* **72**, 1448–1457.

Halayko, A. J., Tran, T., Ji, S. Y. *et al.* (2006). Airway smooth muscle phenotype and function: interactions with current asthma therapies. *Curr Drug Targets* **7**, 525–540.

Hall, I. P. and Kotlikoff, M. (1995). Use of cultured airway myocytes for study of airway smooth muscle. *Am J Physiol* **268**, L1–11.

Hirst, S. J., Twort, C. H. and Lee, T. H. (2000). Differential effects of extracellular matrix proteins on human airway smooth muscle cell proliferation and phenotype. *Am J Respir Cell Mol Biol* **23**, 335–344.

Howarth, P. H., Knox, A. J., Amrani, Y. *et al.* (2004). Synthetic responses in airway smooth muscle. *J Allergy Clin Immunol* **114**, S32–50.

Jiang, H., Rao, K., Halayko, A. J. *et al.* (1992). Ragweed sensitization-induced increase of myosin light chain kinase content in canine airway smooth muscle. *Am J Respir Cell Mol Biol* **7**, 567–573.

Johnson, P. R. (2001). Role of human airway smooth muscle in altered extracellular matrix production in asthma. *Clin Exp Pharmacol Physiol* **28**, 233–236.

Johnson, P. R., Black, J. L., Carlin, S. *et al.* (2000). The production of extracellular matrix proteins by human passively sensitized airway smooth-muscle cells in culture: the effect of beclomethasone. *Am J Respir Crit Care Med* **162**, 2145–2151.

Johnson, P. R., Burgess, J. K., Underwood, P. A. *et al.* (2004). Extracellular matrix proteins modulate asthmatic airway smooth muscle cell proliferation via an autocrine mechanism. *J Allergy Clin Immunol* **113**, 690–696.

Johnson, P. R., Roth, M., Tamm, M. *et al.* (2001). Airway smooth muscle cell proliferation is increased in asthma. *Am J Respir Crit Care Med* **164**, 474–477.

Iyemere, V. P., Proudfoot, D., Weissberg, P. L. *et al.* (2006). Vascular smooth muscle cell phenotypic plasticity and the regulation of vascular calcification. *J Intern Med* **260**, 192–210.

Li, S., Sims, S., Jiao, Y. *et al.* (1999). Evidence from a novel human cell clone that adult vascular smooth muscle cells can convert reversibly between noncontractile and contractile phenotypes. *Circ Res* **85**, 338–348.

Liu, H. W., Halayko, A. J., Fernandes, D. J. *et al.* (2003). The RhoA/Rho kinase pathway regulates nuclear localization of serum response factor. *Am J Respir Cell Mol Biol* **29**, 39–47.

Liu, Z. P., Wang, Z., Yanagisawa, H. *et al.* (2005). Phenotypic modulation of smooth muscle cells through interaction of Foxo4 and myocardin. *Dev Cell* **9**, 261–270.

Ma, X., Wang, Y. and Stephens, N. L. (1998). Serum deprivation induces a unique hypercontractile phenotype of cultured smooth muscle cells. *Am J Physiol* **274**, C1206–1214.

Mack, C. P., Somlyo, A. V., Hautmann, M. *et al.* (2001). Smooth muscle differentiation marker gene expression is regulated by RhoA-mediated actin polymerization. *J Biol Chem* **276**, 341–347.

McWhinnie, R., Pechkovsky, D. V., Zhou, D. *et al.* (2006). Endothelin-1 induces hypertrophy and inhibits apoptosis in human airway smooth muscle cells. *Am J Physiol Lung Cell Mol Physiol* **292**, L278–L286.

Miralles, F., Posern, G., Zaromytidou, A. I. *et al.* (2003). Actin dynamics control SRF activity by regulation of its coactivator MAL. *Cell* **113**, 329–342.

Mitchell, R. W., Halayko, A. J., Kahraman, S. *et al.* (2000). Selective restoration of calcium coupling to muscarinic M(3) receptors in contractile cultured airway myocytes. *Am J Physiol Lung Cell Mol Physiol* **278**, L1091–1100.

Nagai, R., Kuro-o, M., Babij, P. *et al.* (1989). Identification of two types of smooth muscle myosin heavy chain isoforms by cDNA cloning and immunoblot analysis. *J Biol Chem* **264**, 9734–9737.

Owens, G. K. (1995). Regulation of differentiation of vascular smooth muscle cells. *Physiol Rev* **75**, 487–517.

Panettieri, R. A., Murray, R. K., DePalo, L. R. *et al.* (1989). A human airway smooth muscle cell line that retains physiological responsiveness. *Am J Physiol* **256**, C329–335.

Parameswaran, K., Radford, K., Zuo, J. *et al.* (2004). Extracellular matrix regulates human airway smooth muscle cell migration. *Eur Respir J* **24**, 545–551.

Pease, D. C. and Paule, W. J. (1960) Electron microscopy of elastic arteries; the thoracic aorta of the rat. *J Ultrastruct Res* **3**, 469–483.

Peng, Q., Lai, D., Nguyen, T. T. *et al.* (2005). Multiple beta 1 integrins mediate enhancement of human airway smooth muscle cytokine secretion by fibronectin and type I collagen. *J Immunol* **174**, 2258–2264.

Roy, J., Kazi, M., Hedin, U. *et al.* (2001). Phenotypic modulation of arterial smooth muscle cells is associated with prolonged activation of ERK1/2. *Differentiation* **67**, 50–58

Sawtell, N. M. and Lessard, J. L. (1989). Cellular distribution of smooth muscle actins during mammalian embryogenesis: expression of the alpha-vascular but not the gamma-enteric isoform in differentiating striated myocytes. *J Cell Biol* **109**, 2929–2937.

Schaafsma, D., McNeill, K. D., Stelmack, G. L. *et al.* (2007). Insulin increases expression of contractile phenotypic markers in airway smooth muscle. *Am J Physiol Cell Physiol* **293**, C429–C439.

Solway, J., Forsythe, S. M., Halayko, A. J. *et al.* (1998). Transcriptional regulation of smooth muscle contractile apparatus expression. *Am J Respir Crit Care Med* **158**, S100–108.

Solway, J., Seltzer, J., Samaha, F. F. *et al.* (1995). Structure and expression of a smooth muscle cell-specific gene, SM22 alpha. *J Biol Chem* **270**, 13460–13469.

Stewart, A. G. (2004). Emigration and immigration of mesenchymal cells: a multicultural airway wall. *Eur Respir J* **24**, 515–517.

Tom-Moy, M., Madison, J. M., Jones, C. A. *et al.* (1987). Morphologic characterization of cultured smooth muscle cells isolated from the tracheas of adult dogs. *Anat Rec* **218**, 313–328.

Tran, T., Gosens, R. and Halayko, A. J. (2007). Effects of extracellular matrix and integrin interactions in airway smooth muscle phenotype and function: it takes two to tango! *Curr Respir Med Rev* (in press).

Tran, T. and Halayko, A. J. (2007). Extracellular matrix and airway smooth muscle interactions: a target for modulating airway wall remodelling and hyperresponsiveness? *Can J Physiol Pharmacol* (in press).

Tran, T., McNeill, K. D., Gerthoffer, W. T. *et al.* (2006). Endogenous laminin is required for human airway smooth muscle cell maturation. *Respir Res* **7**, 117.

Twort, C. H. and van Breemen, C. (1989). Human airway smooth muscle in cell culture: control of the intracellular calcium store. *Pulm Pharmacol* **2**, 45–53.

van der Loop, F. T., Schaart, G., Timmer, E. D. *et al.* (1996). Smoothelin, a novel cytoskeletal protein specific for smooth muscle cells. *J Cell Biol* **134**, 401–411.

Wang, D. Z. and Olson, E. N. (2004). Control of smooth muscle development by the myocardin family of transcriptional coactivators. *Curr Opin Genet Dev* **14**, 558–566.

Wang, L., Liu, H. W., McNeill, K. D. *et al.* (2004). Mechanical strain inhibits airway smooth muscle gene transcription via protein kinase C signalling. *Am J Respir Cell Mol Biol* **31**, 54–61.

Wang, Z., Wang, D. Z., Hockemeyer, D. *et al.* (2004). Myocardin and ternary complex factors compete for SRF to control smooth muscle gene expression. *Nature* **428**, 185–189.

Weaver, M., Batts, L. and Hogan, B. L. (2003). Tissue interactions pattern the mesenchyme of the embryonic mouse lung. *Dev Biol* **258**, 169–184.

Wissler, R. W. (1967). The arterial medial cell: smooth muscle, or multifunctional mesenchyme? *Circulation* **36**, 1–4.

Woodruff, P. G., Dolganov, G. M., Ferrando, R. E. *et al.* (2004). Hyperplasia of smooth muscle in mild to moderate asthma without changes in cell size or gene expression. *Am J Respir Crit Care Med* **169**, 1001–1006.

Zhou, L., Goldsmith, A. M., Bentley, J. K. *et al.* (2005). 4E-binding protein phosphorylation and eukaryotic initiation factor-4E release are required for airway smooth muscle hypertrophy. *Am J Respir Cell Mol Biol* **33**, 195–202.

Zhou, D., Zheng, X., Wang, L. *et al.* (2003). Expression and effects of cardiotrophin-1 (CT-1) in human airway smooth muscle cells. *Br J Pharmacol* **140**, 1237–1244.

5

Airway smooth muscle proliferation: insights into mechanisms regulating airway smooth muscle mass

Reynold A. Panettieri, Jr.

University of Pennsylvania School of Medicine, Philadelphia, PA, USA

5.1 Increases in airway smooth muscle (ASM) mass and the functional consequences

Increases in airway smooth muscle (ASM) mass were first described in fatal asthma (Huber and Koessler, 1922); however, new evidence suggests that subjects with mild persistent asthma may also exhibit increased smooth muscle mass (Woodruff *et al.*, 2004). Although the histological determination of ASM mass in asthma is well established, the functional consequences of the increased number and size of myocytes remain unclear. Potentially, alterations in mass could generate more shortening, leading to increased airway narrowing. Investigators using mathematical modelling and actual measurements of airway wall thickness from subjects with asthma showed that increased smooth muscle mass, rather than

Airway Smooth Muscle   Edited by Kian Fan Chung
© 2008 John Wiley & Sons, Ltd

submucosal or adventitial thickening, was the sole structural change likely to increase airways resistance in response to contractile agonists (Lambert *et al.*, 1993). Despite these sophisticated modelling systems, ASM mass and the functional consequences are difficult to study *in vivo*. As a consequence, investigators correlate ASM mass and asthma severity. High-resolution CT scans of subjects with stable asthma have shown an increased wall thickness that inversely correlated with airway reactivity to methacholine, suggesting that airway remodelling potentially attenuates airway narrowing (Niimi *et al.*, 2003). Unfortunately, such macroscopic approaches cannot discriminate epithelial, submucosal, smooth muscle and adventitial layers, and thus the relative contribution of alterations in ASM mass remains unclear. Several studies have compared ASM derived from subjects with asthma to that derived from healthy volunteers (Bai, 1991; Bjorck *et al.*, 1992; Black, 1991; de Jongste *et al.*, 1987; Goldie *et al.*, 1986; Whicker *et al.*, 1988). Although some studies have shown enhanced force generation in asthma, others have shown normal or reduced stimulated shortening (Bai, 1991). Since data derived from subjects with asthma are lacking, investigators use animal models of allergen-induced airway inflammation and hyperresponsiveness to correlate ASM thickness and function. In some studies, ASM thickness and methacholine responsiveness increased with ovalbumin sensitization and challenge (Henderson *et al.*, 2002). Other studies, however, show an uncoupling of ASM proliferation and responsiveness. After repeated exposures of rats and mice to allergen, airways hyperresponsiveness and allergic inflammation resolved but ASM proliferation persisted (Leung *et al.*, 2004; McMillan and Lloyd, 2004). Consistent with human airways, canine and murine airways showed increased ASM shortening, velocity and capacity following passive sensitization (Fan *et al.*, 1997; Jiang *et al.*, 1992). To date, however, only one study has directly examined the effects of smooth muscle hypertrophy and hyperplasia on *ex vivo* airway function. These investigators measured mechanical properties of guinea-pig airway explants treated with cardiotropin, a member of the interleukin (IL)-6 family (Zheng *et al.*, 2004). Cardiotropin increased the size and protein synthesis of cultured human bronchial smooth muscle cells (Zhou *et al.*, 2003) and increased ASM content in guinea-pig airway explants. Maximum isometric stress, however, was decreased, suggesting that the contractile apparatus of the enlarged ASM cells may not be completely functional. Overall, the lack of definitive studies comparing alterations in ASM mass to functional consequences is a substantial gap in our understanding and should foster research interest in the future.

Another approach to determining the functional consequences of ASM mass is to utilize cultured airway myocytes from subjects with asthma *in vitro*. Investigators have demonstrated that cultured ASM from subjects with asthma proliferates at a faster rate than that from nonasthmatic individuals (Johnson *et al.*, 2001). While corticosteroids inhibited the proliferation of normal airway myocytes by reducing cyclin D1 expression and retinoblastoma protein phosphorylation (Fernandes *et al.*, 1999), steroids apparently failed to modulate ASM proliferation in cells derived from subjects with asthma. Further, the investigators suggested that a dysfunctional interaction between C/EBPα and the glucocorticoid receptor was responsible for the lack of steroid effects (Roth *et al.*, 2002). Although these results are provocative, further studies are needed to demonstrate that such processes actually occur *in vivo*. Despite the limitations of studying ASM cell proliferation *in vitro*, substantial progress has been made in studying the molecular processes that modulate myocyte growth.

5.2 Growth factors, inflammatory mediators and cytokines modulate ASM proliferation

ASM proliferation

Inflammatory mediators are increased in bronchoalveolar lavage (BAL) from subjects with asthma, and some induce ASM mitogenesis *in vitro*. The known mitogenic stimuli include growth factors, such as epidermal growth factor (EGF), insulin-like growth factors, platelet-derived growth factor (PDGF) isoforms BB and AB, and basic fibroblast growth factor; plasma- or inflammatory cell-derived mediators, such as lysosomal hydrolases (β-hexosaminidases and β-glucuronidase), α-thrombin, tryptase, and sphingosine 1-phosphate (SPP); and contractile agonists, such as histamine, endothelin-1, substance P, phenylephrine, serotonin, thromboxane A_2, and leukotriene D_4 (Panettieri *et al.*, 1998) (reviewed in (Ammit and Panettieri, 2001)).

Although the cytokines IL-1β, IL-6, and tumor necrosis factor α (TNFα) are also increased in the BAL of asthmatics (Broide *et al.*, 1992), whether these cytokines stimulate ASM proliferation *in vitro* remains controversial. De *et al.* (1995) reported that IL-1β and IL-6 cause hyperplasia and hypertrophy of cultured guinea-pig ASM cells; however, other studies have shown that IL-1β (Belvisi *et al.*, 1998) and IL-6 (McKay *et al.*, 2000) are not mitogenic for human ASM cells. McKay *et al.* (2000) also reported that TNFα (~30 pM) has no immediate mitogenic effect on human ASM cells. These results are in contrast to those of Stewart *et al.* (1995); there the proliferative effect of TNFα on human ASM cells appeared to be biphasic; low concentrations of TNFα (0.3–30 pM) were promitogenic, while at higher concentrations (300 pM), the mitogenic effect was abolished. Such conflicting reports may be due to cytokine-induced cyclo-oxygenase 2-dependent prostanoid production (Belvisi *et al.*, 1998). Recent evidence suggests that TNFα induces interferon β secretion, which in turn inhibits ASM mitogenesis. Thus, autocrine secreton of interferon β regulates cytokine effects on ASM cell growth (Tliba *et al.*, 2003). Cyclo-oxygenase products, such as prostaglandin E$_2$, inhibit DNA synthesis (Belvisi *et al.*, 1998). Therefore, cytokine-induced proliferative responses in ASM may be greater under conditions of cyclo-oxygenase inhibition, in which the expression of growth inhibitory prostanoids, such as prostaglandin E$_2$, is limited (Belvisi *et al.*, 1998; De *et al.*, 1995; Stewart *et al.*, 1995).

Airway remodelling, a key feature of persistent asthma, is also characterized by the deposition of extracellular matrix (ECM) proteins in the airways (Laitinen and Laitinen, 1995; Roberts, 1995). ECM proteins (collagen I, III, and V; fibronectin; tenascin; hyaluronan; versican; and laminin α2/β2) are increased in asthmatic airways (Altraja *et al.*, 1996; Bousquet *et al.*, 1995; Laitinen and Laitinen, 1995; Roberts and Burke, 1998). Components of the ECM also modulate mitogen-induced ASM growth. Fibronectin and collagen I increase human ASM cell mitogenesis in response to PDGF-BB or α-thrombin, whereas laminin inhibits proliferation (Hirst *et al.*, 2000a). In this study, the increase in cell proliferation was accompanied by a decrease in expression of smooth muscle cell contractile proteins, such as α-actin, calponin and myosin heavy chain, suggesting that the matrix may also modulate smooth muscle phenotype. Recently, human ASM cells were shown to secrete ECM proteins in response to asthmatic sera (Johnson *et al.*, 2000), suggesting a cellular source for ECM deposition in airways, and implicating a novel mechanism in which ASM cells may modulate autocrine proliferative responses.

Cell-cycle regulation

Extracellular stimuli transduce proliferative responses that move the cell through the cell cycle, which comprises distinct phases termed G$_1$, S (DNA synthesis), G$_2$ and M (mitosis) (Figure 5.1). ASM growth appears to occur by activating cell-cycle events similar to those described in other cell types. Hence, the following section provides an overview of the mammalian cell cycle (reviewed in (Sherr, 1994; Sherr and Roberts, 1999)) with particular emphasis on G$_1$-to-S transition, the most widely studied cell-cycle phase in ASM biology, shown schematically in Figure 5.1. While many of the studies cited refer to ASM, the general phenomena are typical of other normal and malignant cell types.

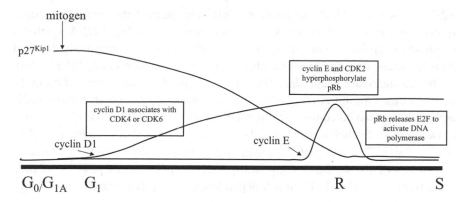

Figure 5.1 Schematic representation of the G_1-to-S transition phase in the cell cycle. In response to mitogens, cells enter the cell cycle from the G_0/G_{1A} phase. D-type cyclins (D1 shown here) are expressed, while the levels of the CKI p27[Kip1], usually high in quiescent cells, fall in response to mitogenic stimulation. Progression through the G_1 phase initially depends on holoenzymes composed of the D-type cyclins in association with cyclin-dependent kinases, CDK4 or CDK6. Most p27[Kip1] becomes complexed with cyclin D-CDK, allowing activation of the cyclin E-CDK2 complex. Together, cyclin E and CDK2 act in a cascade to hyperphosphorylate pRb, which then releases the elongation factor E2F, which activates DNA polymerase. Cell commitment to traverse completely through to mitosis is achieved on or near this point, termed the restriction point (R), in the cell cycle. Subsequently, cells initiate DNA synthesis (S-phase) (from Ammit, A. J. and Panettieri, Jr., R. P. (2001). *J Appl Physiol* **91**, 1431 (Ammit and Panettieri, 2001)). (Reproduced with permission from the Journal of Applied Physiology, Vol. 91, pp. 1431–1437, 2001. © 2005 The American Physiological Society)

To date, few studies have examined ASM cell growth *in vivo*. Proliferative responses in ASM cells are often studied in cell-culture models. ASM cells are grown to confluence, and then growth-arrested in low-serum media or serum-free conditions for 24–48 h (Panettieri *et al.*, 1989; 1990). This experimental design synchronizes ASM cells in the G_0 or early G_1 phase (G_{1A}) of the cell cycle where ASM minimally incorporates [35S]methionine and [3H]thymidine (Panettieri *et al.*, 1989; 1990). As cells enter the cycle from G_0/G_{1A}, one or more D-type cyclins (D1, D2, and D3) are expressed as part of the delayed early response to mitogen stimulation, as shown in Figure 5.1. Progression through the G_1 phase initially depends on holoenzymes composed of one or more of the D type cyclins (D1, D2, and/or D3) in association with cyclin-dependent kinases, CDK4 or CDK6. This is followed by activation of the cyclin E-CDK2 complex as cells approach the G_1/S transition. Together, cyclin E and CDK2 act to hyperphospho-rylate retinoblastoma protein (pRb), which then releases the elongation factor

E2F that activates DNA polymerase. This step, termed the restriction point, represents the *point of no return*; cell commitment to undergo DNA synthesis (S phase) and mitosis is inevitable. In ASM cells, S phase is commonly detected by using incorporation of radiolabeled thymidine (Panettieri *et al.*, 1989; 1990), or by immunofluorescent detection of the thymidine analogue 5-bromo-2'-deoxyuridine (Ammit *et al.*, 1999). At each phase of G_1-to-S transition, CDK activities can also be constrained by CDK inhibitors (CKIs). CKIs are assigned to two families by their structures and CDK targets: i) the INK4 family (p16^{INK4a}, p15^{INK4b} p18^{INK4c}, and p19^{INK4d}) specifically inhibits the catalytic subunits of CDK4 and CDK6; ii) the Cip/Kip family (p21^{Cip1}, p27^{Kip1}, and p57^{Kip2}) inhibits the activities of cyclin D-, E-, and A-dependent kinases (Sherr and Roberts, 1995).

Regulation of cell cycle in ASM cell proliferation

ASM mitogens may act via different receptor-operated mechanisms (reviewed in (Hirst *et al.*, 2000b)), as shown in Figure 5.2. While growth factors induce ASM cell mitogenesis by activating receptors with intrinsic protein tyrosine kinase (RTK) activity, contractile agonists released from inflammatory cells mediate their effects via activation of seven transmembrane G protein-coupled receptors (GPCRs). Cytokines signal through cell-surface glycoprotein receptors that function as oligomeric complexes consisting typically of two to four receptor chains (Bagley *et al.*, 1997), coupled to Src family non-receptor tyrosine kinases, such as lyn (Bolen and Brugge, 1997).

Despite disparate receptor-operated mechanisms, recent evidence suggests that the small guanidine triphosphatase (GTPase), p21ras, acts as a point of convergence for diverse extracellular signal-stimulated pathways in ASM cells, as shown in Figure 5.2 (Ammit *et al.*, 1999). Interestingly, synergy can occur between RTK and GPCRs, promoting human ASM mitogenesis and p21ras activation (Krymskaya *et al.*, 2000). In their GTP-bound active state, p21ras proteins interact with downstream effectors, namely, Raf-1 and phosphatidylinositol 3-kinase (PI3K). By recruiting Raf-1, a 74-kDa cytoplasmic serine/threonine kinase, to the plasma membrane, GTP-bound p21ras activates the extracellular signal-regulated kinase (ERK) pathway, although Raf-1-independent signalling to ERK also has been shown (Kartha *et al.*, 1999). P21ras also binds and activates PI3K by using specific regions termed switch I (Asp^{30}–Asp^{38}) and switch II (Gly^{60}–Glu^{76}) (Pacold *et al.*, 2000; Rodriguez-Viciana *et al.*, 1994). Although alternative pathways do exist, such as protein kinase C-dependent pathways (reviewed in (Hirst *et al.*, 2000b)) or reactive oxygen-dependent pathways (Brar *et al.*, 1999), PI3K and ERK activation appears to be the predominant signal transduction pathway for RTK-, GPCR-, or cytokine-stimulated growth of ASM cells.

Extracellular signal-regulated kinase pathway

Raf-1 activation induces phosphorylation and activation of mitogen-activated protein (MAP) kinase/extracellular signal-regulated kinase (ERK) kinase (MEK1). Activated MEK1 then directly phosphorylates (on both tyrosine and threonine residues) and activates the 42-kDa ERK2 and 44-kDa ERK1, also collectively referred to as p42/p44 MAP kinases, as shown in Figure 5.2. In bovine ASM,

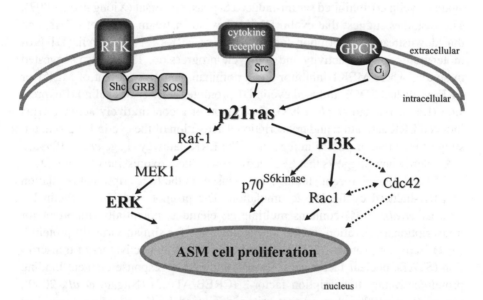

Figure 5.2 Schematic representation of signal transduction mechanisms that regulate ASM cell proliferation. ASM mitogens act via RTKs, cytokine receptors, or GPCRs to activate the small GTPase p21ras. p21ras proteins then interact with the downstream effectors, Raf-1 and PI3K. Raf-1 activates MEK1, which then phosphorylates ERK. PI3K activates the downstream effectors, p70^{S6k} or members of the Rho family GTPases, Rac1 and Cdc42 (although whether Cdc42 acts upstream of Rac1, or cross-talk exists is unknown [indicated by the dashed lines]). ERK, PI3K, and the downstream effectors of PI3K regulate cell-cycle proteins, and thus the ERK and PI3K pathways are considered to be two major independent signaling pathways regulating ASM cell growth (from Ammit, A. J. and Panettieri, Jr., R. P. (2001). *J Appl Physiol* **91**, 1431 (Ammit and Panettieri, 2001)). (Reproduced with permission from the Journal of Applied Physiology, Vol. 91, pp. 1431–1437, 2001. © The American Physiological Society)

inhibition of MEK1 and ERK activity attenuates PDGF-induced DNA synthesis, suggesting that activation of MEK1 and ERKs is required for proliferation (Karpova *et al.*, 1997). In human ASM (Orsini *et al.*, 1999), mitogens, including EGF, PDGF-BB, and thrombin, produced a robust and sustained activation of ERK1 and ERK2 that was correlated with ASM growth responses and was inhibited by MEK1 inhibition. Studies such as these suggest that the ERK pathway is a key signalling event mediating mitogen-induced ASM proliferation.

D-type cyclins (cyclins D1, D2, and D3) are key regulators of G_1 progression in mammalian cells; consequently, cyclin D1 has been the most widely studied cyclin in ASM biology. In bovine ASM, mitogenic stimulation with PDGF induced cyclin D1 transcriptional activation and protein synthesis, with consequent hyperphosphorylation of pRb, while microinjection with a neutralizing antibody against cyclin D1 inhibited serum-induced S-phase traversal (Xiong *et al.*, 1997). These studies suggest that cyclin D1 is a key downstream target of ERKs, and that downstream transcription factor targets of ERKs regulate cyclin D1 promoter transcriptional activity and cell-cycle progression. This is also suggested in studies where MEK1 inhibitor and a dominant negative mutant of MEK1 or ERK abolished PDGF-induced cyclin D1 promoter activity or cyclin D1 expression (Ramakrishnan *et al.*, 1998). Expression of a constitutively active p21ras induced ERK activation and transcriptional activation of the cyclin D1 promoter, suggesting a role of p21ras in regulating the ERK pathway (Page *et al.*, 1999a).

Evidence now suggests that ERK activation induces expression of cyclin D1 in ASM cells. Hence, recent studies have focused on the transcriptional regulation of ERK-induced cyclin D1 accumulation. The promoter region of cyclin D1 (Herber *et al.*, 1994) contains multiple *cis*-elements potentially important for transcriptional activation, including binding sites for simian virus 40 protein 1 (Sp1), activator protein-1 (AP-1), signal transducers and activators of transcription (STAT), nuclear factor κB (NF-κB), and cAMP response element-binding protein/activating transcription factor-2 (CREB/ATF-2) (Nagata *et al.*, 2000). Orsini *et al.* (1999) showed that mitogen-induced ERK activation, thymidine incorporation, and Elk-1 and AP-1 reporter activity were similarly abrogated by MEK1 inhibition. Such studies suggest a link between ERK activation, transcription factor activation, cyclin D1 expression, and ASM proliferation. Similarly, MEK1 inhibition also attenuated expression of c-Fos (Lee *et al.*, 2001), suggesting that c-Fos may be one or both of the dimer pairs in the AP-1 transcription factor complex responsible for cyclin D1 expression in ASM cells. Whether ERK-dependent transcriptional regulation of cyclin D1 gene expression is via direct *cis*-activation with AP-1 dimers (composed of c-Fos), or via Elk-1 mediated *trans*-activation, requires further investigation. In addition, cyclin D1 protein, but not mRNA levels, was affected by MEK1 inhibition

(Ravenhall *et al.*, 2000), suggesting that post-transcriptional control of cyclin D1 protein levels may also occur independently of the MEK1/ERK signalling pathways.

Another critical cell-cycle protein is p27^{Kip1} (Sherr and Roberts, 1999), as shown in Figure 5.1. In quiescent cells, the cytosolic protein levels of p27^{Kip1} remain high. A coordinated increase of cyclin D1 expression promotes complexing of unbound p27^{Kip1} molecules with cyclin D-dependent kinases, relieving cyclin E-CDK2 from CKI constraint, and thereby facilitating cyclin E-CDK2 activation later in the G$_1$ phase (Sherr and Roberts, 1999). In human ASM cells (Ammit *et al.*, 2001), SPP, an agonist that activates multiple GPCRs, was shown to increase cyclin D1 levels and decrease p27^{Kip1}, possibly via an ERK-mediated pathway (Pyne and Pyne, 1996). SPP also appeared to augment EGF- and thrombin-induced DNA proliferation by increasing G$_1$/S progression (Ammit *et al.*, 2001). This was due to an enhancement of the stimulatory/inhibitory effect of EGF and thrombin on cyclin D1/p27^{Kip1} expression by SPP (Ammit *et al.*, 2001). A summary of the signal transduction pathways that modulate cell-cycle events in ASM is shown in Figure 5.2.

Phosphatidylinositol 3-kinase pathway

PI3K isoforms are divided into three classes by their structure and substrate specificity (Rameh and Cantley, 1999). Class IA PI3Ks are cytoplasmic heterodimers composed of a 110-kDa (p110α, -β, or -δ) catalytic subunit and an 85-kDa (p85, p55, or p50) adaptor protein. Class IA isoforms can be activated by RTKs and nonreceptor tyrosine kinases, whereas class IB p110γ is activated by G$\beta\gamma$ subunits of GPCRs. Class II isoforms are mainly associated with the phospholipid membranes, and are concentrated in the *trans*-Golgi network and present in clathrin-coated vesicles (Domin *et al.*, 2000). Class III isoforms are structurally related to the yeast vesicular sorting protein Vps34p (Volinia *et al.*, 1995). Recent data (Krymskaya *et al.*, 2001) show that human ASM cells express class IA, II, and III PI3K, but not the class IB p110γ isoform.

PI3K phosphorylates membrane phosphoinositides on the D3 hydroxyl of the inositol ring to form the phosphoinositides (PI) 3-phosphate, PI 3,4-diphosphate, and PI 3,4,5-triphosphate. These D3 phosphoinositides function as second messengers and activate downstream effector molecules, such as the 70-kDa, ribosomal S6 kinase (p70^{S6k}) (Krymskaya *et al.*, 1999; Scott *et al.*, 1996) or members of the Rho family GTPases (Rac1 (Page *et al.*, 1999b) and Cdc42 (Bauerfeld *et al.*, 2001), but not RhoA (Bauerfeld *et al.*, 2001)) to regulate cell-cycle protein expression and thus modulate cell-cycle traversal in ASM cells.

The use of PI3K inhibitors has shown that activation of PI3K is critical for ASM cell-cycle progression in human (Krymskaya *et al.*, 1999) and bovine ASM (Scott *et al.*, 1996; Walker *et al.*, 1998). Transfection or microinjection of cells with constitutively active class IA PI3K alone markedly increased DNA synthesis (Krymskaya *et al.*, 2001). This is the first study to show that a constitutively active signalling molecule can induce DNA synthesis in human ASM cells. The extent of DNA synthesis stimulated in cells microinjected with constitutively active PI3K, however, was substantially less than that induced by receptor-mediated pathways. These data suggest that although active PI3K is sufficient to stimulate ASM DNA synthesis, other signalling events are also necessary to promote maximal ASM growth responses. Interestingly, PI3K inhibition does not alter ERK activation (Krymskaya *et al.*, 1999), confirming that PI3K regulates DNA synthesis in an ERK-independent or possibly parallel manner.

In bovine (Scott *et al.*, 1996) and human ASM (Krymskaya *et al.*, 1999), rapamycin, an inhibitor of $p70^{S6k}$, attenuates growth factor-induced DNA synthesis, showing that $p70^{S6k}$ is an essential step in the pathway to ASM cellular proliferation, as shown in Figure 5.2. Through the phosphorylation of the 40S ribosomal protein, $p70^{S6k}$ upregulates the translation of mRNAs that contain an oligopyrimidine tract at their transcriptional start site. Such mRNA moieties encode proteins required for cell-cycle progression in the G_1 phase, such as the elongation factor E2F (Brennan *et al.*, 1999).

Studies have also examined the involvement of the Rho GTPases, Rac1 (Page *et al.*, 1999b), Cdc42 (Bauerfeld *et al.*, 2001), and RhoA (Bauerfeld *et al.*, 2001) in cyclin D1 upregulation and bovine ASM proliferation. Rac1 overexpression induced transcription of a cyclin D1 promoter construct, while a dominant-negative allele of Rac1 inhibited PDGF-induced cyclin D1 transcription (Page *et al.*, 1999b). Rac1-induced cyclin D1 promoter activation was also independent of ERK, since inhibition of MEK1 had little effect (Page *et al.*, 1999b). In other studies, overexpression of the catalytically active subunit of PI3K ($p110^{PI3K}$CAAX) was sufficient to activate the cyclin D1 promoter, and cyclin D1 promoter activation could be attenuated by inhibitors of Rac1 signalling (Page *et al.*, 2000). These results suggest that Rac1 may be downstream of PI3K; however, further study is necessary to confirm this observation. Other studies using overexpression constructs of Cdc42 and RhoA also showed that overexpression of Cdc42, but not RhoA, induced transcription from the cyclin D1 promoter in an ERK-independent manner (Bauerfeld *et al.*, 2001). In addition, $p110^{PI3K}$CAAX (Page *et al.*, 2000), Rac1 (Page *et al.*, 2000), and Cdc42 (Bauerfeld *et al.*, 2001) were shown to activate the cyclin D1 promoter via the CREB/ATF-2 binding site. These results led the investigators to speculate that Cdc42 acts upstream of Rac1 (Bauerfeld *et al.*, 2001). Whether this implicates PI3K in a linked signalling cascade remains unknown.

Inhibition of ASM cell proliferation by anti-asthma therapies

The most widely used therapies for the control of asthma symptoms are the corticosteroids and the β_2-agonists. Inhaled corticosteroids inhibit inflammatory cell activation, while β_2-agonists are effective bronchodilators. In addition, these anti-asthma therapies are potent inhibitors of ASM cell proliferation.

β_2-agonists activate the β_2-adrenergic receptor G_s-adenylyl cyclase pathway to elevate $3',5'$-cyclic adenosine monophosphate (cAMP) in ASM cells. Because of their cAMP-elevating ability (Tomlinson *et al.*, 1995), albuterol (Tomlinson *et al.*, 1994) and fenoterol (Stewart *et al.*, 1997) have been shown to inhibit mitogen-induced proliferation of human ASM cells. β_2-Adrenergic receptor agonists and other cAMP-elevating agents are thought to induce G_1 arrest by post-transcriptionally inhibiting cyclin D1 protein levels by action on a proteasome-dependent degradation pathway (Stewart *et al.*, 1999). Musa *et al.* (1999) examined the effects of forskolin, an activator of adenylate cyclase, on DNA synthesis, cyclin D1 expression, cAMP response element-binding protein (CREB) phosphorylation, and DNA binding in bovine ASM. By increasing cAMP in ASM cells, Musa *et al.* (1999) showed that forskolin suppressed cyclin D_1 gene expression by phosphorylation and transactivation of CREB, suggesting that the effect of cAMP on cyclin D1 gene expression is by *cis*-repression of the cyclin D1 promoter.

In human ASM (Fernandes *et al.*, 1999), the corticosteroids dexamethasone and fluticasone propionate were shown to arrest ASM cells in the G_1 phase of the cell cycle. In this study, corticosteroids reduced thrombin-stimulated increases in cyclin D1 protein and mRNA levels, and attenuated pRb phosphorylation, via a pathway either downstream of, or parallel to, the ERK cascade (Fernandes *et al.*, 1999). The difference between the relative inhibition of RTK- and GPCR-mediated ASM cell growth by dexamethasone suggests that steroid effects on ASM mitogenesis are complex. Further elucidation of the signalling and transcriptional targets to inhibit cell-cycle progression by corticosteroids and β_2-agonists may indicate how these anti-asthma therapies could be used optimally, and possibly in combination, to modulate airway wall remodelling in asthma.

References

Altraja, A., Laitinen, A., Virtanen, I. *et al.* (1996). Expression of laminins in the airways in various types of asthmatic patients: a morphometric study. *Am J Respir Cell Mol Biol* **15**, 482–488.

Ammit, A. J., Kane, S. A. and Panettieri, R. A., Jr. (1999). Activation of K-p21ras and N-p21ras, but not H-p21ras, is necessary for mitogen-induced human airway smooth muscle proliferation. *Am J Respir Cell Mol Biol* **21**, 719–727.

Ammit, A. J., Hastie, A. T., Edsall, L. C. *et al.* (2001). Sphingosine 1-phosphate modulates human airway smooth muscle cell functions that promote inflammation and airway remodeling in asthma. *FASEB J* **15**, 1212–1214.

Ammit, A. J. and Panettieri, R. A., Jr. (2001). Signal transduction in smooth muscle. Invited review: The circle of life – cell cycle regulation in airway smooth muscle. *J Appl Physiol* **91**, 1431–1437.

Bagley, C. J., Woodcock, J. M., Stomski, F. C. *et al.* (1997). The structural and functional basis of cytokine receptor activation: lessons from the common beta subunit of the granulocyte-macrophage colony-stimulating factor, interleukin-3 (IL-3), and IL-5 receptors. *Blood* **89**, 1471–1482.

Bai, T. R. (1991). Abnormalities in airway smooth muscle in fatal asthma. A comparison between trachea and bronchus. *Am Rev Respir Dis* **143**, 441–443.

Bauerfeld, C. P., Hershenson, M. B. and Page, K. (2001). Cdc42, but not RhoA, regulates cyclin D1 expression in bovine tracheal myocytes. *Am J Physiol Lung Cell Mol Physiol* **280**, L974–L982.

Belvisi, M. G., Saunders, M., Yacoub, M. *et al.* (1998). Expression of cyclo-oxygenase-2 in human airway smooth muscle is associated with profound reductions in cell growth. *Br J Pharmacol* **125**, 1102–1108.

Bjorck, T., Gustafsson, L. E. and Dahlen, S. E. (1992). Isolated bronchi from asthmatics are hyperresponsive to adenosine, which apparently acts indirectly by liberation of leukotrienes and histamine. *Am Rev Respir Dis* **145**, 1087–1091.

Black, J. L. (1991). Pharmacology of airway smooth muscle in chronic obstructive pulmonary disease and in asthma. *Am Rev Respir Dis* **143**, 1177–1181.

Bolen, J. B. and Brugge, J. S. (1997). Leukocyte protein tyrosine kinases: Potential targets for drug discovery. *Annu Rev Pharmacol Toxicol* **15**, 371–404.

Bousquet, J., Vignola, A. M., Chanez, P. *et al.* (1995). Airways remodelling in asthma: no doubt, no more? *Int Arch Allergy Immunol* **107**, 211–214.

Brar, S. S., Kennedy, T. P., Whorton, A. R. *et al.* (1999). Requirement for reactive oxygen species in serum-induced and platelet-derived growth factor-induced growth of airway smooth muscle. *J Biol Chem* **274**, 20017–20026.

Brennan, P., Babbage, J. W., Thomas, G. *et al.* (1999). p70s6k integrates phosphatidylinositol 3-kinase and rapamycin-regulated signals for E2F regulation in T lymphocytes. *Mol Cell Biol* **19**, 4729–4738.

Broide, D. H., Lotz, M., Cuomo, A. J. *et al.* (1992). Cytokines in symptomatic asthma airways. *J Allergy Clin Immunol* **89**, 958–967.

De, S., Zelazny, E. T., Souhrada, J. F. *et al.* (1995). IL-1β and IL-6 induce hyperplasia and hypertrophy of cultured guinea pig airway smooth muscle cells. *J Appl Physiol* **78**, 1555–1563.

de Jongste, J. C., Mons, H., Van Strik, R. *et al.* (1987). Comparison of human bronchiolar smooth muscle responsiveness in vitro with histological signs of inflammation. *Thorax* **42**, 870–876.

Domin, J., Gaidarov, I., Smith, M. E. K. *et al.* (2000). The class II phosphoinositide 3-kinase PI3K-C2α is concentrated in the trans-golgi network and present in clathrin-coated vesicles. *J Biol Chem* **275**, 11943–11950.

Fan, T., Yang, M., Halayko, A. J. *et al.* (1997). Airway responsiveness in two inbred strains of mouse disparate in IgE and IL-4 production. *Am J Respir Cell Mol Biol* **17**, 156–163.

Fernandes, D., Guida, E., Koutsoubos, V. *et al.* (1999). Glucocorticoids inhibit proliferation, cyclin D1 expression, and retinoblastoma protein phosphorylation, but not activity of the extracellular-regulated kinases in human cultured airway smooth muscle. *Am J Respir Cell Mol Biol* **21**, 77–88.

Goldie, R. G., Spina, D., Henry, P. J. *et al.* (1986). In vitro responsiveness of human asthmatic bronchus to carbachol, histamine, beta-adrenoceptor agonists and theophylline. *Br J Clin Pharmacol* **22**, 669–676.

Henderson, W. R., Jr., Tang, L. O., Chu, S. J. *et al.* (2002). A role for cysteinyl leukotrienes in airway remodeling in a mouse asthma model. *Am J Respir Crit Care Med* **165**, 108–116.

Herber, B., Truss, M., Beato, M. *et al.* (1994). Inducible regulatory elements in the human cyclin D1 promoter. *Oncogene* **9**, 2105–2107.

Hirst, S. J., Twort, C. H. C. and Lee, T. H. (2000a). Differential effects of extracellular matrix proteins on human airway smooth muscle cell proliferation and phenotype. *Am J Respir Cell Mol Biol* **23**, 335–344.

Hirst, S. J., Walker, T. R. and Chilvers, E. R. (2000b). Phenotypic diversity and molecular mechanisms of airway smooth muscle proliferation in asthma. *Eur Respir J* **16**, 159–177.

Huber, H. L. and Koessler, K. K. (1922). The pathology of bronchial asthma. *Arch Intern Med* **30**, 690–760.

Jiang, H., Rao, K., Halayko, A. J. *et al.* (1992). Bronchial smooth muscle mechanics of a canine model of allergic airway hyperresponsiveness. *J Appl Physiol* **72**, 39–45.

Johnson, P. R., Roth, M., Tamm, M. *et al.* (2001). Airway smooth muscle cell proliferation is increased in asthma. *Am J Respir Crit Care Med* **164**, 474–477.

Johnson, P. R. A., Black, J. L., Cralin, S. *et al.* (2000). The production of extracellular matrix proteins by human passively sensitized airway smooth-muscle cells in culture: The effect of beclomethasone. *Am J Respir Crit Care Med* **162**, 2145–2151.

Karpova, A. K., Abe, M. K., Li, J. *et al.* (1997). MEK1 is required for PDGF-induced ERK activation and DNA synthesis in tracheal monocytes. *Am J Physiol Lung Cell Mol Physiol* **272**, L558–L565.

Kartha, S., Naureckas, E. T., Li, J. *et al.* (1999). Partial characterization of a novel mitogen-activated protein kinase/extracellular signal-regulated kinase activator in airway smooth-muscle cells. *Am J Respir Cell Mol Biol* **20**, 1041–1048.

Krymskaya, V. P., Penn, R. B., Orsini, M. J. *et al.* (1999). Phosphatidylinositol 3-kinase mediates mitogen-induced human airways smooth muscle cell proliferation. *Am J Physiol Lung Cell Mol Physiol* **277/21**, L65–L78.

Krymskaya, V. P., Orsini, M. J., Eszterhas, A. J. *et al.* (2000). Mechanisms of proliferation synergy by receptor tyrosine kinase and G protein-coupled receptor activation in human airway smooth muscle. *Am J Respir Cell Mol Biol* **23**, 546–554.

Krymskaya, V. P., Ammit, A. J., Hoffman, R. K. *et al.* (2001). Activation of class IA phosphatidylinositol 3-kinase stimulates DNA synthesis in human airway smooth muscle cells. *Am J Physiol Lung Cell Mol Physiol* **280**, L1009–L1018.

Laitinen, L. A. and Laitinen, A. (1995). Inhaled corticosteroid treatment and extracellular matrix in the airways in asthma. *Int Arch Allergy Immunol* **107**, 215–216.

Lambert, B. R., Wiggs, B. R., Kuwano, K. *et al.* (1993). Functional significance of increased airway smooth muscle in asthma and COPD. *J Appl Physiol* **74**, 2771–2781.

Lee, J.-H., Johnson, P. R. A., Roth, M. *et al.* (2001). ERK activation and mitogenesis in human airway smooth muscle cells. *Am J Physiol Lung Cell Mol Physiol* **280**, L1019–L1029.

Leung, S.-Y., Eynott, P. R., Noble, A. *et al.* (2004). Resolution of allergic airways inflammation but persistence of airway smooth muscle proliferation after repeated allergen exposures. *Clin Exp Allergy* **34**, 213–220.

McKay, S., Hirst, S. J., Bertrand-de Haas, M. *et al.* (2000). Tumor necrosis factor-α enhances mRNA expression and secretion of interleukin-6 in cultured human airway smooth muscle cells. *Am J Respir Cell Mol Biol* **23**, 103–111.

McMillan, S. J. and Lloyd, C. M. (2004). Prolonged allergen challenge in mice leads to persistent airway remodelling. *Clin Exp Allergy* **34**, 497–507.

Musa, N. L., Ramakrishnan, M., Li, J. *et al.* (1999). Forskolin inhibits cyclin D1 expression in cultured airway smooth-muscle cells. *Am J Respir Cell Mol Biol* **20**, 352–358.

Nagata, D., Suzuki, E., Nishimatsu, H. *et al.* (2000). Transcriptional activation of the cyclin D1 gene is mediated by multiple *cis*-elements, including SP1 sites and a cAMP-responsive element in vascular endothelial cells. *J Biol Chem* **276**, 662–669.

Niimi, A., Matsumoto, H., Takemura, M. *et al.* (2003). Relationship of airway wall thickness to airway sensitivity and airway reactivity in asthma. *Am J Respir Crit Care Med* **168**, 983–988.

Orsini, M. J., Krymskaya, V. P., Eszterhas, A. J. *et al.* (1999). MAPK superfamily activation in human airway smooth muscle: mitogenesis requires prolonged p42/p44 activation. *Am J Physiol Lung Cell Mol Physiol* **277/21**, L479–L488.

Pacold, M. E., Suire, S., Perisic, O. *et al.* (2000). Crystal structure and functional analysis of Ras binding to its effector phosphoinositide 3-kinase gamma. *Cell* **103**, 931–943.

Page, K., Li, J. and Hershenson, M. B. (1999a). Platelet-derived growth factor stimulation of mitogen-activated protein kinases and cyclin D_1 promoter activity in cultured airway smooth-muscle cells. *Am J Respir Cell Mol Biol* **20**, 1294–1302.

Page, K., Li, J., Hodge, J. A. *et al.* (1999b). Characterization of a Rac1 signaling pathway to cyclin D(1) expression in airway smooth muscle cells. *J Biol Chem* **274**, 22065–22071.

Page, K., Li, J., Wang, Y. *et al.* (2000). Regulation of cyclin D(1)' expression and DNA synthesis by phosphatidylinositol 3-kinase in airway smooth muscle cells. *Am J Respir Cell Mol Biol* **23**, 436–443.

Panettieri, R. A., Jr., Murray, R. K., DePalo, L. R. *et al.* (1989). A human airway smooth muscle cell line that retains physiological responsiveness. *Am J Physiol Cell Physiol* **256/25**, C329–C335.

Panettieri, R. A., Jr., Yadvish, P. A., Kelly, A. M. *et al.* (1990). Histamine stimulates proliferation of airway smooth muscle and induces c-fos expression. *Am J Physiol Lung Cell Mol Physiol* **259/3**, L365–L371.

Panettieri, R. A., Jr., Tan, E. M. L., Ciocca, V. *et al.* (1998). Effects of LTD_4 on human airways smooth muscle cell proliferation, matrix expression, and contraction in vitro: Differential sensitivity to cysteinyl leukotriene receptor antagonists. *Am J Respir Cell Mol Biol* **19**, 453–461.

Pyne, S. and Pyne, N. J. (1996). The differential regulation of cyclic AMP by sphingomyelin-derived lipids and the modulation of sphingolipid-stimulated extracellular signal regulated kinase-2 in airway smooth muscle. *Biochemical Journal*, **315**, 917–923.

Ramakrishnan, M., Musa, N. L., Li, J. *et al.* (1998). Catalytic activation of extracellular signal-regulated kinases induces cyclin D1 expression in primary tracheal myocytes. *Am J Respir Cell Mol Biol* **18**, 736–740.

Rameh, L. E. and Cantley, L. C. (1999). The role of phosphoinositide 3-kinase lipid products in cell function. *J Biol Chem* **274**, 8347–8350.

Ravenhall, C., Guida, E., Harris, T. *et al.* (2000). The importance of ERK activity in the regulation of cyclin D1 levels and DNA synthesis in human cultured airway smooth muscle. *Br J Pharmacol* **131**, 17–28.

Roberts, C. R. (1995). Is asthma a fibrotic disease? *Chest* **107**, 111S–117S.

Roberts, C. R. and Burke, A. (1998). Remodelling of the extracellular matrix in asthma: proteoglycan synthesis and degradation. *Can Respir J* **5**, 48–50.

Rodriguez-Viciana, P., Warne, P. H., Dhand, R. *et al.* (1994). Phosphatidylinositol-3-OH kinase as a direct target of ras. *Nature* **370**, 527–532.

Roth, M., Johnson, P. R., Rudiger, J. J. *et al.* (2002). Interaction between glucocorticoids and beta2 agonists on bronchial airway smooth muscle cells through synchronised cellular signalling. *Lancet* **360**, 1293–1299.

Scott, P. H., Belham, C. M., Al-Hafidh, J. *et al.* (1996). A regulatory role for cAMP in phosphatidylinositol 3-kinase/p70 ribosomal S6 kinase-mediated DNA synthesis in platelet-derived-growth-factor-stimulated bovine airway smooth-muscle cells. *Biochem J* **318**, 965–971.

Sherr, C. J. (1994). G1 phase progression: cycling on cue. *Cell* **79**, 551–555.

Sherr, C. J. and Roberts, J. M. (1995). Inhibitors of mammalian G1 cyclin-dependent kinases. *Genes Dev* **9**, 1149–1163.

Sherr, C. J. and Roberts, J. M. (1999). CDK inhibitors: positive and negative regulators of G1-phase progression. *Genes Dev* **13**, 1501–1512.

Stewart, A. G., Tomlinson, P. R., Fernandes, D. J. *et al.* (1995). Tumor necrosis factor α modulates mitogenic responses of human cultured airway smooth muscle. *Am J Respir Cell Mol Biol* **12**, 110–119.

Stewart, A. G., Tomlinson, P. R. and Wilson, J. W. (1997). Beta 2-adrenoceptor agonist-mediated inhibition of human airway smooth muscle cell proliferation: importance of the duration of beta 2-adrenoceptor stimulation. *Br J Pharmacol* **121**, 361–368.

Stewart, A. G., Harris, T., Fernandes, D. J. *et al.* (1999). β2-adrenergic receptor agonists and cAMP arrest human cultured airway smooth muscle cells in the G1 phase of the cell cycle: Role of proteasome degradation of cyclin D1. *Mol Pharmacol* **56**, 1079–1086.

Tliba, O., Tliba, S., Huang, C. D. *et al.* (2003). TNFα modulates airway smooth muscle function via the autocrine action of IFNβ. *J Biol Chem* **278**, 50615–50623.

Tomlinson, P. R., Wilson, J. W. and Stewart, A. G. (1994). Inhibition by salbutamol of the proliferation of human airway smooth muscle cells grown in culture. *Br J Pharmacol* **111**, 641–647.

Tomlinson, P. R., Wilson, J. W. and Stewart, A. G. (1995). Salbutamol inhibits the proliferation of human airway smooth muscle cells grown in culture: relationship to elevated cAMP levels. *Biochem Pharmacol* **49**, 1809–1819.

Volinia, S., Dhand, R., Vanhaesebroeck, B. *et al.* (1995). A human phosphatidylinositol 3-kinase complex related to the yeast Vps34p–Vps15p protein sorting system. *EMBO J* **14**, 3339–3348.

Walker, T. R., Moore, S. M., Lawson, M. F. *et al.* (1998). Platelet-derived growth factor-BB and thrombin activate phosphoinositide 3-kinase and protein kinase B: Role in mediating airway smooth muscle proliferation. *Mol Pharmacol* **54**, 1007–1015.

Whicker, S. D., Armour, C. L. and Black, J. L. (1988). Responsiveness of bronchial smooth muscle from asthmatic patients to relaxant and contractile agonists. *Pulm Pharmacol* **1**, 25–31.

Woodruff, P. G., Dolganov, G. M., Ferrando, R. E. *et al.* (2004). Hyperplasia of smooth muscle in mild to moderate asthma without changes in cell size or gene expression. *Am J Respir Crit Care Med* **169**, 1001–1006.

Xiong, W., Pestell, R. G., Watanabe, G. *et al.* (1997). Cyclin D1 is required for S phase traversal in bovine tracheal myocytes. *Annu Rev Physiol* **272**, L1205–L1210.

Zheng, X., Zhou, D., Seow, C. Y. *et al.* (2004). Cardiotrophin-1 alters airway smooth muscle structure and mechanical properties in airway explants. *Am J Physiol Lung Cell Mol Physiol*, **287**, L1165–1171. Epub 2004 Jul 1123.

Zhou, D., Zheng, X., Wang, L. *et al.* (2003). Expression and effects of cardiotrophin-1 in human airway smooth muscle cells. *Br J Pharmacol* **140**, 1237–1244.

6

Airway smooth muscle bidirectional interactions with extracellular matrix

Jane E. Ward[1] and Stuart J. Hirst[2]

[1] *Department of Pharmacology, University of Melbourne, Victoria, Australia*

[2] *King's College London, MRC and Asthma UK Centre in Allergic Mechanisms of Asthma, London, UK*

6.1 Overview

Inflammation and subepithelial layer fibrosis are characteristic features of asthmatic airways. The fibrotic response includes an increase in volume occupied by extracellular matrix (ECM) tissue, as well as changes in its composition favouring wound type collagens, fibronectin, and several glycoproteins and proteoglycans normally associated with lung development and injury repair. Airway smooth muscle (ASM) secretes multiple ECM proteins and ECM-degrading enzymes in response to pro-fibrogenic stimuli, identifying it as a putative cellular source and regulator of ECM composition in the airways. ASM cells also express variable levels of integrins, the major receptors for the ECM, and so have the capacity to interact with their immediate ECM environment. ASM cells from asthmatic subjects secrete higher levels of ECM and emerging data strongly implicate the

Airway Smooth Muscle   Edited by Kian Fan Chung
© 2008 John Wiley & Sons, Ltd

ECM as a critical regulator of all the known functions of ASM including contraction, attachment and migration, apoptosis, proliferation and cytokine production. Alterations in the ECM composition may also determine responses of ASM to frontline anti-asthma therapy. Currently available frontline therapies such as glucocorticoids and β_2-adrenoceptor agonists are poorly effective in preventing ECM protein accumulation, and contact with the ECM appears to regulate the efficacy of these drugs on ASM function. In the case of glucocorticoids, these drugs may even contribute to local ECM changes. Further understanding of the role of the ECM and integrins in modulating ASM function in asthma and other chronic lung diseases may provide novel therapeutic targets that prevent or reverse the remodelling process as well as offering improvements in the efficacy of existing therapies.

6.2 Introduction

Tissue remodelling and persistent inflammation are key features of the airway wall of various lung disorders, including asthma (Bousquet *et al.*, 1995; Davies *et al.*, 2003), chronic obstructive pulmonary disease (Jeffery, 2001), cystic fibrosis (Hays *et al.*, 2005) and chronic lung disease of immaturity (bronchopulmonary dysplasia). Two prominent components of the remodelling process in patients with asthma include accumulation of airway smooth muscle (ASM) and alterations in the amount and composition of extracellular matrix (ECM) proteins (Roberts *et al.*, 2002; Roche, 1991; Roche *et al.*, 1989). Together, these changes are thought to underlie the cause of increased airway narrowing and hyperresponsiveness, and the exaggerated bronchoconstrictor response to various unrelated stimuli, as well as the progressive loss of reversibility of airflow obstruction seen in patients with asthma (Jeffery, 2001).

The ECM forms the supporting structure for the airway wall, and ASM cells are found surrounded by a bed of ECM (Figure 6.1) (James, 2005). The ECM in the airways and in all tissues of the body is a complex ordered aggregate comprising several classes of macromolecules. These are the collagen superfamily, where approximately 28 different forms of collagen are known, of which half are found in the lung (Thickett *et al.*, 2001), the structural glycoproteins (including fibronectin, laminins, tenascins, thrombospondin, osteonectin, fibrillin and vitronectin), the proteogylcans (such as bigylcan, decorin, versican and perlecan), the glycosaminoglycans (hyaluronan, heparin sulphate and chondroitin sulphate) and elastin. With the exception of elastin and the glycosaminoglycans, each class of matrix macromolecule comprises families of structurally related proteins, each member being a unique gene product (Labat-Robert and Robert, 2005). It is now

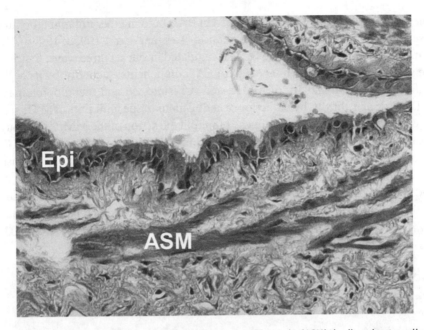

Figure 6.1 Photomicrograph (×200) of airway smooth muscle (ASM) (red) and extracellular matrix (ECM) (blue) seen in a transverse section of a normal human airway stained with Masson's trichrome. ECM is present throughout the airway wall, and ASM bundles are embedded within and surrounded by the ECM. Also shown is the ciliated epithelium (Epi). (For a colour reproduction of this figure, please see the colour section, located towards the centre of the book).

widely acknowledged that the complex matrix resulting from this macromolecular diversity goes beyond being a structurally stable material that provides support for cells and tissues. Instead, the ECM also provides direct signals to cells, often via integrins expressed at the cell surface, to modulate many pivotal cellular functions, including proliferation, differentiation and migration. The ECM also acts as a reservoir for stored inflammatory mediators and growth factors, which can be activated or released via matrix metalloproteinases (MMPs). The physical properties of the various elements that comprise the ECM can profoundly alter the fluid balance and elasticity of the airway wall.

The ECM is not a static structure; rather, it is in a state of continuous turnover. In the lung, total collagen turnover has been estimated to be as high as 10–15% per day (McAnulty *et al.*, 1988). In the bronchoalveolar lavage (BAL) fluid of asthmatics, increased levels of fibronectin, hyaluronan and laminin breakdown products (reflecting increased ECM turnover) are found, correlating with asthma severity (Bousquet *et al.*, 1991; Meerschaert *et al.*, 1999). Vitronectin, fibronectin and collagen type III levels are increased in BAL fluid in other inflammatory lung diseases such as extrinsic allergic alveolitis (Teschler *et al.*, 1993a; 1993b). The composition and turnover of the ECM is regulated in at least three ways: firstly, by the synthesis of new proteins and other ECM components; secondly, by its degradation through the activity of MMPs and A disintegrin and metalloproteinases (ADAMs); and, thirdly, by the action of endogenous tissue inhibitors of MMPs, the so-called TIMPs, which act to oppose or inhibit the action of the MMPs. Although not the only cell in the airway wall involved in the dynamic regulation of the ECM, ASM contributes to the turnover of ECM by producing ECM proteins (Johnson *et al.*, 2000), MMPs and TIMPs (Elshaw *et al.*, 2004), ADAMs (Haitchi *et al.*, 2005) and other key molecules intimately associated with ECM regulation, such as the pro-fibrotic growth factors transforming growth factor (TGF)-β (Coutts *et al.*, 2001) and connective tissue growth factor (CTGF) (Burgess *et al.*, 2003).

6.3 Airway extracellular matrix (ECM) in health and disease

Accumulating evidence strongly supports key changes in the ECM in chronic airway diseases following repeated cycles of injury and repair, particularly in asthma, chronic obstructive pulmonary disease (COPD) and cystic fibrosis. Most is known of the changes occurring in asthma whereby the ECM contributes to thickening of the subepithelial basement membrane (Jeffery, 2001), increased ASM accumulation (Thomson *et al.*, 1996) and overall increased adventitial thickness. The effect of this remodelling of the ECM on airway function in asthma is uncertain (reviewed in MacParland *et al.*, 2003). It has been proposed that thickening of the subepithelial basement membrane ECM is a protective response that can limit the magnitude of airway narrowing. Likewise, increases in ECM within ASM bundles may protect against excess bronchoconstriction by providing internal resistance to shortening (Pare *et al.*, 1997; Thomson *et al.*, 1996). However, mathematical modelling studies predict that airway wall remodelling underlies the development of airway hyperresponsiveness in asthma (James *et al.*, 1989). In keeping with this, ECM remodelling in the airway adventitia may uncouple ASM tissue from the parenchymal elastic recoil, and thus exaggerate narrowing during ASM shortening (Macklem, 1996).

As well as changes in the quantity of ECM in chronic lung disease, studies of airway secretions and airway histology have revealed marked changes in its composition. A key histopathologic feature of the asthmatic airway is the thickened subepithelial basement membrane, which has a hyaline appearance

(Jeffery, 2001). This is associated with increased deposition of collagens I, III and V; laminin α2 and β2 chains; fibronectin; and tenascin (Altraja *et al.*, 1996; Flood-Page *et al.*, 2003; Laitinen *et al.*, 1997; Roche *et al.*, 1989), as well as a potential contribution from oedema (Jeffery, 2001). Some of these changes appear very early; in particular, tenascin, lumican and procollagen type III levels are increased in asthmatic airways as early as 24 h after allergen challenge (Flood-Page *et al.*, 2003). In contrast to the thickening of the subepithelial basement membrane in the airways seen in asthma, there is usually no change in this thickness in smokers with chronic bronchitis or COPD (Jeffery, 2001). A study examining ECM differences in endobronchial biopsies found more fibroblasts and collagen III staining in biopsies from patients with persistent severe asthma than in normal subjects. In contrast, there were fewer fibroblasts and no difference in total collagen III staining in COPD patients, indicating a relative lack of fibrotic changes in the proximal large airways in COPD (Benayoun *et al.*, 2003).

Total proteoglycan levels are increased in asthmatic airways (Huang *et al.*, 1999; Roberts and Burke, 1998). Versican, a large chondroitin sulphate proteoglycan, was expressed in the airways in fatal asthma (Roberts, 1995). Although decorin was poorly expressed in the airways and ASM bundles of healthy subjects and those with mild asthma (Huang *et al.*, 1999), levels of two other small proteoglycans, biglycan and lumican, and the glycosaminoglycan, hyaluronan, were elevated in the airways of asthmatics (Bousquet *et al.*, 1991; Huang *et al.*, 1999). In a recent report, immunolabelling for biglycan, lumican and versican was prominent in the ASM layer in moderate asthmatics. Staining for these proteoglycans within the ASM layer was relatively sparse in severe asthmatics, although it was still greater than that observed in control subjects (Pini *et al.*, 2007). In contrast, levels of biglycan, decorin and interstitial proteoglycans were reported to be reduced in COPD, but only when the condition was severe (van Straaten *et al.*, 1999). These changes were noted in the peribronchial area of small airways, but without changes in expression of collagen types I, III and IV; laminin; or fibronectin.

The role of elastic fibre changes in asthmatic airways is unclear. Bousquet *et al.* (1996) showed an abnormal distribution of elastic fibres in the submucosa of asthmatics, but another study suggested no difference in amount of elastic fibres compared with healthy subjects (Godfrey *et al.*, 1995). Deeper within the mucosa, increased collagen type I, hyaluronan and versican have been found localized within and surrounding ASM bundles from individuals with atopic asthma (Huang *et al.*, 1999; Roberts *et al.*, 2002; Wilson and Li, 1997). In adventitial areas, there was an increase in ECM deposition in small airways (Cosio *et al.*, 1978; de Boer *et al.*, 1998), perhaps indicating that the fibrotic process is more important in the small airways. Moreover, remodelling of the airway wall differed between moderate and severe asthmatics in terms of proteoglycan expression (Pini *et al.*, 2007). Although deposition in the subepithelial layer was similar in the two groups of asthmatic patients, moderate asthmatics had greater proteoglycan deposition within the ASM layer than severe asthmatics and healthy controls. Overall, the pattern of ECM deposition in COPD differs subtly from that found in asthma, where changes are localized predominantly in the subepithelial layers. In COPD, this deposition may be more diffuse throughout the airway wall, but the extent of changes in ECM in COPD has yet to be fully investigated.

6.4 Integrins

Integrins are the principal cell-surface receptors for ECM proteins such as collagen, fibronectin and laminin, and comprise two non-covalently linked transmembrane heterodimers, α and β, the combination of which confers selectivity for the multitude of ECM protein ligands. Eighteen α and eight β subunits form the 24 known αβ-heterodimers depending on cell type and cellular function (Staunton *et al.*, 2006). Their extracellular domains recognize the short tripeptide Arg–Gly–Asp (RGD) sequence present in many ECM proteins, including collagens and fibronectin. Integrins mediate signalling from the extracellular space into the cell through integrin-associated signalling and adaptor molecules such as FAK (focal adhesion kinase), ILK (integrin-linked kinase), PINCH (a particularly interesting, new, cysteine-histidine-rich protein) and Nck2 (non-catalytic (region of) tyrosine kinase adaptor protein 2). Via these molecules, integrin signalling interacts tightly and cooperatively with receptor tyrosine kinase signalling to regulate survival, proliferation and cell shape as well as polarity, adhesion, migration and differentiation. Integrins are also known as bidirectional receptors. Thus, integrin activation can also be utilized to signal from within the cell out to the ECM. In this way, cells can modulate integrin ligand-binding affinity and control cell adhesion (reviewed by Hynes, 2002), but this aspect of integrin biology has not been studied in detail in ASM.

Little is known of the expression of integrins in intact ASM in the airway wall or in disease. Histologic analysis has demonstrated that healthy human ASM cells *in situ* express integrins α1, α3, β1 and β3, but not α2, α6, or β4 (Damjanovich *et al.*, 1992; Pilewski *et al.*, 1997; Virtanen *et al.*, 1996; Weinacker *et al.*, 1995). In culture, human ASM cells expressed varying levels of integrins α1-5, αv, β1 and β3, with α2, α5 and β1 being universally expressed (Freyer *et al.*, 2001; Nguyen *et al.*, 2005) (Figure 6.2). Integrin β2, β4 and β5 subunits were not expressed in cultured cells (Freyer *et al.*, 2001; Nguyen *et al.*, 2005). The reasons for the discrepancy in integrin α subunit expression between cultured ASM cells and

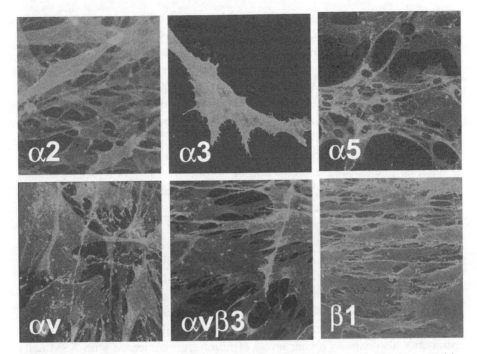

Figure 6.2 Fluorescent immunocytochemical localization of major integrins expressed by cultured human airway smooth muscle cells. Cell nuclei (blue) are labelled with the nucleic acid stain, Hoechst 33342 (×400). (For a colour reproduction of this figure, please see the colour section, located towards the centre of the book).

ASM in the intact airway wall are unclear but may relate to removal of these cells from their native environment or their proliferative status when in culture (Pickering *et al.*, 2000). The expression pattern in ASM represents less than half of the known heterodimers. No information exists for other fibronectin, and laminin receptors belonging to the $\beta1$ subfamily ($\alpha7-\alpha11$) or for $\beta6-\beta8$ integrins, and there are no data for their expression on ASM from asthmatics or patients with other lung diseases such as COPD.

6.5 Airway smooth muscle (ASM) as a modulator of airway ECM

As well as having the capacity to respond to ECM signals via expression of multiple integrin receptors, ASM cells synthesize a variety of ECM substrates *in vitro*, including collagens type I and IV, fibronectin, elastin, laminins, and the chrondroitin sulphate proteoglycans, biglycan, decorin and perlecan (Coutts *et al.*, 2001; Johnson *et al.*, 2000; Panettieri *et al.*, 1998). Many of these ECM components are upregulated in ASM by stimulation with growth factors such as TGFβ (Coutts *et al.*, 2001; Johnson *et al.*, 2006; Panettieri *et al.*, 1998), vascular endothelial growth factor (Kazi *et al.*, 2004) and CTGF (Johnson *et al.*, 2006), although gene expression for decorin is downregulated following TGFβ stimulation (Burgess *et al.*, 2003), and the cysteinyl leukotrienes appear to have a varying role (Panettieri *et al.*, 1998).

Exposure of human ASM in culture to serum from an asthmatic individual is reported to enhance the production of fibronectin, laminin α1, perlecan and chondroitin sulphate compared with serum from a healthy individual (Johnson *et al.*, 2000). Interestingly, production of these ECM substrates was not prevented by glucocorticoids but was enhanced in the case of fibronectin, perlecan and chrondroitin sulphate. In a subsequent study, Johnson *et al.* (2004) demonstrated increased ECM production by asthmatic cells in culture. Collagen type I and perlecan were found to be elevated in asthmatic ASM cells compared with non-asthmatic ASM cells. In contrast, production of collagen IV and laminin α1 was decreased, and chondroitin sulphate was only detectable in non-asthmatic cells (Johnson *et al.*, 2004). Recently, Chan *et al.* (2006) reported that ASM cells from asthmatics also express higher levels of fibronectin compared with ASM cells cultured from healthy individuals.

Elsewhere, conflicting data exist on the ability of leukotriene D_4 (LTD_4) to stimulate ECM secretion by human ASM. One study reported that LTD_4 alone or in combination with TGFβ had no effect on gene expression for collagen I, collagen IV, fibronectin, elastin, or biglycan (Panettieri *et al.*, 1998). No gene expression for the proteoglycan, versican, was detected in ASM at baseline or after TGFβ stimulation. Another study investigated the effect of epidermal growth factor (EGF) in combination with LTD_4 on ECM synthesis, and examined gene expression in human ASM for collagen I and fibronectin and for the proteoglycans, versican, biglycan, decorin, and perlecan (Potter-Perigo *et al.*, 2004). Versican and fibronectin expression was increased in human ASM cells treated with LTD_4 and EGF, while collagen I and decorin gene expression was reduced. Overall, total proteoglycan levels were elevated by LTD_4 and EGF, decorin being the major proteoglycan expressed in cultured human ASM.

In addition to their role in ECM deposition, ASM cells also have the capacity to degrade the ECM by secreting multiple MMPs (Elshaw *et al.*, 2004; Foda *et al.*, 1999; Johnson and Knox 1999) and may have a role in the anchoring of VEGF to the ECM derived from ASM (Burgess *et al.*, 2006a). The MMPs secreted by ASM comprise the collagenases, gelatinases, stromelysins, elastases, and membrane-bound forms, the most abundant MMPs being pro-MMP-2, pro-MMP-3, active MMP-3 and MT1-MMP. Human ASM cells also produce MMP-1 (Rajah *et al.*, 1996) and MMP-9, the levels of which are altered when the cells are exposed to allergic serum (Johnson *et al.*, 1999). The autocrine production of MMP-2 by human ASM has also been identified as a regulatory event in the ASM proliferative response (Johnson and Knox, 1999) and more recently in cellular migration (Henderson *et al.*, 2007).

Little is known of the release and activity of TIMPs in ASM (Black *et al.*, 2003). TIMPs counteract the proteolytic activity of the MMPs. They are 20–30-kDa proteins that bind to the catalytic domain of the MMP in a 1:1 stoichiometry, thereby inhibiting enzymatic activity (Black *et al.*, 2003). Knowledge of the specificity of the TIMPs is limited, although TIMP-1 is believed to inhibit MMP-1, -2, -3 and -9, whereas TIMP-2 has a greater selectivity for MMP-2. Johnson *et al.* (1999) reported production of TIMP-1, but not of TIMP-2 or -3, from human ASM cells in culture. In contrast, Elshaw *et al.* (2004) demonstrated that ASM-derived MMP-2 was closely controlled by autocrine TIMP-2, and in a later study collagen type I was shown to reduce TIMP-2 protein expression by ASM and potentiate MMP-2-dependent cellular migration (Henderson *et al.*, 2007).

6.6 Airway ECM as modulator of ASM function

Emerging data strongly implicate the ECM as a critical regulator of all the known functions of ASM (reviewed in Fernandes *et al.*, 2006) (Figure 6.3). These include contraction (Dekkers *et al.*, 2007), attachment (Nguyen *et al.*, 2005), migration (Parameswaran *et al.*, 2004), apoptosis (Freyer *et al.*, 2001), proliferation (Bonacci *et al.*, 2003; 2006; Hirst *et al.*, 2000a; Johnson *et al.*, 2004; Nguyen *et al.*, 2005), cytokine release (Chan *et al.*, 2006; Peng *et al.*, 2005) and the response of ASM to anti-asthma therapies such as β_2-adrenoceptor agonists (Freyer *et al.*, 2005) and glucocorticoids (Bonacci *et al.*, 2003, 2006).

The majority of studies have examined ASM cells on ECM beds in a two-dimensional culture in which cells are plated onto a single ECM substrate coating the bottom of a dish. Efforts are being made to improve this model by introducing three-dimensional cultures (Ward *et al.*, 2006) and by plating cells in two dimensions on ECM substrates derived from asthmatic or non-asthmatic ASM (Chan *et al.*, 2006; Johnson *et al.*, 2004). In almost all cases, attempts to mimic the ECM environment in asthma by culture on either collagen type I or fibronectin and their combination, or on ECM beds derived from asthmatic ASM, have resulted in enhancement of the pro-asthmatic function of ASM (Bonacci *et al.*,

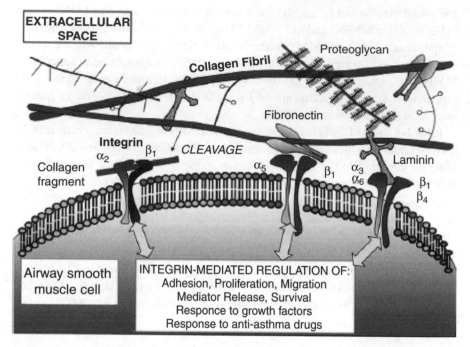

Figure 6.3   Putative bidirectional integrin-dependent interactions between the extracellular matrix and airway smooth muscle as well as other airway mesenchymal cells that result in altered cellular function or responses to anti-asthma therapy. Modified from Fernandes *et al.* (2006) with permission (Adapted from Figure 1 in Curr Drug Targets 7, 567-577 (2006))

2003; Chan *et al.*, 2006; Hirst *et al.*, 2000a; Johnson *et al.*, 2004; Nguyen *et al.*, 2005; Parameswaran *et al.*, 2004; Peng *et al.*, 2004).

Contraction and phenotype

Hirst *et al.* (2000a) reported in cultured ASM that the ECM differentially affects the expression of contractile proteins, including smooth muscle (sm)-α-actin, calponin, and sm-myosin heavy chain (MHC). Thus, in human ASM cells cultured on fibronectin or collagen type I matrices, there was reduced expression of these contractile proteins. In contrast, cells cultured on a laminin bed retained expression of sm-α-actin, calponin, and sm-MHC. Reduced contractile protein expression in smooth muscle is strongly associated with modulation towards a less contractile or less differentiated phenotype, often referred to as the synthetic

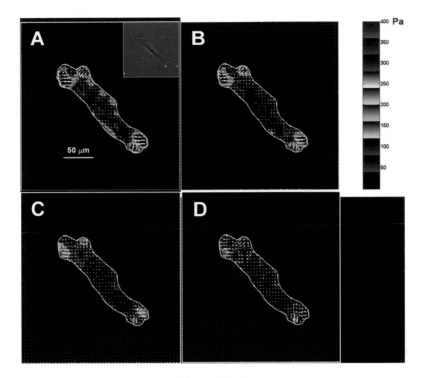

Figure 1.3

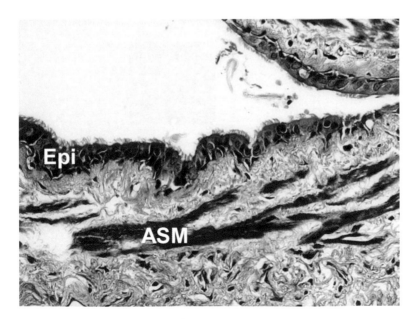

Figure 6.1

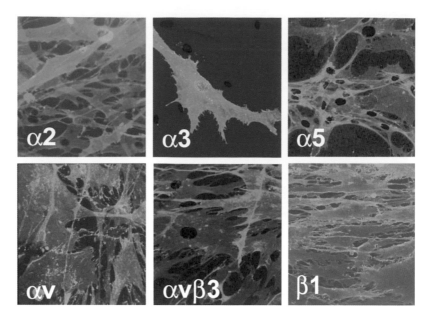

Figure 6.2

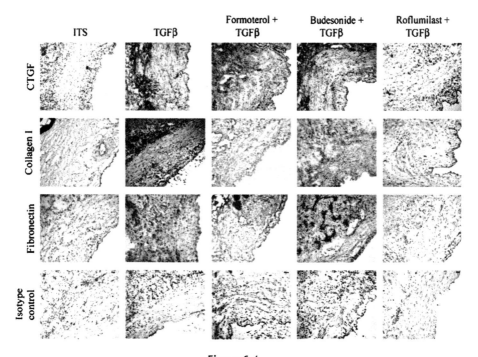

Figure 6.4

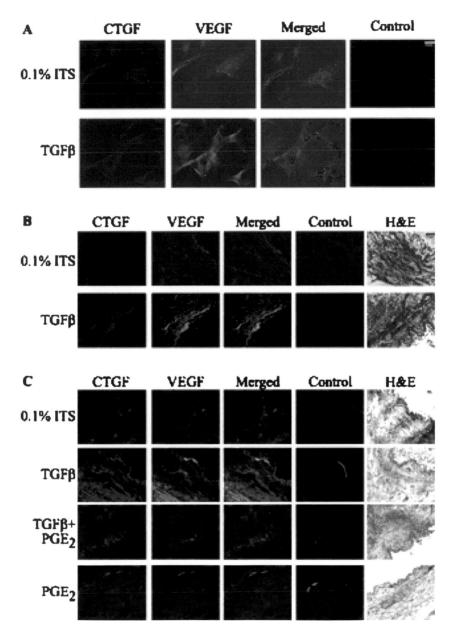

Figure 10.3

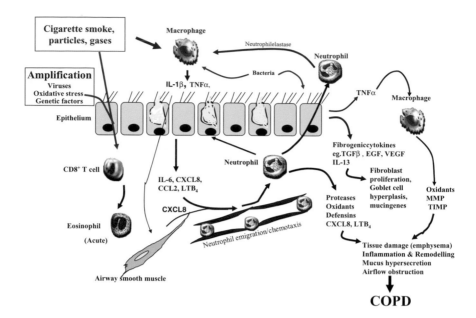

Figure 11.1

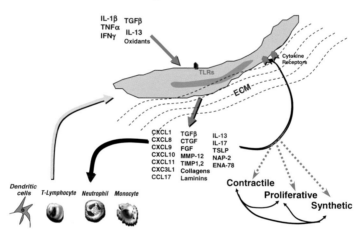

Figure 11.2

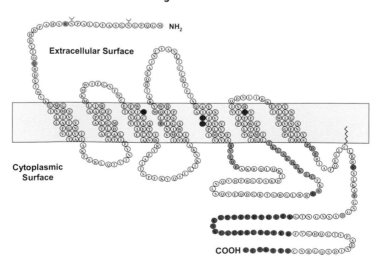

Figure 13.1

phenotype (Hirst *et al.*, 2000b). A recent study examined possible implications of this phenotype switching by the ECM on ASM contractile responses in isolated tracheal smooth muscle preparations (Dekkers *et al.*, 2007). Prolonged (4 days) culture of intact bovine tracheal muscle strips in the presence of fibronectin or collagen type I induced a decline in maximal contraction in response to both a receptor-dependent (e.g., methacholine) and a receptor-independent (e.g., KCl) stimulus. This was associated with a reduction in contractile protein expression. The observed fibronectin-induced suppression of contractility was blocked by the integrin RGD-containing antagonist Arg–Gly–Asp–Ser (RGDS), but not by the non-RGD control peptide, Gly–Arg–Ala–Asp–Ser–Pro (GRADSP). In contrast, culture with laminin maintained both the levels of sm-α-actin, calponin, and sm-MHC and contractility to methacholine and KCl, suggesting that laminin may be involved in maintaining a (normo)contractile phenotype (Dekkers *et al.*, 2007).

Cell adhesion and migration

Cell attachment to the ECM is an important event in migration of cells between tissue compartments. Recent endobronchial biopsy studies provide evidence for hypertrophy (Benayoun *et al.*, 2003; Ebina *et al.*, 1993); however, detailed stereologic studies have also emphasized the importance of hyperplasia (Woodruff *et al.*, 2004). While the *in vitro* proliferative capacity of human ASM from asthmatics is reported to be increased (Johnson *et al.*, 2001), other mechanisms could account for the accumulation. These could include reduced apoptosis (see next section) of resident smooth muscle cells, or migration of interstitial mesenchymal cells or circulating stem cells that then differentiate into fibroblasts or muscle precursor cells to form the additional muscle bulk (Stewart, 2004).

Nguyen *et al.* (2005) characterized attachment of human ASM cells in culture to varying ECM substrates including monomeric or fibrillar type I collagen, fibronectin, fibrillar type III collagen, vitronectin, and tenascin-C, and found that integrins α2β1 (major collagen type I receptor) and αvβ3 (vitronectin receptor) mediated attachment to collagen type I, while attachment to fibronectin required integrin α5β1 (major fibronectin receptor). Parameswaran *et al.* (2004) investigated the effect of different ECM proteins on leukotriene E_4-primed ASM migration towards platelet-derived growth (PDGF). Migration was greater across membranes coated with collagen types III and V and fibronectin, than with collagen type I, laminin or elastin. Blocking antibodies directed against a combination of integrin α5, αv and β1 subunits each attenuated ASM migration and adhesion on a collagen type I ECM bed. More recently, Henderson *et al.* (2007) examined the role of MMPs, in particular MMP-2, in relation to the functions of ASM in airway remodelling and found that both migration and proliferation were MMP dependent, but adhesion and apoptosis were not.

Apoptosis

The ECM provides a strong survival signal to cultured ASM cells. This was demonstrated by Freyer *et al.* (2001), who found that multiple ECM beds, such as collagen types I and IV, fibronectin and laminin, each induced the survival of ASM cells, while collagen type V, vitronectin and elastin provided no survival signal. Furthermore, collagen types I and IV, fibronectin, laminin, and vitronectin all produced marked inhibition of apoptosis of ASM seeded in the presence of cycloheximide, whereas elastin did not have this effect. Intriguingly, the anti-apoptotic signal for each of the survival-inducing ECM substrates appeared to require a common integrin heterodimer, the fibronectin receptor, $\alpha5\beta1$ (Freyer *et al.*, 2001).

Proliferation

Proliferation of human ASM cells in response to diverse stimuli such as PDGF, thrombin and fibroblast growth factor (FGF)-2 is markedly enhanced on monomeric collagen type I or fibronectin relative to cells plated on a laminin ECM bed or on fibrillar collagen type I, collagen type III or tenascin (Bonacci *et al.*, 2003; Hirst *et al.*, 2000a; Nguyen *et al.*, 2005). Furthermore, when fibronectin and monomeric collagen type I were combined to mimic the dominance of these substrates in the asthmatic airway ECM environment, there was a marked synergistic enhancement in PDGF-induced proliferation (Nguyen *et al.*, 2005). Interrogation of these responses with integrin function-blocking antibodies implicated $\alpha2\beta1$, $\alpha4\beta1$ (vascular cell adhesion molecule-1 receptor and a secondary fibronectin receptor), and $\alpha5\beta1$ integrins mediating the enhanced proliferative effects of fibronectin and monomeric collagen type I (Bonacci *et al.*, 2006; Nguyen *et al.*, 2005). In another report, proliferation of non-asthmatic ASM cells was enhanced by culture on an ECM bed from asthmatic ASM cells compared with similar beds from non-asthmatic ASM cells (Johnson *et al.*, 2004).

Of note, other components of the ECM that are non-integrin dependent may also modulate human ASM proliferation. Kanabar *et al.* (2005) demonstrated the importance of the degree of sulphation, ionic charge and polymer size in the antiproliferative effects of heparin, and showed that chondroitin sulphates, heparan sulphate and hyaluron each inhibited serum-stimulated proliferation to varying degrees.

Cytokine secretion

Peng *et al.* (2005) examined production of eotaxin, RANTES (regulated on activation, normal T-cell expressed and secreted), and granulocyte-macrophage colony-stimulating factor (GM-CSF) in human ASM cells cultured on varying ECM substrates. Interleukin (IL)-1β-stimulated levels of all three mediators were increased in cells plated on a collagen type I or fibronectin compared with cells plated on plastic or collagen types III or V. Enhancement of cytokine release by type I or fibronectin was blocked by the RGD-sequence-containing integrin antagonists Gly–Arg–Gly–Asp–Ser (GRGDS) and Gly–Arg–Gly–Asp–Thr–Pro (GRGDTP), but not by a non-RGD negative control Gly–Arg–Ala–Asp–Ser–Pro (GRADSP). Targeted integrin-blocking experiments confirmed a requirement for integrin ligation and demonstrated that the RGD-binding integrins $\alpha2\beta1$, $\alpha5\beta1$, $\alpha v\beta1$ (fibronectin receptor) and $\alpha v\beta3$ (vitronectin/fibronectin receptor) were each important for mediating increased cytokine expression in response to fibronectin, while integrin $\alpha2\beta1$ was the major transducer for collagen type I.

A subsequent report showed similar enhancement of IL-13- or IL-4-dependent eotaxin release from ASM cells by fibronectin or collagen type I (Chan *et al.*, 2006). In the same study, when they were seeded on ECM beds from asthmatic ASM, IL-13-dependent eotaxin release from healthy or asthmatic ASM was enhanced compared with culture on healthy ECM. This was associated with increased autocrine fibronectin expression by ASM cultured from asthmatics and was partially prevented by integrin $\alpha5\beta1$ neutralization. Given that asthmatic airways are characterized by upregulation of collagen I and fibronectin, it seems likely that these ECM proteins contribute to the increased synthetic function (Freyer *et al.*, 2001; Nguyen *et al.*, 2005; Peng *et al.*, 2005). Importantly, asthmatic ASM cells in culture produce collagen type I and fibronectin to a greater extent than non-asthmatic cells, leading to the possibility that ASM cells themselves might regulate their synthetic activity in an autocrine manner (Chan *et al.*, 2006; Johnson *et al.*, 2004). Interest is therefore growing in the potential for synergism/antagonism between ECM exposure and anti-asthma therapy.

6.7 Impact of anti-asthma therapy on ASM–ECM interactions

The therapeutic actions of the major classes of current anti-asthma agents on ASM are modified or impaired by contact with an altered ECM environment (Figure 6.3). Glucocorticoids, which are anti-proliferative in ASM cultured from non-asthmatic subjects, are poorly effective in asthmatic ASM (Roth *et al.*, 2004). The antiproliferative actions of glucocorticoids in ASM are also lost when ASM cells are plated on monomeric collagen type I, but not when seeded on plastic or a laminin substrate (Bonacci *et al.*, 2003). Given that asthmatic ASM synthesize more collagen I than non-asthmatic cells (Johnson *et al.*, 2004), it is possible that the insensitivity to glucocorticoids might be mediated in part by an altered ECM environment. Consistent with this notion, Johnson *et al.* (2000) reported that the glucocorticoid beclomethasone *enhanced* rather than decreased the synthesis of several ECM substrates induced by atopic asthmatic serum, including fibronectin, perlecan and chrondroitin sulphate.

Impaired relaxation of ASM and incomplete bronchodilatation are character- istic features of asthma. Freyer *et al.* (2004) noted that a fibronectin-dominated environment increased responses to β_2-adrenoceptor activation, while exposure to collagen type V or laminin reduced β_2-adrenoceptor signalling compared with collagens types I or IV. This effect was accompanied by a reduction in intracellu- lar cyclic AMP levels that was associated with modulation of Giα. More recently, it was reported that neither glucocorticoids nor long-acting β_2-adreonoceptor agonists reduced TGFβ-induced collagen type I or fibronectin in ASM from individuals with or without asthma (Burgess *et al.*, 2006b). However, the phos- phodiesterase 4 (PDE4) inhibitor roflumilast prevented upregulation of induced collagen type I, fibronectin or CTGF by TGFβ in intact bronchial rings from non-asthmatics (Figure 6.4). In contrast, the PDE inhibitor was less effective against TGFβ-induced ECM production from ASM cells cultured from healthy or asthmatics donors (Burgess *et al.*, 2006b). Recently, treatment of allergen- exposed atopic asthmatics with a humanized, anti-IL-5 monoclonal antibody (mepolizumab) was found to suppress both airway eosinophil numbers and the subepithelial basement membrane expression of tenascin, lumican and procolla- gen type III (Flood-Page *et al.*, 2003). More studies of this type are needed to understand the impact of novel anti-asthma therapies on ECM deposition in the airways, particularly in the tissue compartments surrounding ASM.

Increasing evidence implicates key roles for multiple transcription factors in the inflammatory and remodelling processes in asthma (reviewed in Roth and Black,

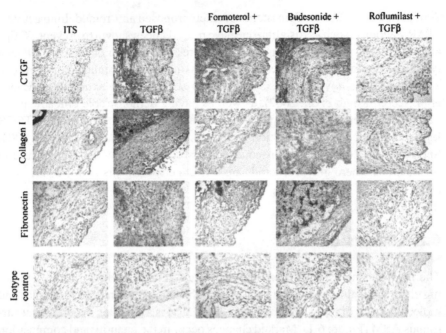

Figure 6.4 Roflumilast (a phosphodiesterase inhibitor), but not budesonide (a glucocorticoid) or formoterol (β_2-adreoceptor agonist), attenuates transforming growth factor (TGF)-β-induced connective tissue growth factor (CTGF), collagen type I and fibronectin upregulation (brown labelling) in non-asthmatic bronchial rings. Modified from Burgess *et al.* (2006b) with permission. (Reprinted from Journal of Allergy & Clinical Immunology, 118(3), with permission from the American Academy of Allergy, Asthma, and Immunology.) (For a colour reproduction of this figure, please see the colour section, located towards the centre of the book).

2006), including the glucocorticoid receptor (GR), nuclear factor κB (NF-κB), activator protein-1 (AP-1), nuclear factor of activated T-cells (NF-AT), cyclic AMP response element-binding protein (CREB), CCAAT/enhancer binding protein (C/EBP), and, more recently, peroxisome proliferator-activated receptor (PPAR). PPARs are members of the nuclear hormone receptor superfamily that regulate the expression of genes involved in a variety of biological processes, including lipid metabolism and insulin sensitivity. PPARγ is expressed in both infiltrating inflammatory and structural resident cells of the lung, and ligand-dependent activation of this receptor results in suppression of effector leukocyte responses, including cytokine production, cellular migration and proliferation (reviewed in Spears *et al.*, 2006).

In addition to its reported anti-inflammatory effects, PPARγ activation suppresses proliferation and cytokine secretion from cultured ASM cells (Patel *et al.*,

2003; Ward *et al.*, 2004). Consistent with their proposed anti-remodelling activity, PPARγ agonists such as rosiglitazone have recently been shown to prevent TGFβ- and CTGF-dependent signalling and to suppress expression of MMPs, collagen type I and fibronectin in a variety of cell types, including vascular smooth muscle and parenchymal myofibroblasts (Burgess *et al.*, 2005; Law *et al.*, 2000; Maeda *et al.*, 2005; Routh *et al.*, 2002), as well as having demonstrable efficacy in targeting the wider remodelling process in asthma, including inhibition of TGFβ levels and collagen III deposition (Benayoun *et al.*, 2001; Honda *et al.*, 2004; Spears *et al.*, 2006). As a consequence, this class of drugs may prove useful in situations where glucocorticoids are less effective (Spears *et al.*, 2006; Ward *et al.*, 2007).

6.8 Conclusions

The mechanisms by which remodelling occurs in chronic lung disease are being elucidated, although major gaps exist in the understanding of the cellular and molecular basis of many of the changes. The ECM is a complex structure that surrounds ASM (Figure 6.1). Marked changes occur in the quantity and composition of the ECM in chronic lung disease, and there is wide speculation as to whether the changes are ultimately harmful or protective in the remodelled airway. In addition to providing structural support to cells in the airway wall, the ECM actively transduces signals, and there are bidirectional interactions, many of which are mediated by multiple cell-surface integrins expressed on ASM (Figure 6.2). Emerging evidence suggests that ASM cells have a substantial capacity to synthesize, secrete and degrade the array of ECM proteins found in the airway wall, and, in doing so, they can regulate their immediate ECM environment and the signals to which they are exposed. New evidence also indicates that many airway wall ECM components affect ASM function, including its contractile, proliferative, migratory and secretory properties. Changes in the ECM environment that reflect the process of chronic airway inflammation generally enhance the pro-asthmatic function of ASM.

Currently available therapies such as glucocorticoids and β$_2$-adrenoceptor agonists are ineffective in preventing ECM protein accumulation, and contact with the ECM appears to regulate the efficacy of these drugs (Figures 6.3 and 6.4). In the case of glucocorticoids, these drugs may even induce secretion of ECM from ASM. It is therefore necessary to consider the effects of novel anti-inflammatory and anti-remodelling therapies that target the ECM and its complex interplay with airway structural cells such as ASM.

Acknowledgements

The authors are grateful to Dr Lyn M. Moir and Dr Trang T. B. Nguyen for providing the photomicrographs in Figures 6.1 and 6.2, respectively.

References

Altraja, A., Laitinen, A., Virtanen, I. *et al.* (1996). Expression of laminins in the airways in various types of asthmatic patients: a morphometric study. *Am J Respir Cell Mol Biol* **15**, 482–488.

Benayoun, L., Druilhe, A., Dombret, M. C. *et al.* (2003). Airway structural alterations selectively associated with severe asthma. *Am J Respir Crit Care Med* **167**, 1360–1368.

Benayoun, L., Letuve, S., Druilhe, A. *et al.* (2001). Regulation of peroxisome proliferator-activated receptor gamma expression in human asthmatic airways: relationship with proliferation, apoptosis, and airway remodeling. *Am J Respir Crit Care Med* **164**, 1487–1494.

Black, J. L., Burgess, J. K. and Johnson, P. R. (2003). Airway smooth muscle – its relationship to the extracellular matrix. *Respir Physiol Neurobiol* **137**, 339–346.

Bonacci, J. V., Harris, T., Wilson, J. W. *et al.* (2003). Collagen-induced resistance to gluco-corticoid anti-mitogenic actions: a potential explanation of smooth muscle hyperplasia in the asthmatic remodelled airway. *Br J Pharmacol* **138**, 1203–1206.

Bonacci, J.V., Schuliga, M., Harris, T. *et al.* (2006). Collagen impairs glucocorticoid actions in airway smooth muscle through integrin signalling. *Br J Pharmacol* **149**, 365–373.

Bousquet, J., Chanez, P., Lacoste, J. Y.*et al.* (1991). Indirect evidence of bronchial inflammation assessed by titration of inflammatory mediators in BAL fluid of patients with asthma. *J Allergy Clin Immunol* **88**, 649–660.

Bousquet, J., Vignola, A. M., Chanez, P. *et al.* (1995). Airways remodelling in asthma: no doubt, no more? *Int Arch Allergy Immunol* **107**, 211–214.

Burgess, H. A., Daugherty, L. E., Thatcher, T. H. *et al.* (2005). PPARgamma agonists inhibit TGF-beta induced pulmonary myofibroblast differentiation and collagen production: implications for therapy of lung fibrosis. *Am J Physiol Lung Cell Mol Physiol* **288**, L1146–L1153.

Burgess, J. K., Ge, Q., Poniris, M. H. *et al.* (2006a). Connective tissue growth factor and vascular endothelial growth factor from airway smooth muscle interact with the extracellular matrix. *Am J Physiol Lung Cell Mol Physiol* **290**, L153–L161.

Burgess, J. K., Johnson, P. R., Ge, Q. *et al.* (2003). Expression of connective tissue growth factor in asthmatic airway smooth muscle cells. *Am J Respir Crit Care Med* **167**, 71–77.

Burgess, J. K., Oliver, B. G., Poniris, M. H. *et al.* (2006b). A phosphodiesterase 4 inhibitor inhibits matrix protein deposition in airways in vitro. *J Allergy Clin Immunol* **118**, 649–657.

Chan, V., Burgess, J. K., Ratoff, J. C. *et al.* (2006). Extracellular matrix regulates enhanced eotaxin expression in asthmatic airway smooth muscle cells. *Am J Respir Crit Care Med* **174**, 379–385.

Chung, K. F. (2005). The role of airway smooth muscle in the pathogenesis of airway wall remodeling in chronic obstructive pulmonary disease. *Proc Am Thorac Soc* **2**, 347–354.

Cosio, M., Ghezzo, H., Hogg, J. C. *et al.* (1978). The relations between structural changes in small airways and pulmonary-function tests. *N Engl J Med* **298**, 1277–1281.

Coutts, A., Chen, G., Stephens, N. *et al.* (2001). Release of biologically active TGF-beta from airway smooth muscle cells induces autocrine synthesis of collagen. *Am J Physiol Lung Cell Mol Physiol* **280**, L999–L1008.

Damjanovich, L., Albelda, S. M., Mette, S. A. *et al.* (1992). Distribution of integrin cell adhesion receptors in normal and malignant lung tissue. *Am J Respir Cell Mol Biol* **6**, 197–206.

Davies D. E., Wicks, J., Powell, R. M. *et al.* (2003). Airway remodeling in asthma: new insights. *J Allergy Clin Immunol* **111**, 215–225.

de Boer, W. I., van Schadewijk, A., Sont, J. K. *et al.* (1998). Transforming growth factor beta1 and recruitment of macrophages and mast cells in airways in chronic obstructive pulmonary disease. *Am J Respir Crit Care Med* **158**, 1951–1957.

Dekkers, B. G., Schaafsma, D., Nelemans, S. A. *et al.* (2007). Extracellular matrix proteins differentially regulate airway smooth muscle phenotype and function. *Am J Physiol Lung Cell Mol Physiol* (in press).

Ebina, M., Takahashi, T., Chiba, T. *et al.* (1993). Cellular hypertrophy and hyperplasia of airway smooth muscles underlying bronchial asthma. A 3-D morphometric study. *Am Rev Respir Dis* **148**, 720–726.

Elshaw, S. R., Henderson, N., Knox, A. J. *et al.* (2004). Matrix metalloproteinase expression and activity in human airway smooth muscle cells. *Br J Pharmacol* **142**, 1318–1324.

Fernandes, D. J., Bonacci, J. V. and Stewart, A. G. (2006). Extracellular matrix, integrins, and mesenchymal cell function in the airways. *Curr Drug Targets* **7**, 567–577.

Flood-Page, P., Menzies-Gow, A., Phipps, S. *et al.* (2003). Anti-IL-5 treatment reduces deposition of ECM proteins in the bronchial subepithelial basement membrane of mild atopic asthmatics. *J Clin Invest* **112**, 1029–1036.

Foda, H. D., George, S., Rollo, E. *et al.* (1999). Regulation of gelatinases in human airway smooth muscle cells: mechanism of progelatinase A activation. *Am J Physiol* **277**, L174–L182.

Freyer, A. M., Billington, C. K., Penn, R. B. *et al.* (2004). Extracellular matrix modulates beta2-adrenergic receptor signaling in human airway smooth muscle cells. *Am J Respir Cell Mol Biol* **31**, 440–445.

Freyer, A.M., Johnson, S.R. and Hall, I.P. (2001). Effects of growth factors and extracellular matrix on survival of human airway smooth muscle cells. *Am J Respir Cell Mol Biol* **25**, 569–576.

Godfrey, R.W., Lorimer, S., Majumdar, S. *et al.* (1995). Airway and lung elastic fibre is not reduced in asthma nor in asthmatics following corticosteroid treatment. *Eur Respir J* **8**, 922–927.

Haitchi, H. M., Powell, R. M., Shaw, T. J. *et al.* (2005). ADAM33 expression in asthmatic airways and human embryonic lungs. *Am J Respir Crit Care Med* **171**, 958–965.

Hays, S. R., Ferrando, R. E., Carter, R. *et al.* (2005). Structural changes to airway smooth muscle in cystic fibrosis. *Thorax* **60**, 226–228.

Henderson, N., Markwick, L. J., Elshaw, S. R. *et al.* (2007). Collagen I and thrombin activate MMP-2 by MMP-14 dependent and independent pathways: implications for airway smooth muscle migration. *Am J Physiol Lung Cell Mol Physiol* **292**, L1030–L1038.

Hirst, S. J., Twort, C. H. and Lee, T. H. (2000a). Differential effects of extracellular matrix proteins on human airway smooth muscle cell proliferation and phenotype. *Am J Respir Cell Mol Biol* **23**, 335–344.

Hirst, S. J., Walker, T. R. and Chilvers, E. R. (2000b). Phenotypic diversity and molecular mechanisms of airway smooth muscle proliferation in asthma. *Eur Respir J* **16**, 159–177.

Honda, K., Marquillies, P., Capron, M. *et al.* (2004). Peroxisome proliferator-activated receptor gamma is expressed in airways and inhibits features of airway remodeling in a mouse asthma model. *J Allergy Clin Immunol* **113**, 882–888.

Huang, J., Olivenstein, R., Taha, R. *et al.* (1999). Enhanced proteoglycan deposition in the airway wall of atopic asthmatics. *Am J Respir Crit Care Med* **160**, 725–729.

Hynes, R. O. (2002). Integrins: bidirectional, allosteric signaling machines. *Cell* **110**, 673–687.

James, A. (2005). Remodelling of airway smooth muscle in asthma: what sort do you have? *Clin Exp Allergy* **35**, 703–707.

James, A. L., Pare, P. D. and Hogg, J. C. (1989). The mechanics of airway narrowing in asthma. *Am Rev Respir Dis* **139**, 242–246.

Jeffery, P. K. (2001). Remodeling in asthma and chronic obstructive lung disease. *Am J Respir Crit Care Med* **164**, S28–S38.

Johnson, P. R., Black, J. L., Carlin, S. *et al.* (2000). The production of extracellular matrix proteins by human passively sensitized airway smooth-muscle cells in culture: the effect of beclomethasone. *Am J Respir Crit Care Med* **162**, 2145–2151.

Johnson, P. R., Burgess, J. K., Ge, Q. *et al.* (2006). Connective tissue growth factor induces extracellular matrix in asthmatic airway smooth muscle. *Am J Respir Crit Care Med* **173**, 32–41.

Johnson, P. R., Burgess, J. K., Underwood, P. A. *et al.* (2004). Extracellular matrix proteins modulate asthmatic airway smooth muscle cell proliferation via an autocrine mechanism. *J Allergy Clin Immunol* **113**, 690–696.

Johnson, P. R., Roth, M., Tamm, M. *et al.* (2001). Airway smooth muscle cell proliferation is increased in asthma. *Am J Respir Crit Care Med* **164**, 474–477.

Johnson, S. and Knox, A. (1999). Autocrine production of matrix metalloproteinase-2 is required for human airway smooth muscle proliferation. *Am J Physiol* **277**, L1109–L1117.

Kanabar, V., Hirst, S. J., O'Connor, B. J. *et al.* (2005). Some structural determinants of the antiproliferative effect of heparin-like molecules on human airway smooth muscle. *Br J Pharmacol* **146**, 370–377.

Kazi, A. S., Lotfi, S., Goncharova, E. A. *et al.* (2004). Vascular endothelial growth factor-induced secretion of fibronectin is ERK dependent. *Am J Physiol Lung Cell Mol Physiol* **286**, L539–L545.

Labat-Robert, J. and Robert, L. (2005). The extracellular matrix during normal development and neoplastic growth. *Prog Mol Subcell Biol* **40**, 79–106.

Laitinen, A., Altraja, A., Kampe, M. *et al.* (1997). Tenascin is increased in airway basement membrane of asthmatics and decreased by an inhaled steroid. *Am J Respir Crit Care Med* **156**, 951–958.

Law, R. E., Goetze, S., Xi, X. P. *et al.* (2000). Expression and function of PPARgamma in rat and human vascular smooth muscle cells. *Circulation* **101**, 1311–1318.

Macklem, P. T. (1996). Can airway function be predicted? *Am J Respir Crit Care Med* **153**, S19–S20.

Maeda, A., Horikoshi, S., Gohda, T. *et al.* (2005). Pioglitazone attenuates TGF-beta(1)-induction of fibronectin synthesis and its splicing variant in human mesangial cells via activation of peroxisome proliferator-activated receptor (PPAR)gamma. *Cell Biol Int* **29**, 422–428.

McAnulty, R. J., Staple, L. H., Guerreiro, D. *et al.* (1988). Extensive changes in collagen synthesis and degradation during compensatory lung growth. *Am J Physiol* **255**, C754–C759.

McParland, B. E., Macklem, P. T. and Pare, P. D. (2003). Airway wall remodeling: friend or foe? *J Appl Physiol* **95**, 426–434.

Meerschaert, J., Kelly, E. A., Mosher, D. F. *et al.* (1999). Segmental antigen challenge increases fibronectin in bronchoalveolar lavage fluid. *Am J Respir Crit Care Med* **159**, 619–625.

Nguyen, T. T., Ward, J. P. and Hirst, S. J. (2005). Beta1-integrins mediate enhancement of airway smooth muscle proliferation by collagen and fibronectin. *Am J Respir Crit Care Med* **171**, 217–223.

Panettieri, R. A., Tan, E. M., Ciocca, V. *et al.* (1998). Effects of LTD4 on human airway smooth muscle cell proliferation, matrix expression, and contraction in vitro: differential sensitivity to cysteinyl leukotriene receptor antagonists. *Am J Respir Cell Mol Biol* **19**, 453–461.

Parameswaran, K., Radford, K., Zuo, J. *et al.* (2004). Extracellular matrix regulates human airway smooth muscle cell migration. *Eur Respir J* **24**, 545–551.

Pare, P. D., Roberts, C. R., Bai, T. R. *et al.* (1997). The functional consequences of airway remodeling in asthma. *Monaldi Arch Chest Dis* **52**, 589–596.

Patel, H. J., Belvisi, M. G., Bishop-Bailey, D. *et al.* (2003). Activation of peroxisome proliferator-activated receptors in human airway smooth muscle cells has a superior anti-inflammatory profile to corticosteroids: relevance for chronic obstructive pulmonary disease therapy. *J Immunol* **170**, 2663–2669.

Peng, Q., Lai, D., Nguyen, T. T. *et al.* (2005). Multiple beta 1 integrins mediate enhancement of human airway smooth muscle cytokine secretion by fibronectin and type I collagen. *J Immunol* **174**, 2258–2264.

Pickering, J. G., Chow, L. H., Li, S. *et al.* (2000). Alpha 5 beta 1 integrin expression and luminal edge fibronectin matrix assembly by smooth muscle cells after arterial injury. *Am J Pathol* **156**, 453–465.

Pilewski, J. M., Latoche, J. D., Arcasoy, S. M. *et al.* (1997). Expression of integrin cell adhesion receptors during human airway epithelial repair in vivo. *Am J Physiol* **273**, L256–L263.

Pini, L., Hamid, Q., Shannon, J. *et al.* (2007). Differences in proteoglycan deposition in the airways of moderate and severe asthmatics. *Eur Respir J* **29**, 71–77.

Potter-Perigo, S., Baker, C., Tsoi, C. *et al.* (2004). Regulation of proteoglycan synthesis by leukotriene D_4 and epidermal growth factor in bronchial smooth muscle cells. *Am J Respir Cell Mol Biol* **30**, 101–108.

Rajah, R., Nunn, S. E., Herrick, D. J. *et al.* (1996). Leukotriene D_4 induces MMP-1, which functions as an IGFBP protease in human airway smooth muscle cells. *Am J Physiol* **271**, L1014–L1022.

Roberts, C. R. (1995). Is asthma a fibrotic disease? *Chest* **107**, 111S–117S.

Roberts, C. R. and Burke, A. K. (1998). Remodelling of the extracellular matrix in asthma: proteoglycan synthesis and degradation. *Can Respir J* **5**, 48–50.

Roberts, C. R., Walker, D. C. and Schellenberg, R. R. (2002). Extracellular matrix. *Clin Allergy Immunol* **16**, 143–178.

Roche, W. R. (1991). Fibroblasts and asthma. *Clin Exp Allergy* **21**, 545–548.

Roche, W. R., Beasley, R., Williams, J. H. *et al.* (1989). Subepithelial fibrosis in bronchi of asthmatics. *Lancet* **1**(8637), 520–524.

Roth, M. and Black, J. L. (2006). Transcription factors in asthma: are transcription factors a new target for asthma therapy? *Curr Drug Targets* **7**, 589–595.

Roth, M., Johnson, P. R., Borger, P. *et al.* (2004). Dysfunctional interaction of C/EBPalpha and the glucocorticoid receptor in asthmatic bronchial smooth-muscle cells. *N Engl J Med* **351**, 560–574.

Routh, R. E., Johnson, J. H. and McCarthy, K. J. (2002). Troglitazone suppresses the secretion of type I collagen by mesangial cells in vitro. *Kidney Int* **61**, 1365–1376.

Spears, M., McSharry, C. and Thomson, N. C. (2206). Peroxisome proliferator-activated receptor-gamma agonists as potential anti-inflammatory agents in asthma and chronic obstructive pulmonary disease. *Clin Exp Allergy* **36**, 1494–1504.

Staunton, D. E., Lupher, M. L., Liddington, R. *et al.* (2006). Targeting integrin structure and function in disease. *Adv Immunol* **91**, 111–157.

Stewart, A. G. (2004). Emigration and immigration of mesenchymal cells: a multicultural airway wall. *Eur Respir J* **24**, 515–517.

Teschler, H., Pohl, W. R., Thompson, A. B. *et al.* (1993a). Elevated levels of bronchoalveolar lavage vitronectin in hypersensitivity pneumonitis. *Am Rev Respir Dis* **147**, 332–337.

Teschler, H., Thompson, A. B., Pohl, W. R. *et al.* (1993b). Bronchoalveolar lavage procollagen-III-peptide in recent onset hypersensitivity pneumonitis: correlation with extracellular matrix components. *Eur Respir J* **6**, 709–714.

Thickett, D. R., Poole, A. R. and Millar, A. B. (2001). The balance between collagen synthesis and degradation in diffuse lung disease. *Sarcoidosis Vasc Diffuse Lung Dis* **18**, 27–33.

Thomson, R. J., Bramley, A. M. and Schellenberg, R. R. (1996). Airway muscle stereology: implications for increased shortening in asthma. *Am J Respir Crit Care Med* **154**, 749–757.

van Straaten, J. F., Coers, W., Noordhoek, J. A. *et al.* (1999). Proteoglycan changes in the extracellular matrix of lung tissue from patients with pulmonary emphysema. *Mod Pathol* **12**, 697–705.

Virtanen, I., Laitinen, A., Tani, T. *et al.* (1996). Differential expression of laminins and their integrin receptors in developing and adult human lung. *Am J Respir Cell Mol Biol* **15**, 184–196.

Ward, J. E. and Tan, X. (2007). The role of peroxisome proliferator activated receptors in the regulation of airway inflammation and remodelling in chronic lung disease. *PPAR Res* **2007**, 14983.

Ward, J. E., Gould, H., Harris, T. *et al.* (2004). PPAR gamma ligands, 15-deoxy-delta12, 14-prostaglandin J$_2$ and rosiglitazone regulate human cultured airway smooth muscle proliferation through different mechanisms. *Br J Pharmacol* **141**, 517–525.

Ward. J. E., Wee, E. P., Bonacci, J. V. *et al.* (2006). Collagen matrix remodeling by human airway smooth muscle is regulated by metalloproteinase activity and endothelin. *Proc Am Thoracic Soc* **3**, A262.

Weinacker, A., Ferrando, R., Elliott, M. *et al.* (1995). Distribution of integrins alpha v beta 6 and alpha 9 beta 1 and their known ligands, fibronectin and tenascin, in human airways. *Am J Respir Cell Mol Biol* **12**, 547–556.

Wilson, J. W. and Li, X. (1997). The measurement of reticular basement membrane and submucosal collagen in the asthmatic airway. *Clin Exp Allergy* **27**, 363–371.

Woodruff, P. G., Dolganov, G. M., Ferrando, R. E. *et al.* (2003). Hyperplasia of smooth muscle in mild to moderate asthma without changes in cell size or gene expression. *Am J Respir Crit Care Med* **169**, 1001–1006.

7
Airway smooth muscle interaction with mast cells

Roger Marthan[1-3], Patrick Berger[1-3], Pierre-Olivier Girodet[1-3] and J. Manuel Tunon-de-Lara[1-3]

[1]*Université Bordeaux, Laboratoire de Physiologie Cellulaire Respiratoire, Bordeaux, France*

[2]*Inserm, U 885 Bordeaux, France*

[3]*CHU de Bordeaux, Pôle Cardio-Thoracique, Pessac, Hôpital du Haut-Lévêque, France*

7.1 Introduction

Mast cells have long been recognized as tissue-resident cells playing an important role in both allergic and non-allergic asthma (Bradding *et al.*, 2006). Several studies performed in humans have documented an increase in mast cell numbers in the bronchial mucosa of asthmatic patients. Mast cells can release preformed, stored mediators and synthesize other factors *de novo* upon activation by both IgE-dependent and non-immunological stimuli. Such activation results in the release of histamine, and serine proteases such as tryptase and chymase and heparin, and in the synthesis of lipid mediators, including leukotriene C4 (LTC4) and

Airway Smooth Muscle   Edited by Kian Fan Chung
© 2008 John Wiley & Sons, Ltd

prostaglandin D_2, which can all interact with airway smooth muscle cells and alter airway smooth muscle function.

Mast cell infiltration within the smooth muscle layer, on the other hand, is only a recent observation that is not fully understood. Ammit and colleagues were the first to point out, in 1997, that patients sensitized to aeroallergens had a significant infiltration of the airway smooth muscle by inflammatory cells, including mast cells (Ammit *et al.*, 1997). Investigating the mechanism involved in the passively sensitized human airway *in vitro* model (Black *et al.*, 1989), we also found that a high proportion of IgE-bearing mast cells were localized to the smooth muscle layer following passive sensitization (Berger *et al.*, 1998). More recently, two groups have convincingly reported that mast cells specifically infiltrate the smooth muscle layer in asthma. Brightling *et al.* demonstrated a specific smooth muscle micro-localization of mast cells in asthmatic bronchi when compared with that of healthy subjects or with bronchi from patients suffering from eosinophilic bronchitis, that is, a bronchial eosinophilic inflammation that was not associated with bronchial hyperresponsiveness (Brightling *et al.*, 2002). Caroll and coworkers reported a significant mast cell infiltration of the airway smooth muscle layer from both fatal and non-fatal asthma (Carroll *et al.*, 2002). Interestingly, these authors reported a high degree of mast cell degranulation within the airway smooth muscle of lung specimens obtained from fatal asthma cases. Moreover, mast cells infiltrating airway smooth muscle in asthma demonstrate ultrastructural feature of activation (Begueret *et al.*, 2007). Finally, additional potential roles have been identifiedfor the mast cell mediators, histamine and tryptase, in the regulationof airway inflammation in the vicinity of the ASM in asthma (Chhabra *et al.*, 2007).

The mechanism of such infiltration remains to be understood. On the one hand, it has been demonstrated that human airway smooth muscle cells (HASMC) produce, upon stimulation, a variety of cytokines and chemokines that may attract mast cell to the smooth muscle layer (Berger *et al.*, 2003, Brightling *et al.*, 2005a). On the other hand, it has been recently demonstrated that cell–cell contact between mast cells and smooth muscle cells enhances mast cell degranulation (Thangam *et al.*, 2005). Moreover, HASMC and mast cells express a variety of adhesion molecules and their counter receptors, such as VCAM-1 and ICAM-1 (Amrani *et al.*, 1999; Lazaar *et al.*, 2001), that may facilitate cell–cell contact, and, as a consequence, induce *in situ* liberation of mast cell products. Mast cells also express transmembrane glycoproteins, such as CD44 (Fukui *et al.*, 2000) and CD51 (Columbo and Bochner, 2001), that recognize extracellular matrix proteins and thus may play a role in maintaining mast cell infiltration.

In this chapter, we review data supporting the hypothesis that mast cell-derived products activate airway smooth muscle cells that, in turn, attract mast cells to airway smooth muscle cells, thus generating an auto-activation loop that may represent a new therapeutic target.

7.2 Mast cell mediators alter smooth muscle function

Mast cells release preformed, stored mediators and synthesize other factors *de novo* on activation. This results in release of histamine, and serine proteases such as tryptase, chymase and heparin, and in the synthesis of lipid mediators, including LTC_4 and prostaglandin D_2, that may interact with airway smooth muscle cells and alter airway smooth muscle function.

The cysteinyl-leukotrienes are potent bronchoconstrictors: LTC4 and LTD4 are 1000-fold more potent than histamine in contracting human airway smooth muscle (Barnes *et al.*, 1984). The effects of the cysteinyl-leukotrienes are mediated by activation of two G-protein-coupled receptor subtypes, CysLT1 and CysLT2 expressed by airway smooth muscle cells (Gorenne *et al.*, 1996). LTD4 markedly induces shortening of smooth muscle in a variety of tissues, including human airways. This response results from the activation of a receptor-mediated pathway that increases phosphoinositide turnover and cytosolic calcium (Hyvelin *et al.*, 2000). *In vitro*, LTD4 contracts human bronchial smooth muscle strips to a level comparable with histamine, but it is far less effective in mobilizing cytosolic Ca^{2+} (Accomazzo *et al.*, 2001). This discrepancy between force generation by the muscle and increase in Ca^{2+} suggests that LTD4-induced contraction of airway smooth muscle is partly calcium-independent (Savineau and Marthan, 1994) and may involve the activation of a protein kinase C isoform, PKC-ε (Accomazzo *et al.*, 2001).

The major product of the mast cell, the trypsin-like serine proteinase, tryptase, has been shown to induce airway hyperresponsiveness and to stimulate airway smooth muscle cell responses in terms of both calcium signalling and cell prolif-eration (Berger *et al.*, 1999; 2001b; 2001a; Johnson *et al.*, 1997). Exogenously administered tryptase induces bronchoconstriction and bronchial hyperrespon-siveness in dogs and sheep (Molinari *et al.*, 1996; Orlowski *et al.*, 1989; Sekizawa *et al.*, 1989), and, *in vitro*, tryptase can potentiate the contractile response of human bronchi to histamine (Berger *et al.*, 1999; Johnson *et al.*, 1997). Most of these effects are likely mediated by proteinase-activated receptor 2, (PAR-2), a member of a family of hepta-helical receptors activated upon the proteolytic activ-ity of enzymes and expressed by airway smooth muscle cells (Figure 7.1). Several serine proteases cleave the amino acids at a specific site of the extracellular NH_2-terminus of the molecule, leading to the exposure of a new NH_2-terminus that acts

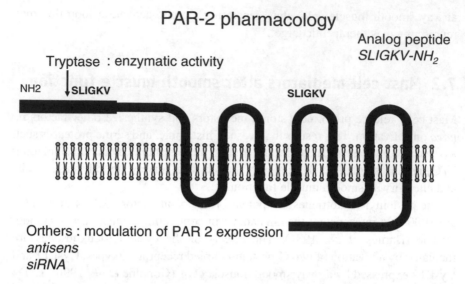

Figure 7.1 Proteinase-activated receptor (PAR)-2 is a member of the hepta-helical receptors family that is activated upon the proteolytic activity of enzymes. Tryptase cleaves the amino acids at a specific site of the extra cellular NH_2-terminus of the molecule, leading to the exposure of a new NH_2-terminus that acts as a tethered ligand. The peptide agonist SLIGKV corresponds to the tethered ligand in human PAR-2. Alternative tools to inhibit PAR2 function include RNAi strategy

as a tethered ligand that binds and activates the cleaved receptor molecule. The peptide agonist, SLIGKV, corresponding to the tethered ligand in human PAR-2, mimics the effects of tryptase on human airway smooth muscle cells, including a rise in intracellular calcium (Berger *et al.*, 2001b), proliferation (Berger *et al.*, 2001a), and cytokine synthesis (Berger *et al.*, 2003). Such studies indicate that human mast cell tryptase has a direct effect on isolated human airway smooth muscle cells. Tryptase activates a $[Ca^{2+}]_i$ response that appears to be mediated through PAR-2. In agreement with the above-described mechanism of activation of PAR, tryptase-induced $[Ca^{2+}]_i$ rise is delayed compared with conventional agonists such as acetylcholine (ACh) or histamine, which also act at G-protein-coupled receptors. Nevertheless, signal transduction mechanisms activated by tryptase and acetylcholine (ACh) or histamine, downstream of the receptor, are similar. PAR-2 activation mobilizes the intracellular Ca^{2+} store (presumably in the sarcoplasmic reticulum) via phosphoinositide phospholipase C (PLC) activation and thus via the inositol triphosphate (IP_3) pathway (Berger *et al.*, 2001b).

Since conventional PAR-2 antagonists are not available, alternative techniques, including RNAi strategy for gene silencing in human airway smooth muscle cells, have been developed to examine the function of PAR-2 (Trian *et al.*, 2006). As a consequence of its effect on PAR-2 expression, RNAi impairs PAR-2-mediated functional effects. Regarding the most selective PAR-2 agonist, the activating peptide SLIGKV, RNAi-mediated inhibition of the receptor decreased the mean $[Ca^{2+}]_i$ response by approximately 50 per cent, in agreement with the amplitude of the decrease in PAR-2 expression (Trian *et al.*, 2006).

In addition to its ability to alter the contractile response of airway smooth muscle, tryptase also acts as a potent mitogen *in vitro* and probably actively contributes to airway remodelling *in vivo*. It has been demonstrated that human tryptase can provide a potent stimulus for both DNA synthesis and cell proliferation in human airway smooth muscle cells. Similar responses can be evoked with trypsin and the peptide agonist of PAR-2, SLIGKV, again supporting the idea that tryptase may act through this receptor (Berger *et al.*, 2001a). Furthermore, signal transduction mechanisms involved in tryptase-induced proliferation of smooth muscle cells have been examined (Berger *et al.*, 2001a). However, the precise mechanism whereby tryptase may interact with these cells remains unclear (Brown *et al.*, 2006). Several reports have suggested the need for an intact catalytic site (Berger *et al.*, 2001a), although other studies indicate that non-proteolytic actions may be involved in mitogenesis (Brown *et al.*, 2002).

Mast cells also interact with airway smooth muscle cells through the release of cytokines, such as interleukin (IL)-4, IL-13 or tumour necrosis factor (TNF)-α, that may contribute to airway hyperresponsiveness (Holgate *et al.*, 2003). IL-4 and IL-13 are believed to play a key role in the development of bronchial hyperresponsiveness. This is supported by *in vivo* data, obtained in mice, showing that instillation of T-helper type 2 (TH2) cytokines to the airways of naive mice induces airway hyperresponsiveness within 6 h. This requires expression of the IL-4 receptor α subunit and signal transducer and activator of transcription 6 (STAT-6), suggesting a critical role for IL-4 and/or IL-13, and each of these cytokines produced similar effects; that is, STAT-6 activation when administered individually. Moreover, IL-4 and IL-13 also enhance the magnitude of agonist-induced intracellular calcium responses in cultured human airway smooth muscle cells. TNF-α, a proinflammatory cytokine strongly implicated in the pathogenesis of asthma, induces both airway hyperresponsiveness and sputum neutrophilia. It also enhances bronchial responsiveness in asthmatic patients when administered by inhalation to animals and humans, (Thomas and Heywood, 2002).

7.3 Smooth muscle cells induce mast cell chemotaxis

It has been demonstrated that the number of mast cells present in the smooth muscle layer of the bronchial wall is significantly higher in asthmatic patients than in normal subjects and in patients suffering from eosiniphilic bronchitis (Brightling *et al.*, 2002). This observation agreed with previous results obtained in human isolated bronchi sensitized to aeroallergens (Ammit *et al.*, 1997) or in patients who died from an asthma attack (Carroll *et al.*, 2002). This co-localization of both cell types prompted the hypothesis that the smooth muscle itself could exert a chemotactic activity for mast cells through the production of various factors. This hypothesis has recently been tested, and it has been demonstrated that HASMC induce mast cell chemotaxis (Berger *et al.*, 2003). *In vitro* experiments indicated that the mechanism of the chemotactic activity is related to the synthesis of transforming growth factor (TGF)-β1 by smooth muscle cells and, to a lesser extent, to that of stem cell factor (SCF) (Berger *et al.*, 2003). Furthermore, *ex vivo* data collected in bronchial biopsies from asthmatic patients indicated that the number of mast cells within the smooth muscle layer is related to TGF-β expression. These findings thus highlighted a novel role for smooth muscle in recruiting mast cells in asthma. However, inhibition of both TGF-β1 and SCF by blocking antibodies did not fully block mast cell migration, suggesting the involvement of additional mechanisms implicating alternative HASMC-derived chemokines. For instance, CCL5/RANTES accounted for such migration only when HASMC were stimulated by the proinflammatory cytokine, TNF-α (Berger *et al.*, 2003), although mast cells express the chemokines receptor CCR1 and CCR3, which can both bind CCL5/RANTES (Brightling *et al.*, 2005b). However, mast cells also express the chemokine receptors CXCR1, CXCR3 and CXCR4, which can bind CXCL8/IL-8, IP-10 and CXCL12/SDF-1α, respectively (Brightling *et al.*, 2005b). Among these chemokines, CXCL10/IP-10 seems to play an important role since, on the one hand, CXCR3 is the most abundantly expressed chemokine receptor found on human lung mast cells in the airway smooth muscle bundle in asthma, and, on the other hand, CXCL1/IP-10 is expressed preferentially by asthmatic airway smooth muscle (Brightling *et al.*, 2005a). Indeed, it is an important inducer of mast cell migration (Brightling *et al.*, 2005a; El-Shazly *et al.*, 2006).

Finally, another chemokine, CX$_3$CL1/fractalkine, has also been implicated in the interaction of mast with smooth muscle cells. This chemokine is produced by HASMC (Sukkar *et al.*, 2004) including those from asthmatic patients (El-Shazly *et al.*, 2006). Mast cells also express the fractalkine receptor CX$_3$CR1. However, fractalkine attracts mast cells in the presence of the peptide, vasointestinal peptide (VIP), a neurotransmitter that is upregulated in the asthmatic smooth muscle (El-Shazly *et al.*, 2006).

The reverse phenomenon, that is, the attraction of HASMC by mast cells, has very recently been investigated (Kaur *et al.*, 2006). On the one hand, HASMC, and also myofibroblasts and fibroblasts, express the chemokine receptor CCR7. On the other hand, one of the CCR7 ligands, the chemokine CCL19, is highly expressed in mast cells and blood vessels in bronchial biopsies of patients with asthma. Indeed, activation of smooth muscle CCR7 by CCL19 derived from mast cells mediates HASMC migration. It has therefore been suggested that CCL19/MIP-3β and its receptor CCR7 may contribute to the development of increased smooth muscle mass in asthma (Kaur *et al.*, 2006).

Collectively, these data provide a mechanistic basis for understanding the microlocalization of mast cells within the asthmatic bronchial smooth muscle. A variety of chemokines produced by stimulated HASMC from asthmatic patients act on their receptors on mast cells to induce chemotaxis, thus generating an auto-activation loop (Figure 7.2). Identification of all the factors involved in such a loop is far from complete and requires further studies.

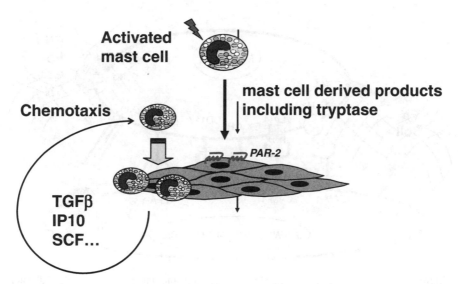

Figure 7.2 Auto-activation loop involving mast cell and smooth muscle cell. Upon activation, mast cell-derived products, including tryptase acting at PAR-2, trigger cytokine secretion by airway smooth muscle cells, which in turns induces the attraction of new mast cells

7.4 Mast cells can adhere to airway smooth muscle

Once they are located at the site of the smooth muscle layer, the question then arises as to how mast cells adhere to bronchial smooth muscle (Figure 7.3). Adhesion mechanism between both cell types could explain long-term effects leading to airway hyperresponsiveness and airway remodelling.

A direct cell–cell contact between mast cells and HASMC has been suggested as an enhancing factor for mast cell degranulation (Thangam *et al.*, 2005). In this study, incubation of human mast cells with HASMC, but not its culture super-natant, caused a significant enhancement of anaphylatoxin C3a-induced mast cell degranulation, via a SCF-c-kit-independent but dexamethasone-sensitive mechanism (Thangam *et al.*, 2005). More recently, Yang *et al.* have shown that human mast cells could adhere to HASMC through a calcium-dependent mechanism (Yang *et al.*, 2006). Blocking the adhesion molecule, tumour sup-pressor in lung cancer-1 (TSLC-1), reduced the mast cell adhesion by less than 40 per cent, suggesting the involvement of additional mechanisms. However, var-ious antibodies blocking ICAM-1, VCAM-1, or CD18 had no significant effects (Yang *et al.*, 2006).

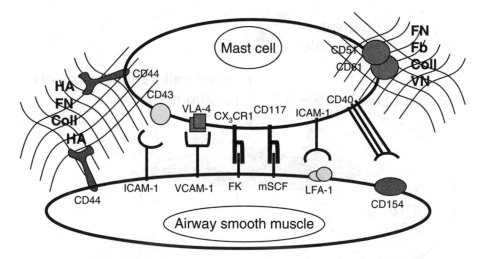

Figure 7.3　Adhesion mechanisms between mast cell and smooth muscle cell. Adhesion mechanisms may involve cell–cell interactions or interactions with the extracellular matrix. ICAM-1: intercellular adhesion molecule-1; VCAM-1: vascular-cell adhesion molecule-1; FK: fractalkine or CX3CL1; mSCF membrane stem cell factor; LFA-1: lymphocyte function-associated antigen-1; VN: vitronectin; Coll: collagen; Fb: fibrinogen; FN: fibronectin; HA: hyaluronic acid

As suggested by recent ultrastructural analysis of the asthmatic airway smooth muscle layer, an alternative mechanism of mast cell adhesion to HASMC could involve the extracellular matrix (Begueret *et al.*, 2007). In asthmatic subjects, the extracellular matrix was frequently noted to be increased between HASMC. Smooth muscle basal lamina was thicker in asthmatic patients (1.4 ± 0.2 μm) than in controls (0.7 ± 0.1 μm). In this study, mast cells infiltrating the smooth muscle layer were not in direct contact with HASMC but were in a very close relationship with the extracellular matrix of HASMC (Begueret *et al.*, 2007). Adhesion molecules involved in such interactions remain to be elucidated. However, recent findings from animal models using knockout mice suggest that CD44 may be implicated in mast cell adhesion to the bronchial smooth muscle (Girodet *et al.*, 2006).

7.5 Conclusion

The localization of mast cells within the airway smooth muscle layer is a recent finding that is likely to be important in the pathophysiology of asthma. On the one hand, mast cells release preformed stored mediators and synthesize other factors *de novo* upon activation by both IgE-dependent and non-immunological stimuli. Among these factors, enzymes such as tryptase, via an interaction with PAR-2 at the site of ASMC, modulate a variety of airway smooth muscle functions. On the other hand, the microlocalization of mast cells within the asthmatic bronchial smooth muscle results from complex interactions between both cells. A variety of chemokines produced by stimulated HASMC, including those from asthmatic patients, act on their respective receptors to induce chemotaxis of mast cells to the ASM bundle, thus generating an auto-activation loop. In this context, an amplification process may thus contribute to both bronchial hyperresponsiveness and airway remodelling. Moreover, once located at the site of the smooth muscle layer, mast cells adhere to bronchial smooth muscle, thus reenforcing long-term effects leading to hyperreactivity and remodelling. The clinical consequences of mast cell/smooth muscle cell interaction require further investigation. It is noteworthy that a study by Slats *et al.* has pointed out an association between the airway response to deep inspiration and mast infiltration in airway smooth muscle in asthma (Slats *et al.*, 2007). Lack of deep-inspiration-induced bronchodilation is often considered a critical phenomenon in asthma-induced alteration in pulmonary mechanics, although its mechanism remains controversial (Fairshter, 1985; Marthan and Woolcock, 1989; Skloot *et al.*, 1995). Therefore, the observation that the reduced bronchodilator effect of a deep inspiration in asthma is associated with an increased number of mast cells within the airway smooth muscle (Slats *et al.*, 2007) further supports the concept that such infiltration is clinically important.

Finally, mast cell myositis may represent an innovative therapeutic target in asthma. Although some of the molecular mechanisms underlying both chemotaxis and adhesion have already been described, a comprehensive identification and evaluation of these mechanisms is required for better definition of relevant therapeutic targets for the treatment of asthma.

Acknowledgements

Work performed in the authors' laboratory was supported by grants from ANR (Agence Nationale de la Recherche) and FRM (Fondation pour la Recherche Medicale).

References

Accomazzo, M. R., Rovati, G. E., Vigano, T. *et al.* (2001). Leukotriene D_4-induced activation of smooth-muscle cells from human bronchi is partly Ca^{2+}-independent. *Am J Respir Crit Care Med* **163**, 266–272.

Ammit, A. J., Bekir, S. S., Johnson, P. R. A. *et al.* (1997). Mast cell numbers are increased in the smooth muscle of human sensitized isolated bronchi. *Am J Respir Crit Care Med* **155**, 1123–1129.

Amrani, Y., Lazaar, A. L. and Panettieri, R. A., Jr. (1999). Up-regulation of ICAM-1 by cytokines in human tracheal smooth muscle cells involves an NF-kappa B-dependent signaling pathway that is only partially sensitive to dexamethasone. *J Immunol* **163**, 2128–2134.

Barnes, N. C., Piper, P. J. and Costello, J. F. (1984). Comparative effects of inhaled leukotriene C_4, leukotriene D_4, and histamine in normal human subjects. *Thorax* **39**, 500–504.

Begueret, H., Berger, P., Vernejoux, J. M. *et al.* (2007). Inflammation of bronchial smooth muscle in allergic asthma. *Thorax* **62**, 8–15.

Berger, P., Compton, S. J., Molimard, M. *et al.* (1999). Mast cell tryptase as a mediator of hyperresponsiveness in human isolated bronchi. *Clin Exp Allergy* **29**, 804–812.

Berger, P., Girodet, P. O., Begueret, H. *et al.* (2003). Tryptase-stimulated human airway smooth muscle cells induce cytokine synthesis and mast cell chemotaxis. *FASEB J* **17**, 2139–2141.

Berger, P., Perng, D. W., Thabrew, H. *et al.* (2001a). Tryptase and agonists of PAR-2 induce the proliferation of human airway smooth muscle cells. *J Appl Physiol* **91**, 1372–1379.

Berger, P., Tunon-De-Lara, J. M., Savineau, J. P. *et al.* (2001b). Tryptase-induced PAR-2-mediated Ca(2+) signaling in human airway smooth muscle cells. *J Appl Physiol* **91**, 995–1003.

Berger, P., Walls, A. F., Marthan, R. *et al.* (1998). Immunoglobulin E-induced passive sensitization of human airways: an immunohistochemical study. *Am J Respir Crit Care Med* **157**, 610–616.

Black, J. L., Marthan, R., Armour, C. L. *et al.* (1989). Sensitization alters contractile responses and calcium influx in human airway smooth muscle. *J Allergy Clin Immunol* **84**, 440–447.

Bradding, P., Walls, A. F. and Holgate, S. T. (2006). The role of the mast cell in the pathophysiology of asthma. *J Allergy Clin Immunol* **117**, 1277–1284.

Brightling, C. E., Ammit, A. J., Kaur, D. *et al.* (2005a). The CXCL10/CXCR3 axis mediates human lung mast cell migration to asthmatic airway smooth muscle. *Am J Respir Crit Care Med* **171**, 1103–1108.

Brightling, C. E., Bradding, P., Symon, F. A. *et al.* (2002). Mast-cell infiltration of airway smooth muscle in asthma. *N Engl J Med* **346**, 1699–1705.

Brightling, C. E., Kaur, D., Berger, P. *et al.* (2005b). Differential expression of CCR3 and CXCR3 by human lung and bone marrow-derived mast cells: implications for tissue mast cell migration. *J Leukoc Biol* **77**, 759–766.

Brown, J. K., Hollenberg, M. D. and Jones, C. A. (2006). Tryptase activates phosphatidylinositol 3-kinases proteolytically independently from proteinase-activated receptor-2 in cultured dog airway smooth muscle cells. *Am J Physiol Lung Cell Mol Physiol* **290**, L259–269.

Brown, J. K., Jones, C. A., Rooney, L. A. *et al.* (2002). Tryptase's potent mitogenic effects in human airway smooth muscle cells are via nonproteolytic actions. *Am J Physiol Lung Cell Mol Physiol* **282**, L197–206.

Carroll, N. G., Mutavdzic, S. and James, A. L. (2002). Distribution and degranulation of airway mast cells in normal and asthmatic subjects. *Eur Respir J* **19**, 879–885.

Chhabra, J., Li, Y. Z., Alkhouri, H. *et al.* (2007). Histamine and tryptase modulate asthmatic airway smooth muscle GM-CSF and RANTES release. *Eur Respir J* **29**, 861–870.

Columbo, M. and Bochner, B. S. (2001). Human skin mast cells adhere to vitronectin via the alphavbeta3 integrin receptor (CD51/CD61). *J Allergy Clin Immunol* **107**, 554.

El-Shazly, A., Berger, P., Girodet, P. O. *et al.* (2006). Fraktalkine produced by airway smooth muscle cells contributes to mast cell recruitment in asthma. *J Immunol* **176**, 1860–1868.

Fairshter, R. D. (1985). Airway hysteresis in normal subjects and individuals with chronic airflow obstruction. *J Appl Physiol* **58**, 1505–1510.

Fukui, M., Whittlesey, K., Metcalfe, D. D. *et al.* (2000). Human mast cells express the hyaluronic-acid-binding isoform of CD44 and adhere to hyaluronic acid. *Clin Immunol* **94**, 173–178.

Girodet, P. O., Berger, P., Begueret, H. *et al.* (2006). Disruption of CD44 variant 6 abolished bronchial hyperresponsiveness, inflammation and mast cell myositis in a murine model of asthma. *Eur Respir J* **28**, 217s (Abstract).

Gorenne, I., Norel, X. and Brink, C. (1996). Cysteinyl leukotriene receptors in the human lung: what's new? *Trends Pharmacol Sci* **17**, 342–345.

Holgate, S. T., Peters-Golden, M., Panettieri, R. A. *et al.* (2003). Roles of cysteinyl leukotrienes in airway inflammation, smooth muscle function, and remodeling. *J Allergy Clin Immunol* **111**, S18-34; discussion S34–36.

Hyvelin, J. M., Martin, C., Roux, E. *et al.* (2000). Human isolated bronchial smooth muscle contains functional ryanodine/caffeine-sensitive Ca-release channels. *Am J Respir Crit Care Med* **162**, 687–694.

Johnson, P. R., Ammit, A. J., Carlin, S. M. *et al.* (1997). Mast cell tryptase potentiates histamine-induced contraction in human sensitized bronchus. *Eur Respir J* **10**, 38–43.

Kaur, D., Saunders, R., Berger, P. *et al.* (2006). Airway smooth muscle and mast cell-derived CC chemokine ligand 19 mediate airway smooth muscle migration in asthma. *Am J Respir Crit Care Med* **174**, 1179–1188.

Lazaar, A. L., Krymskaya, V. P. and Das, S. K. (2001). VCAM-1 activates phosphatidylinositol 3-kinase and induces p120Cbl phosphorylation in human airway smooth muscle cells. *J Immunol* **166**, 155–161.

Marthan, R. and Woolcock, A. J. (1989). Is a myogenic response involved in deep inspiration-induced bronchoconstriction in asthmatics? *Am Rev Respir Dis* **140**, 1354–1358.

Molinari, J. F., Scuri, M., Moore, W. R. *et al.* (1996). Inhaled tryptase causes bronchoconstriction in sheep via histamine release. *Am J Respir Crit Care Med* **154**, 649–653.

Orlowski, M., Lesser, M., Ayala, J. *et al.* (1989). Substrate specificity and inhibitors of a capillary injury-related protease from sheep lung lymph. *Arch Biochem Biophys* **269**, 125–136.

Savineau, J. P. and Marthan, R. (1994). Activation properties of chemically skinned fibres from human isolated bronchial smooth muscle. *J Physiol* **474**, 433–438.

Sekizawa, K., Caughey, G. H., Lazarus, S. C. *et al.* (1989). Mast cell tryptase causes airway smooth muscle hyperresponsiveness in dogs. *J Clin Invest* **83**, 175–179.

Skloot, G., Permutt, S. and Togias, A. (1995). Airway hyperresponsiveness in asthma: a problem of limited smooth muscle relaxation with inspiration. *J Clin Invest* **96**, 2393–2403.

Slats, A. M., Janssen, K., Van Schadewijk, A. *et al.* (2007). Bronchial inflammation and airway responses to deep inspiration in asthma and COPD. *Am J Respir Crit Care Med* **176**, 121–128.

Sukkar, M. B., Issa, R., Xie, S. *et al.* (2004). Fractalkine/CX3CL1 production by human airway smooth muscle cells: induction by IFN-gamma and TNF-alpha and regulation by TGF-beta and corticosteroids. *Am J Physiol Lung Cell Mol Physiol* **287**, L1230–1240.

Thangam, E. B., Venkatesha, R. T., Zaidi, A. K. *et al.* (2005). Airway smooth muscle cells enhance C3a-induced mast cell degranulation following cell-cell contact. *Faseb J* **19**, 798–800.

Thomas, P. S. and Heywood, G. (2002). Effects of inhaled tumour necrosis factor alpha in subjects with mild asthma. *Thorax* **57**, 774–778.

Trian, T., Girodet, P. O., Ousova, O. *et al.* (2006). RNA interference decreases PAR-2 expression and function in human airway smooth muscle cells. *Am J Respir Cell Mol Biol* **34**, 49–55.

Yang, W., Kaur, D., Okayama, Y. *et al.* (2006). Human lung mast cells adhere to human airway smooth muscle, in part, via tumor suppressor in lung cancer-1. *J Immunol* **176**, 1238–1243.

8

Airway smooth muscle synthesis of inflammatory mediators

Alison E. John, Deborah L. Clarke, Alan J. Knox and Karl Deacon

Division of Respiratory Medicine, Centre for Respiratory Research,
City Hospital, University of Nottingham, Nottingham, UK

8.1 Introduction

Our understanding of airway smooth muscle (ASM) functions in asthma has advanced considerably over the last 10–15 years. ASM was originally thought to be a passive partner in airway inflammation, contracting in response to proinflammatory mediators and neurotransmitters, and relaxing in response to endogenous and exogenous bronchodilators. This paradigm was simplistic and it is clear that ASM has several other important properties of relevance to obstructive lung diseases. These functions include the ability to proliferate, undergo hypertrophy and migrate, and thereby contribute to the dysfunctional repair mechanisms that cause airway remodelling and poorly reversible airflow obstruction. Of particular note are its synthetic functions whereby ASM cells synthesize and release a diverse repertoire of biologically active inflammatory mediators. ASM is now

Airway Smooth Muscle Edited by Kian Fan Chung
© 2008 John Wiley & Sons, Ltd

considered to play a central role in orchestrating the inflammatory response within the bronchial wall. Here we discuss some of the key molecules these cells produce and the underlying molecular mechanisms.

8.2 Lipid mediators

For many years, lipids were considered to be solely of dietary or structural importance. This view changed dramatically in the 1970s and 1980s with the discovery of the prostaglandins (PGs), the first lipid-derived intercellular mediators. It is now known that other important mediators are lipid-derived, including leukotrienes and lipoxins, formed via the mobilization of arachidonic acid from cellular phospholipids, via phospholipase (PL) A_2 (Figure 8.1). Arachidonic acid can also undergo non-enzymatic peroxidation, either in its free form or while esterified to membrane phospholipids, by free radicals and reactive oxygen species to produce 8-isoprostanes (Figure 8.1).

Prostanoids

Prostanoids exhibit a wide range of biological activities, inhibition of which underlies the action of one of the world's all-time favourite medicines, aspirin, as well as its more modern counterparts. There are three isoforms of cyclooxygenase (COX). COX-1 is constitutively expressed and a 'housekeeping gene', whereas COX-2 is inducible and responsible for prostanoid formation at sites of inflammation. COX-3 is found predominantly in the brain and is the site of action of paracetamol. COX-1, but not COX-2, is constitutively expressed in ASM, and inflammatory stimuli such as IL-1β and bradykinin induce COX-2 mRNA and protein expression (Belvisi *et al.*, 1997; Pang and Knox, 1997; Petkova *et al.*, 1999).

Prostanoids have varied effects on ASM tone. Prostaglandin (PG)D_2, $PGF_{2\alpha}$ and thromboxane A_2 contract smooth muscle *in vitro* via TP receptors(Gardiner and Collier, 1980), (Armour *et al.*, 1989), and are also potent bronchoconstictors *in vivo* (Hardy *et al.*, 1984). In contrast, PGE_2 relaxes ASM *in vitro* at low concentrations, but contracts at higher concentrations through TP receptors (Armour *et al.*, 1989). PGE_2 potently inhibits induced bronchoconstriction (Melillo *et al.*, 1994; Pavord and Tattersfield, 1995), and is mainly protective in the lung. PGI_2 relaxes isolated precontracted human bronchus via IP receptors (Gardiner and Collier, 1980).

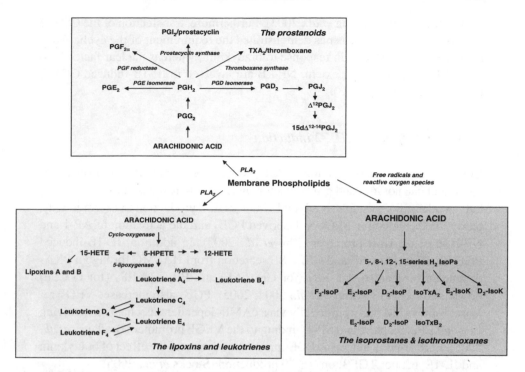

Figure 8.1 Lipid mediators. Arachidonic acid mobilized from membrane phospholipids is enzymatically converted to prostanoids via cyclo-oxygenase, or to leukotrienes and lipoxins via lipoxygenases. Arachidonic acid can undergo non-enzymatic peroxidation, either in its free form or while it is still esterified to membrane phospholipids, by free radicals and reactive oxygen species to produce isoprostanes

A number of pharmacological and molecular studies have looked at the regulation and functional consequence of COX-2 induction in HASM cells.

Molecular mechanisms of COX-2 induction

ERK and p38 MAPK have been implicated in IL-1β-induced PGE$_2$ synthesis (Laporte *et al.*, 1999; 2000) and PKCε mediates bradykinin-induced COX-2 induction in HASM (Pang *et al.*, 2003). The COX-2 promoter has binding sites for several transcription factors including nuclear factor-κB (NF-κB), C/EBP, AP-2, and cAMP response element (CRE). Our group reported that COX-2 expression is increased by NSAIDs in a prostanoid-independent manner involving PPARγ activation and activation of the peroxisome proliferator response element (PPRE) on the COX-2 promoter (Pang *et al.*, 2003). Furthermore, transfection of the COX-2 promoter into HASM cells has demonstrated the requirement of the cyclic AMP response element (CRE) in response to bradykinin, whereas nuclear factor IL-6 and CRE, and, to a lesser extent, NF-κB are involved in IL-1β-induced COX-2 induction (Nie *et al.,* 2003).

Functional effects of COX-2 induction

COX-2 induction and production of PGE_2 is an important autocrine loop in the induction of chemokines and growth factors in HASM in response to some stimuli, but not others. For example, bradykinin induces IL-8 production transcriptionally, partly via the induction of COX-2-derived PGE_2 and the activation of AP-1 and NF-IL-6 at the IL-8 promoter (Zhu *et al.,* 2003). In addition, IL-1β-induced PGE_2 facilitates G-CSF and GM-CSF secretion from HASMC through a PKA-dependent mechanism via EP_2 (for GM-CSF) and EP_2 and EP_4 (for G-CSF) prostanoid receptors (Clarke *et al.,* 2004; 2005). PGE_2 also increases VEGF release from HASM transcriptionally via a cAMP-dependent mechanism involving EP_2 and EP_4 receptors and SP-1 binding to the VEGF promoter (Bradbury *et al.,* 2005). Furthermore, autocrine PGE_2 production mediates the effect of bradykinin and IL-1β, but not TGFβ, on VEGF production (Stocks *et al.,* 2005).

Additionally, induction of COX-2 downregulates adenylyl cyclase (Laporte *et al.,* 1999; Pang and Knox, 1997). This is potentially very important, as it suggests that upregulation of COX-2 as a result of airway inflammation in asthma impairs the action of $β_2$-adrenoceptor agonists, the main class of bronchodilator drug used to treat asthma. By preventing COX-2 induction, glucocorticoids may restore $β_2$-adrenergic responsiveness.

Leukotrienes

Leukotrienes are formed by the enzymatic action of lipoxygenases (LO) on arachidonic acid, and are associated with allergic and inflammatory actions. COX and LO enzymes compete for AA, and inhibition of one of them may shunt the production of lipids to the other. This 'shunting' may partly explain aspirin-sensitive asthma (ASA), whereby aspirin induces asthma attacks in a small percentage of asthmatic patients (Ferreri *et al.,* 1988).

Leukotrienes induce potent, prolonged bronchoconstriction. Leukotriene antagonists, the first new class of anti-asthmatics introduced in over 30 years (Barnes,

2003), competitively block cysteinyl-leukotriene-1 receptors. In clinical practice, they have been relatively disappointing and remain as a weak, second-line therapy.

Our unpublished observations suggest that although HASM cells produce leukotrienes, the levels produced are not as functionally relevant or dynamically regulated as prostanoids.

Isoprostanes

The effects of isoprostanes generated from lipids has been less characterized. 8-iso-$PGF_{2\alpha}$ and 8-iso-PGE_2 contract smooth muscle via prostanoid TP receptors (Janssen *et al.,* 2000; Kawikova *et al.,* 1996). In carbachol-precontracted tissues (pretreated with a TP-receptor antagonist to block any potential contraction), 8-iso-$PGF_{3\alpha}$ completely relaxes HASM, although the receptors through which these actions are mediated were not investigated (Janssen *et al.,* 2000). E-ring 8-isoprostanes regulate the secretion of GM-CSF and G-CSF from HASM cells by a cAMP- and PKA-dependent mechanism. Moreover, antagonist studies revealed that 8-iso-PGE_1 and 8-iso-PGE_2 act solely via EP_2-receptors to inhibit GM-CSF release, whereas both EP_2- and EP_4-receptor subtypes positively regulate G-CSF output (Clarke *et al.,* 2005).

HASM cells release 8-isoprostanes in response to IL-17, and this can be suppressed by macrolides (erythromycin and azithromycin), but not by steroids or immunosuppressive agents such as cyclosporin and rapamycin (Vanaudenaerde *et al.,* 2007).

8.3 Chemokines

Chemokines are small chemoattractant cytokines (8–14 kDa) with diverse effects on cellular recruitment, activation and differentiation via interactions with seven-transmembrane spanning G-protein-coupled chemokine receptors (Murphy *et al.,* 2000). Chemokines are classified into four subfamilies, CXC, CC, C and CX3C on the basis of their sequence homology and the position of conserved cysteine residues within the protein (Zlotnik and Yoshie, 2000). Chemokines play a critical role in the pathogenesis of a number of acute and chronic inflammatory lung diseases (Gerard and Rollins, 2001; John and Lukacs, 2003; Smit and Lukacs, 2006).

Chemokines can recruit and activate a range of leukocyte subtypes including eosinophils, TH_2 lymphocytes, neutrophils and mast cells to the bronchi (Lukacs *et al.,* 2005). Chemokines can also activate or modulate the phenotype of structural cells within the airways (Oyamada *et al.,* 1999; Stellato *et al.,* 2001). Chemokines found in the airways of patients with inflammatory lung disease include CCL5/RANTES, CCL11/eotaxin, IP-10, and CXCL8/IL-8 (John and Lukacs, 2003; Pease and Sabroe, 2002; Smit and Lukacs, 2006).

ASM cells release a wide variety of chemokines following stimulation with inflammatory mediators, including the CC chemokine family members, CCL11 (Chung *et al.,* 1999; Ghaffar *et al.,* 1999), CCL5 (Ammit *et al.,* 2000; John *et al.,* 1997; Pype *et al.,* 1999), CCL2, CCL7, CCL8 (Pype *et al.,* 1999; Watson *et al.,* 1998), CCL19 (Kaur *et al.,* 2006) and CCL17 (Faffe *et al.,* 2003); the CXC chemokines CXCL8 (John *et al.,* 1998; Pang and Knox, 1998; Watson *et al.,* 1998), CXCL10 (Brightling *et al.,* 2005; Hardaker *et al.,* 2004), CXCL1, CXCL2 and CXCL3 (Jarai *et al.,* 2004); and, finally, CX3CL1 (El-Shazly *et al.,* 2006; Sukkar *et al.,* 2004).

The ability of ASM to synthesize and release chemokines during an inflammatory response has profound implications for the regulation of airway inflammation, and these cells can therefore be considered a target for anti-inflammatory therapy with agents such as glucocorticoids and β_2-agonists. The molecular mechanisms regulating chemokine expression appear to be chemokine- and stimulus-specific, and most of the data are available for the CC chemokines CCL5/RANTES and CCL11/eotaxin and the CXC chemokines CXCL/IL-8 and CXCL10/IP-10, although some data exist for CCL2/MCP-1 and CCL17TARC.

CCL5/RANTES

RANTES mRNA and protein are induced in ASM by TNFα, IL-1β or platelet-activating factor (PAF), and can be augmented by IFNγ (Ammit *et al.,* 2000; John *et al.,* 1997; Pype *et al.,* 1999). The RANTES promoter contains several key response elements, including a CD28-responsive element (CD28RE), two AP-1-binding sites, binding sites for signal transducer and activator of transcription (STAT) protein, nuclear factor of activated T cells (NF-AT), and C/EBP, and two NF-κB -binding elements (Nelson *et al.,* 1993). Using a series of site-directed mutations within the human RANTES promoter, Ammit and colleagues determined that AP-1 and NF-AT, but not NF-κB, regulate the RANTES promoter (Ammit *et al.,* 2002). Furthermore TNFα-induced AP-1 DNA binding was attenuated by dexamethasone. Elevated cAMP levels also inhibit cytokine-induced RANTES secretion from ASM cells (Ammit *et al.,* 2000; Hallsworth *et al.,* 2001b), but not by inhibiting AP-1 DNA binding (Ammit *et al.,* 2002). IL-1β-stimulated RANTES release is dependent on JNK (Oltmanns *et al.,* 2003) and p42/p44 ERK activation, but not p38 MAPK (Hallsworth *et al.,* 2001a).

CCL11/eotaxin

Eotaxin is released constitutively and increased by TNFα, IL-1β, IL-4 and IL-13 (Chung *et al.,* 1999; Ghaffar *et al.,* 1999; Pang and Knox, 2001) The human eotaxin promoter contains C/EBP, AP-1, STAT6, and NF-κB binding sites (Matsukura *et al.,* 1999; Ponath *et al.,* 1996). IL-1β-stimulated eotaxin release involves p38 MAP kinase, JNK kinase and p42/p44 ERK (Hallsworth *et al.,* 2001b; Wuyts *et al.,* 2003a), and ERK activation is involved in IL-13-, IL-4- or TNFα-stimulated eotaxin release (Moore *et al.,* 2002). STAT6 is involved in IL-4- and IL-13-mediated eotaxin production, whereas TNFα-induced transcription is dependent on NF-κB, but not STAT-6 (Moore *et al.,* 2002; Nie *et al.,* 2005b).

Inflammatory cytokines also regulate chemokine gene expression by chromatin remodelling. In quiescent cells, basal transcription factors and RNA polymerase II are unable to bind to recognition sequences in gene promoters due to the tight packaging of DNA in chromatin. However, the four core histones, an H3-H4 tetramer, and two H2A-H2B dimers that form each chromatin subunit are susceptible to post-transcriptional covalent modifications, including acetylation, phosphorylation and methylation. These processes are tightly controlled by a series of enzymes and include the histone acetyltransferases (HATs) and histone deacetylases (HDACs) (De Ruijter 2003). Histone modifications regulate the unravelling of DNA; as a result, transcription factor-binding sites are exposed and gene transcription is initiated (Wolfe and Hayes 1999; Pazin and Kadonaga 1997). Our recent studies have shown that histone H4 acetylation following TNFα stimulation is a key event in regulating binding of NF-κB p65 to the eotaxin promoter.

β$_2$-Agonists and glucocorticoids partially inhibit TNFα-induced eotaxin production from HASM, and, as with IL-8, their combined use results in greater inhibition of chemokine gene transcription (Pang and Knox, 2001). These compounds inhibit histone H4 acetylation, and the conformational change in chromatin inhibits NF-κB p65 binding to the eotaxin promoter (Nie *et al.,* 2005b) (Figure 8.2).

CXCL8/IL-8

ASM release IL-8 in large quantities after stimulation by TNFα, IL-1β (Pang and Knox, 2000; Watson *et al.,* 1998), TGFβ (Fong *et al.,* 2000) or bradykinin (Zhu *et al.,* 2003). The IL-8 promoter contains several transcription factor-binding motifs, and in many other cell types NF-κB is the key regulator of IL-8 transcription (Kunsch *et al.,* 1994; Mukaida *et al.,* 1994). Mutation analysis of the IL-8

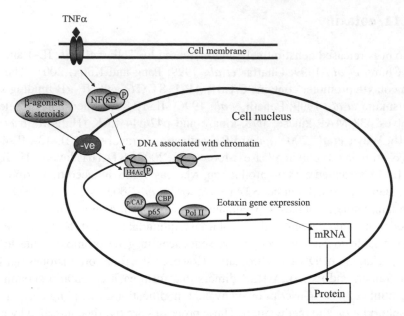

Figure 8.2 The molecular mechanism of action of β₂-agonists and steroids on eotaxin gene transcription. TNFα-induced eotaxin gene transcription is mediated mainly by the transcription factor NF-κB (p65/p50), as analysed by luciferase reporter gene assay, Western blotting, EMSA, and electrophoretic mobility supershift assay. Chromatin immunoprecipitation assays on these cells have demonstrated that TNFα induces selective histone H4 acetylation on lysines 5 and 12 at the eotaxin promoter site and p65 binding to the eotaxin promoter, resulting in eotaxin gene transcription. The inhibition of eotaxin production by β₂-agonists and glucocorticoids is due not to altered NF-κB nuclear translocation or *in vitro* promoter binding capability, but to their inhibition of TNFα-induced histone H4 acetylation and p65 by *in vivo* binding to the promoter, as depicted by the diagram

promoter determined that NF-κB is the major transcription factor in bradykinin-stimulated IL-8 gene transcription, but that AP-1 and NF-IL-6 also contribute (Zhu *et al.*, 2003). Glucocorticoids suppress TNFα-stimulated IL-8 production by HASM, and β₂ agonists augment their inhibition (Pang and Knox, 2000). Dexamethasone inhibits bradykinin-induced IL-8 production via the inhibition of NF-κB, AP-1, and NF-IL-6 binding.

IFNγ abrogates TNFα-induced NF-κB-dependent gene expression by impairing NF-κB transactivation (Keslacy *et al.*, 2007) via an increase in HDAC activity. Cytokine-mediated IL-8 release has also been shown to require activation of ERK, p38 MAPK (Hedges *et al.*, 2000) and JNK (Oltmanns *et al.*, 2003).

CXCL10/IP-10

Interferon (IFN)-γ-inducible protein-10 (IP-10) is preferentially expressed in ASM in bronchial biopsies from asthmatic and COPD patients and in *ex vivo* ASM cultures from asthmatic patients (Brightling *et al.*, 2005; Hardaker *et al.*, 2004). TNFα and IFNγ synergistically enhanced IP-10 mRNA and protein accumulation (Hardaker *et al.*, 2004). Furthermore, the NF-κB inhibitor, salicylanilide, blocked TNFα-, but not IFNγ-, tinduced IP-10 expression.

CCL2/MCP-1

IL-1β and TNFα increase MCP-1mRNA and protein (Nie *et al.*, 2005a; Pype *et al.*, 1999; Watson *et al.*, 1998). IL-1β-induced MCP-1 involves p38 MAPK, JNK kinase, ERK and NF-κB (Wuyts *et al.*, 2003a). Increasing cAMP reduced MCP-1, but not p38 MAPK, ERK or JNK inhibition (Wuyts *et al.*, 2003b).

CCL17/TARC

The TH$_2$ cell chemoattractant, TARC, is released by HASM by combined stimulation with either IL-4 or IL-13 and TNFα (Faffe *et al.*, 2003). IL-4, IL-13, IL-1β, TNFα and IFNγ alone do not cause TARC release. Little is known about the molecular mechanisms regulating TARC gene expression in ASM, although in bronchial epithelial cells NF-κB is required for TNFα-induced TARC release (Sekiya *et al.*, 2000). IL-4- or IL-13- and TNFα-induced TARC release is inhibited by isoproterenol, cAMP analogues or forskolin (Faffe *et al.*, 2003), but not by dexamethasone.

8.4 Growth and remodelling factors

Remodelled airways in asthma show increases in ASM mass, extracellular matrix deposition and vascularization, and ASM cells themselves can produce several factors which contribute to these changes.

ASM can synthesize several mitogens. Protein array analysis of ASM supernatants demonstrated that human ASM can produce EGF, IGF and SCF, but not PDGF (Joubert and Hamid, 2005), and although vascular smooth muscle can synthesize PDGF there is no concomitant evidence in ASM (Ikeda *et al.*, 1991). Both bFGF (Rodel *et al.*, 2000) and NGF (Pons *et al.*, 2001) are synthesized by ASM, the former during *Chlamydia pneumoniae* infection of cultured human ASM and the latter in response to IL-1β.

TGFβ

TGFβ has multiple roles in the remodelling process. TGFβ_1-can act as a mitogen for mesenchymal cells such as fibroblasts and ASM (Battegay *et al.*, 1990; Okona-Mensah *et al.*, 1998). Thickening of the airway wall may also be the result of mesenchymal cell differentiation and migration in response to increased levels of TGFβ_1 in the lung. Genes expressed in differentiated smooth muscle cells, such as smoothelin, calponin and smooth muscle myosin heavy chain, are upregulated by TGFβ_1 treatment of human fetal lung fibroblasts (Chambers *et al.*, 2003). TGFβ_1 is produced by ASM under inflammatory conditions. Neutrophil elastase (Lee *et al.*, 2006) and mast cell tryptase (Berger *et al.*, 2003) can both induce the release of TGFβ_1. Neutrophil elastase induces TGFβ_1 mRNA synthesis and consequent protein production through TLR4, MyD88, IRAK signal transduction, leading to NF-κB activation, but, although this was inferred from the action of 'NF-κB, inhibitors', the actual mechanism at the TGFβ_1 promoter was not investigated.

Extracellular matrix components

ASM expresses fibronectin, laminin, perlecan and chondroitin in response to serum, and this ECM synthetic potential is greater in response to serum derived from asthmatic patients than to 'healthy patient' serum. Treatment of ASM cells with TGFβ_1 causes synthesis of collagen, fibronectin and elastin.

Connective tissue growth factor (CTGF)

TGFβ can release CTGF (Burgess *et al.*, 2006), which then induces fibronectin and collagen synthesis in lung fibroblasts (Johnson *et al.*, 2006). Asthmatic lung-derived ASM secretes more CTGF in response to TGFβ. The MEK inhibitor PD98059 and the JNK1 inhibitor SP600125 (Xie *et al.*, 2005) reduce TGFβ_1-induced CTGF expression and SMAD2/3 phosphorylation. In contrast, PI3-K or p38-MAPK inhibitors are without effect. CTGF can induce the synthesis of fibronectin and collagen I by ASM itself (Johnson *et al.*, 2006), although there was no increase in the response in ASM cells derived from asthmatic patients.

Matrix metalloproteinases (MMPs)

ASM cells produce MMP-2 (Elshaw *et al.*, 2004; Johnson and Knox, 1999), MMP-9 (Liang *et al.*, 2007) and MMP-12 (Chetta *et al.*, 2005). ASM migration may also be enhanced by the interaction of matrix-metalloproteases with the ECM in response to growth factors. Collagen I and thrombin synergistically increase MMP-2 activity in ASM cell culture, and ASM migration towards PDGF is MMP-dependent (Henderson *et al.*, 2007). MMP-9 and MMP-12 expression are both induced by IL-1β..(Liang *et al.*, 2007; Xie *et al.*, 2005). MMP-9 expression is regulated by ERK1/2, JNK, and p38 MAPK, and the blocking of IL-1β-induced movement of NF-κB to the ASM nucleus prevents MMP-9 promoter-reporter construct activity (Liang *et al.*, 2007).

Vascular endothelial growth factor (VEGF) and other angiogenic factors

A growing body of evidence correlates increased vasculature and enhanced angiogenic factors with the symptoms of asthma. Asthmatic lung contains increased vascular branching and elevated VEGF levels (Abdel-Rahman *et al.*, 2006; Asai *et al.*, 2003; Chetta *et al.*, 2005; Feltis *et al.*, 2006; Hoshino *et al.*, 2001a; 2001b; Lee *et al.*, 2006). ASM produces VEGF in addition to several other angiogenic factors. TH2 cytokines (IL-4, IL-5 and IL-13) induce VEGF expression (Faffe *et al.*, 2006; Wen *et al.*, 2003); IL-1β, bradykinin and TGFβ also induce VEGF expression. In the case of IL1-β, bradykinin involves an autocrine loop with COX-2 and PGE$_2$ (Bradbury *et al.*, 2005; Burgess *et al.*, 2006; Stocks *et al.*, 2005). We have shown that the SP-1 transcription factor binding site in the VEGF promoter is essential for PGE$_2$-induced VEGF release in ASM. Unpublished observations from our laboratory suggest that angiopoeitin 1 and 2 are expressed constitutively in ASM but are less dynamically regulated than VEGF.

Several chemokines and cytokines produced by ASM also have angiogenic properties. For example, ASM produces GM-CSF (Saunders *et al.*, 1997), IL-8 and eotaxin, all of which have angiogenic potential (Bussolino *et al.*, 1989; Hu *et al.*, 1993; Koch *et al.*, 1992; Szekanecz *et al.*, 1994; Vanaudenaerde *et al.*, 2007). Furthermore matrix metalloproteinases have an essential role in remodelling the ECM to allow endothelial cell invasion and association to form nascent blood vessels (Asai *et al.*, 2003).

8.5 Conclusions

It is clear that ASM has a myriad of synthetic functions, the spectrum of which is constantly being updated as new publications emerge. As the bulk of ASM is markedly increased in chronic asthma, it can contribute large amounts of different mediators, chemokines and growth factors to the inflammatory mileu and thereby contribute to the perpetuation of chronic inflammation and remodelling in chronic asthma. We are starting to understand the molecular mechanisms involved in these processes, particularly in terms of the key transcription factors that drive these changes. The mainstay of drug treatment for asthma consists of glucocorticoids and long- and short-acting β_2-adrenoceptor agonists. These agents can inhibit the production of a number of molecules produced by ASM cells, but not others. A greater understanding of these processes should lead to the identification of new targets for the treatment of chronic asthma.

References

Abdel-Rahman, A. M., el-Sahrigy, S. A. and Bakr, S. I. (2006). A comparative study of two angiogenic factors: vascular endothelial growth factor and angiogenin in induced sputum from asthmatic children in acute attack. *Chest* **129**, 266–271.

Ammit, A. J., Hoffman, R. K., Amrani, Y. *et al.* (2000). Tumor necrosis factor-alpha-induced secretion of RANTES and interleukin-6 from human airway smooth-muscle cells. Modulation by cyclic adenosine monophosphate. *Am J Respir Cell Mol Biol* **23**, 794–802.

Ammit, A. J., Lazaar, A. L., Irani, C. *et al.* (2002). Tumor necrosis factor-alpha-induced secretion of RANTES and interleukin-6 from human airway smooth muscle cells: modulation by glucocorticoids and beta-agonists. *Am J Respir Cell Mol Biol* **26**, 465–474.

Armour, C. L., Johnson, P. R., Alfredson, M. L. *et al.* (1989). Characterization of contractile prostanoid receptors on human airway smooth muscle. *Eur J Pharmacol* **165**(2-3), 215–222.

Asai, K., Kanazawa, H., Kamoi, H. *et al.* (2003). Increased levels of vascular endothelial growth factor in induced sputum in asthmatic patients. *Clin Exp Allergy* **33**, 595–599.

Barnes, P. J. (2003). Anti-leukotrienes: here to stay? *Curr Opin Pharmacol* **3**, 257–263.

Battegay, E. J., Raines, E. W., Seifert, R. A. *et al.* (1990). TGF-beta induces bimodal proliferation of connective tissue cells via complex control of an autocrine PDGF loop. *Cell* **63**, 515–524.

Belvisi, M. G., Saunders, M. A., El Haddad, B. *et al.* (1997). Induction of cyclo-oxygenase-2 by cytokines in human cultured airway smooth muscle cells: novel inflammatory role of this cell type. *Br J Pharmacol* **120**, 910–916.

Berger, P., Girodet, P. O., Begueret, H. *et al.* (2003). Tryptase-stimulated human airway smooth muscle cells induce cytokine synthesis and mast cell chemotaxis. *FASEB J* **17**, 2139–2141.

Bradbury, D., Clarke, D., Seedhouse, C. *et al.* (2005). Vascular endothelial growth factor induction by prostaglandin E_2 in human airway smooth muscle cells is mediated by E prostanoid EP2/EP4 receptors and SP-1 transcription factor binding sites. *J Biol Chem* **280**, 29993–30000.

Brightling, C. E., Ammit, A. J., Kaur, D. *et al.* (2005). The CXCL10/CXCR3 axis mediates human lung mast cell migration to asthmatic airway smooth muscle. *Am J Respir Crit Care Med* **171**, 1103–1108.

Burgess, J. K., Ge, Q., Poniris, M. H. *et al.* (2006). Connective tissue growth factor and vascular endothelial growth factor from airway smooth muscle interact with the extracellular matrix. *Am J Physiol* **290**, L153–161.

Bussolino, F., Wang, J. M., Defilippi, P. *et al.* (1989). Granulocyte- and granulocyte-macrophage-colony stimulating factors induce human endothelial cells to migrate and proliferate. *Nature* **337**(6206), 471–473.

Chambers, R. C., Leoni, P., Kaminski, N. *et al.* (2003). Global expression profiling of fibroblast responses to transforming growth factor-beta1 reveals the induction of inhibitor of differentiation-1 and provides evidence of smooth muscle cell phenotypic switching. *Am J Pathol* **162**, 533–546.

Chetta, A., Zanini, A., Foresi, A. *et al.* (2005). Vascular endothelial growth factor up-regulation and bronchial wall remodelling in asthma. *Clin Exp Allergy* **35**, 1437–1442.

Chung, K. F., Patel, H. J., Fadlon, E. J. *et al.* (1999). Induction of eotaxin expression and release from human airway smooth muscle cells by IL-1beta and TNFalpha: effects of IL-10 and corticosteroids. *Br J Pharmacol* **127**, 1145–1150.

Clarke, D. L., Belvisi, M. G., Catley, M. C. *et al.* (2004). Identification in human airways smooth muscle cells of the prostanoid receptor and signalling pathway through which PGE_2 inhibits the release of GM-CSF. *Br J Pharmacol* **141**, 1141–1150.

Clarke, D. L., Belvisi, M. G., Hardaker, E. *et al.* (2005). E-ring 8-isoprostanes are agonists at EP2- and EP4-prostanoid receptors on human airway smooth muscle cells and regulate the release of colony-stimulating factors by activating cAMP-dependent protein kinase. *Mol Pharmacol* **67**, 383–393.

El-Shazly, A., Berger, P., Girodet, P. O. *et al.* (2006). Fraktalkine produced by airway smooth muscle cells contributes to mast cell recruitment in asthma. *J Immunol* **176**, 1860–1868.

Elshaw, S. R., Henderson, N., Knox, A. J. *et al.* (2004). Matrix metalloproteinase expression and activity in human airway smooth muscle cells. *Br J Pharmacol* **142**, 1318–1324.

Faffe, D. S., Flynt, L., Bourgeois, K. *et al.* (2006). Interleukin-13 and interleukin-4 induce vascular endothelial growth factor release from airway smooth muscle cells: role of vascular endothelial growth factor genotype. *Am J Respir Cell Mol Biol* **34**, 213–218.

Faffe, D. S., Whitehead, T., Moore, P. E. *et al.* (2003). IL-13 and IL-4 promote TARC release in human airway smooth muscle cells: role of IL-4 receptor genotype. *Am J Physiol* **285**, L907–914.

Feltis, B. N., Wignarajah, D., Zheng, L. *et al.* (2006). Increased vascular endothelial growth factor and receptors: relationship to angiogenesis in asthma. *Am J Respir Crit Care Med* **173**, 1201–1207.

Ferreri, N. R., Howland, W. C., Stevenson, D. D. *et al.* (1988). Release of leukotrienes, prostaglandins, and histamine into nasal secretions of aspirin-sensitive asthmatics during reaction to aspirin. *Am Rev Respir Dis* **137**, 847–854.

Fong, C. Y., Pang, L., Holland, E. *et al.* (2000). TGF-beta1 stimulates IL-8 release, COX-2 expression, and PGE(2) release in human airway smooth muscle cells. *Am J Physiol* **279**, L201–207.

Gardiner, P. J. and Collier, H. O. (1980). Specific receptors for prostaglandins in airways. *Prostaglandins* **19**, 819–841.

Gerard, C. and Rollins, B. J. (2001). Chemokines and disease. *Nat Immunol* **2**, 108–115.

Ghaffar, O., Hamid, Q., Renzi, P. M. *et al.* (1999). Constitutive and cytokine-stimulated expression of eotaxin by human airway smooth muscle cells. *Am J Respir Crit Care Med* **159**, 1933–1942.

Hallsworth, M. P., Moir, L. M., Lai, D. *et al.* (2001a). Inhibitors of mitogen-activated protein kinases differentially regulate eosinophil-activating cytokine release from human airway smooth muscle. *Am J Respir Crit Care Med* **164**, 688–697.

Hallsworth, M. P., Twort, C. H., Lee, T. H. *et al.* (2001b). beta(2)-Adrenoceptor agonists inhibit release of eosinophil-activating cytokines from human airway smooth muscle cells. *Br J Pharmacol* **132**, 729–741.

Hardaker, E. L., Bacon, A. M., Carlson, K. *et al.* (2004). Regulation of TNF-alpha- and IFN-gamma-induced CXCL10 expression: participation of the airway smooth muscle in the pulmonary inflammatory response in chronic obstructive pulmonary disease. *FASEB J* **18**, 191–193.

Hardy, C. C., Robinson, C., Tattersfield, A. E. *et al.* (1984). The bronchoconstrictor effect of inhaled prostaglandin D_2 in normal and asthmatic men. *N Engl J Med* **311**, 209–213.

Hedges, J. C., Singer, C. A., and Gerthoffer, W. T. (2000). Mitogen-activated protein kinases regulate cytokine gene expression in human airway myocytes. *Am J Respir Cell Mol Biol* **23**, 86–94.

Henderson, N., Markwick, L. J., Elshaw, S. R. *et al.* (2007). Collagen I and thrombin activate MMP-2 by MMP-14-dependent and -independent pathways: implications for airway smooth muscle migration. *Am J Physiol* **292**, L1030–1038.

Hoshino, M., Nakamura, Y. and Hamid, Q. A. (2001a). Gene expression of vascular endothelial growth factor and its receptors and angiogenesis in bronchial asthma. *J Allergy Clin Immunol* **107**, 1034–1038.

Hoshino, M., Takahashi, M., Takai, Y. *et al.* (2001b). Inhaled corticosteroids decrease vascularity of the bronchial mucosa in patients with asthma. *Clin Exp Allergy* **31**, 722–730.

Hu, D. E., Hori, Y. and Fan, T. P. (1993). Interleukin-8 stimulates angiogenesis in rats. *Inflammation* **17**, 135–143.

Ikeda, U., Ikeda, M., Oohara, T. *et al.* (1991). Interleukin 6 stimulates growth of vascular smooth muscle cells in a PDGF-dependent manner. *Am J Physiol* **260**(5 Pt 2), H1713–1717.

Janssen, L. J., Premji, M., Netherton, S. *et al.* (2000). Excitatory and inhibitory actions of isoprostanes in human and canine airway smooth muscle. *J Pharmacol Exp Ther* **295**, 506–511.

Jarai, G., Sukkar, M., Garrett, S. *et al.* (2004). Effects of interleukin-1beta, interleukin-13 and transforming growth factor-beta on gene expression in human airway smooth muscle using gene microarrays. *Eur J Pharmacol* **497**, 255–265.

John, A. E., and Lukacs, N. W. (2003). Chemokines and asthma. *Sarcoidosis Vasc Diffuse Lung Dis* **20**, 180–189.

John, M., Au, B. T., Jose, P. J. *et al.* (1998). Expression and release of interleukin-8 by human airway smooth muscle cells: inhibition by Th-2 cytokines and corticosteroids. *Am J Respir Cell Mol Biol* **18**, 84–90.

John, M., Hirst, S. J., Jose, P. J. *et al.* (1997). Human airway smooth muscle cells express and release RANTES in response to T helper 1 cytokines: regulation by T helper 2 cytokines and corticosteroids. *J Immunol* **158**, 1841–1847.

Johnson, P. R., Burgess, J. K., Ge, Q. *et al.* (2006). Connective tissue growth factor induces extracellular matrix in asthmatic airway smooth muscle. *Am J Respir Crit Care Med* **173**, 32–41.

Johnson, S. and Knox, A. (1999). Autocrine production of matrix metalloproteinase-2 is required for human airway muscle proliferation. *Am J Physiol* **277**(6 Pt 1), L1109–1117.

Joubert, P. and Hamid, Q. (2005). Role of airway smooth muscle in airway remodeling. *J Allergy Clin Immunol* **116**, 713–716.

Kaur, D., Saunders, R., Berger, P. *et al.* (2006). Airway smooth muscle and mast cell-derived CC chemokine ligand 19 mediate airway smooth muscle migration in asthma. *Am J Respir Crit Care Med* **174**, 1179–1188.

Kawikova, I., Barnes, P. J., Takahashi, T. *et al.* (1996). 8-Epi-PGF$_2$ alpha, a novel noncyclooxygenase-derived prostaglandin, constricts airways in vitro. *Am J Respir Crit Care Med* **153**, 590–596.

Keslacy, S., Tliba, O., Baidouri, H. *et al.* (2007). Inhibition of tumor necrosis factor-alpha-inducible inflammatory genes by interferon-gamma is associated with altered nuclear factor-kappaB transactivation and enhanced histone deacetylase activity. *Mol Pharmacol* **71**, 609–618.

Koch, A. E., Polverini, P. J., Kunkel, S. L. *et al.* (1992). Interleukin-8 as a macrophage-derived mediator of angiogenesis. *Science* **258**(5089), 1798–1801.

Kunsch, C., Lang, R. K., Rosen, C. A. *et al.* (1994). Synergistic transcriptional activation of the IL-8 gene by NF-kappa B p65 (RelA) and NF-IL-6. *J Immunol* **153**, 153–164.

Laporte, J. D., Moore, P. E., Abraham, J. H. *et al.* (1999). Role of ERK MAP kinases in responses of cultured human airway smooth muscle cells to IL-1beta. *Am J Physiol* **277**(5 Pt 1), L943–951.

Laporte, J. D., Moore, P. E., Lahiri, T. *et al.* (2000). p38 MAP kinase regulates IL-1 beta responses in cultured airway smooth muscle cells. *Am J Physiol* **279**, L932–941.

Lee, K. Y., Ho, S. C., Lin, H. C. *et al.* (2006). Neutrophil-derived elastase induces TGF-beta1 secretion in human airway smooth muscle via NF-kappaB pathway. *Am J Respir Cell Mol Biol* **35**, 407–414.

Liang, K. C., Lee, C. W., Lin, W. N. *et al.* (2007). Interleukin-1beta induces MMP-9 expression via p42/p44 MAPK, p38 MAPK, JNK, and nuclear factor-kappaB signaling pathways in human tracheal smooth muscle cells. *J Cell Physiol* **211**, 759–770.

Lukacs, N. W., Hogaboam, C. M. and Kunkel, S. L. (2005). Chemokines and their receptors in chronic pulmonary disease. *Curr Drug Targets Inflamm Allergy* **4**, 313–317.

Matsukura, S., Stellato, C., Plitt, J. R. *et al.* (1999). Activation of eotaxin gene transcription by NF-kappa B and STAT6 in human airway epithelial cells. *J Immunol* **163**, 6876–6883.

Melillo, E., Woolley, K. L., Manning, P. J. *et al.* (1994). Effect of inhaled PGE$_2$ on exercise-induced bronchoconstriction in asthmatic subjects. *Am J Respir Crit Care Med* **149**, 1138–1141.

Moore, P. E., Church, T. L., Chism, D. D. *et al.* (2002). IL-13 and IL-4 cause eotaxin release in human airway smooth muscle cells: a role for ERK. *Am J Physiol* **282**, L847–853.

Mukaida, N., Okamoto, S., Ishikawa, Y. *et al.* (1994). Molecular mechanism of interleukin-8 gene expression. *J Leukoc Biol* **56**, 554–558.

Murphy, P. M., Baggiolini, M., Charo, I. F. *et al.* (2000). International union of pharmacology. XXII. Nomenclature for chemokine receptors. *Pharmacol Rev* **52**, 145–176.

Nelson, P. J., Kim, H. T., Manning, W. C. *et al.* (1993). Genomic organization and transcriptional regulation of the RANTES chemokine gene. *J Immunol* **151**, 2601–2612.

Nie, M., Corbett, L., Knox, A. J. *et al.* (2005a). Differential regulation of chemokine expression by peroxisome proliferator-activated receptor gamma agonists: interactions with glucocorticoids and beta2-agonists. *J Biol Chem* **280**, 2550–2561.

Nie, M., Knox, A. J. and Pang, L. (2005b). beta2-Adrenoceptor agonists, like glucocorticoids, repress eotaxin gene transcription by selective inhibition of histone H4 acetylation. *J Immunol* **175**, 478–486.

Nie, M., Pang, L., Inoue, H. *et al.* (2003). Transcriptional regulation of cyclooxygenase 2 by bradykinin and interleukin-1beta in human airway smooth muscle cells: involvement of different promoter elements, transcription factors, and histone h4 acetylation. *Mol Cell Biol* **23**, 9233–9244.

Okona-Mensah, K. B., Shittu, E., Page, C. *et al.* (1998). Inhibition of serum and trans-forming growth factor beta (TGF-beta1)-induced DNA synthesis in confluent airway smooth muscle by heparin. *Br J Pharmacol* **125**, 599–606.

Oltmanns, U., Issa, R., Sukkar, M. B. *et al.* (2003). Role of c-jun N-terminal kinase in the induced release of GM-CSF, RANTES and IL-8 from human airway smooth muscle cells. *Br J Pharmacol* **139**, 1228–1234.

Oyamada, H., Kamada, Y., Kuwasaki, T. *et al.* (1999). CCR3 mRNA expression in bronchial epithelial cells and various cells in allergic inflammation. *Int Arch Allergy Immunol* **120** Suppl 1, 45–47.

Pang, L. and Knox, A. J. (1997). Effect of interleukin-1 beta, tumour necrosis factor-alpha and interferon-gamma on the induction of cyclo-oxygenase-2 in cultured human airway smooth muscle cells. *Br J Pharmacol* **121**, 579–587.

Pang, L. and Knox, A. J. (1998). Bradykinin stimulates IL-8 production in cultured human airway smooth muscle cells: role of cyclooxygenase products. *J Immunol* **161**, 2509–2515.

Pang, L. and Knox, A. J. (2000). Synergistic inhibition by beta(2)-agonists and corticosteroids on tumor necrosis factor-alpha-induced interleukin-8 release from cultured human airway smooth-muscle cells. *Am J Respir Cell Mol Biol* **23**, 79–85.

Pang, L. and Knox, A. J. (2001). Regulation of TNF-alpha-induced eotaxin release from cultured human airway smooth muscle cells by beta2-agonists and corticosteroids. *FASEB J* **15**, 261–269.

Pang, L., Nie, M., Corbett, L. *et al.* (2003). Cyclooxygenase-2 expression by nonsteroidal anti-inflammatory drugs in human airway smooth muscle cells: role of peroxisome proliferator-activated receptors. *J Immunol* **170**, 1043–1051.

Pavord, I. D. and Tattersfield, A. E. (1995). Bronchoprotective role for endogenous prostaglandin E_2. *Lancet* **345**(8947), 436–438.

Pease, J. E. and Sabroe, I. (2002). The role of interleukin-8 and its receptors in inflammatory lung disease: implications for therapy. *Am J Respir Med* **1**, 19–25.

Petkova, D. K., Pang, L., Range, S. P. *et al.* (1999). Immunocytochemical localization of cyclo-oxygenase isoforms in cultured human airway structural cells. *Clin Exp Allergy* **29**, 965–972.

Ponath, P. D., Qin, S., Ringler, D. J. *et al.* (1996). Cloning of the human eosinophil chemoattractant, eotaxin. Expression, receptor binding, and functional properties suggest a mechanism for the selective recruitment of eosinophils. *J Clin Invest* **97**, 604–612.

Pons, F., Freund, V., Kuissu, H. *et al.* (2001). Nerve growth factor secretion by human lung epithelial A549 cells in pro- and anti-inflammatory conditions. *Eur J Pharmacol* **428**, 365–369.

Pype, J. L., Dupont, L. J., Menten, P. *et al.* (1999). Expression of monocyte chemotactic protein (MCP)-1, MCP-2, and MCP-3 by human airway smooth-muscle cells. Modulation by corticosteroids and T-helper 2 cytokines. *Am J Respir Cell Mol Biol* **21**, 528–536.

Rodel, J., Woytas, M., Groh, A. *et al.* (2000). Production of basic fibroblast growth factor and interleukin 6 by human smooth muscle cells following infection with *Chlamydia pneumoniae*. *Infect Immun* **68**, 3635–3641.

Saunders, M. A., Mitchell, J. A., Seldon, P. M. *et al.* (1997). Release of granulocyte-macrophage colony stimulating factor by human cultured airway smooth muscle cells: suppression by dexamethasone. *Br J Pharmacol* **120**, 545–546.

Sekiya, T., Miyamasu, M., Imanishi, M. *et al.* (2000). Inducible expression of a Th2-type CC chemokine thymus- and activation-regulated chemokine by human bronchial epithelial cells. *J Immunol* **165**, 2205–2213.

Smit, J. J. and Lukacs, N. W. (2006). A closer look at chemokines and their role in asthmatic responses. *Eur J Pharmacol* **533**(1-3), 277–288.

Stellato, C., Brummet, M. E., Plitt, J. R. *et al.* (2001). Expression of the C-C chemokine receptor CCR3 in human airway epithelial cells. *J Immunol* **166**, 1457–1461.

Stocks, J., Bradbury, D., Corbett, L. *et al.* (2005). Cytokines upregulate vascular endothelial growth factor secretion by human airway smooth muscle cells: role of endogenous prostanoids. *FEBS Lett* **579**, 2551–2556.

Sukkar, M. B., Issa, R., Xie, S. *et al.* (2004). Fractalkine/CX3CL1 production by human airway smooth muscle cells: induction by IFN-gamma and TNF-alpha and regulation by TGF-beta and corticosteroids. *Am J Physiol* **287**, L1230–1240.

Szekanecz, Z., Shah, M. R., Harlow, L. A. *et al.* (1994). Interleukin-8 and tumor necrosis factor-alpha are involved in human aortic endothelial cell migration. The possible role of these cytokines in human aortic aneurysmal blood vessel growth. *Pathobiology* **62**, 134–139.

Vanaudenaerde, B. M., Wuyts, W. A., Geudens, N. *et al.* (2007). Macrolides inhibit IL17-induced IL8 and 8-isoprostane release from human airway smooth muscle cells. *Am J Transplant* **7**, 76–82.

Watson, M. L., Grix, S. P., Jordan, N. J. *et al.* (1998). Interleukin 8 and monocyte chemoattractant protein 1 production by cultured human airway smooth muscle cells. *Cytokine* **10**, 346–352.

Wen, F. Q., Liu, X., Manda, W. *et al.* (2003). TH2 Cytokine-enhanced and TGF-beta-enhanced vascular endothelial growth factor production by cultured human airway smooth muscle cells is attenuated by IFN-gamma and corticosteroids. *J Allergy Clin Immunol* **111**, 1307–1318.

Wuyts, W. A., Vanaudenaerde, B. M., Dupont, L. J. *et al.* (2003a). Involvement of p38 MAPK, JNK, p42/p44 ERK and NF-kappaB in IL-1beta-induced chemokine release in human airway smooth muscle cells. *Respir Med* **97**, 811–817.

Wuyts, W. A., Vanaudenaerde, B. M., Dupont, L. J. *et al.* (2003b). Modulation by cAMP of IL-1beta-induced eotaxin and MCP-1 expression and release in human airway smooth muscle cells. *Eur Respir J* **22**, 220–226.

Xie, S., Sukkar, M. B., Issa, R. *et al.* (2005). Regulation of TGF-beta 1-induced connective tissue growth factor expression in airway smooth muscle cells. *Am J Physiol* **288**, L68–76.

Zhu, Y. M., Bradbury, D. A., Pang, L. *et al.* (2003). Transcriptional regulation of interleukin (IL)-8 by bradykinin in human airway smooth muscle cells involves prostanoid-dependent activation of AP-1 and nuclear factor (NF)-IL-6 and prostanoid-independent activation of NF-kappaB. *J Biol Chem* **278**, 29366–29375.

Zlotnik, A. and Yoshie, O. (2000). Chemokines: a new classification system and their role in immunity. *Immunity* **12**, 121–127.

9
Airway smooth muscle in experimental models

Anne-Marie Lauzon and James G. Martin

Meakins-Christie Laboratories, Department of Medicine, McGill University, Montréal, Québec, Canada

9.1 Introduction

Research on airway smooth muscle (ASM) has been greatly motivated by the possibility that its function is altered in asthma and other obstructive lung diseases. Although definitive evidence of ASM dysfunction in asthma is lacking, there are many ways by which alterations in ASM might lead to airway hyper-responsiveness (AHR) (Solway and Fredberg, 1997). The interest in ASM has broadened since the recognition of properties beyond its role as a contractile tissue. The demonstration of the secretory properties of ASM has raised interesting questions concerning its place in the inflammation associated with airway disease. The plasticity of ASM is also reflected in the finding of the increased muscle mass in airway disease that is a particularly noteworthy feature of airway remodelling in asthma (Carroll *et al.*, 1993; Dunnill *et al.*, 1969). However, the responsiveness of ASM in the absence of obvious airway disorder is also of considerable interest. It is possible that subjects have innate AHR that may occur as a result of factors intrinsic to the host without apparent relationship to environmental

Airway Smooth Muscle   Edited by Kian Fan Chung
© 2008 John Wiley & Sons, Ltd

insults, whereas acquired AHR follows more clearly defined triggers. Innate AHR may, of course, predispose to the development of acquired AHR, although the experimental data to support this hypothesis are limited and the idea is largely based on epidemiological data. The isolation, from other factors, of the contribution of ASM to measures of responsiveness has not been easy. A number of experimental models, ranging from *in vivo* to *in silico*, have been used to evaluate ASM function; contractile, secretory, proliferative and migratory capacities have all been addressed to a greater or lesser extent. In this chapter, we will review some of the techniques available to assess, indirectly and directly, the properties of ASM, and we will review the implications of current findings for airway disease.

9.2 Methods of assessment of airway smooth muscle (ASM) function

The study of airway responsiveness in human subjects and in animal and tissue models has contributed substantially to our understanding of the potential mechanisms of excessive ASM contraction. The evaluation of ASM function can be approached in a variety of ways. Assessment *in vivo* provides the most relevant information but is also associated with the greatest uncertainty because of the complications induced by the multiple modulating influences. Provocation testing with agonists such as methacholine and histamine provides data concerning the capacity of the airways to respond to the contracting stimulus. An inherent assumption in measuring so-called non-specific responsiveness is that it is predominantly a reflection of the degree of ASM contraction and is therefore a measure of the contractile properties of muscle. The effectiveness of contraction of ASM in causing airway narrowing is affected by the architecture of the ASM and the constitutive properties of the airway wall itself (Bates and Martin, 1990; Moreno *et al.*, 1986; 1993). There is an additional component of the airway response attributable to microvascular leakage from bronchial vessels in mammals greater in size than the mouse (Yager *et al.*, 1995). In the mouse airway, closure from fluid is also an important determinant of the measurement of responsiveness and in particular AHR induced by certain stimuli (Wagers *et al.*, 2004). The impedance to airway narrowing provided by the elastic properties of the lung parenchyma limits the degree of induced bronchoconstriction (Ding *et al.*, 1987). The cyclical stresses imposed in the contracting muscle by the act of breathing are potently bronchodilating in action (Fredberg *et al.*, 1997; Gunst and Stropp, 1995). Modulation of responses by the epithelial barrier and by epithelium-derived mediators may also alter the outcome of non-specific challenge tests.

Classical pharmacological methods of studying tracheal strips or bronchial rings have provided much of our basic understanding of ASM contractile properties. However, the disruption of the lung architecture has motivated the development of other approaches. Lung slices in culture have been used effectively in the assessment of ASM and have complemented other techniques (Dandurand *et al.*, 1993). Lung slices allow preservation of lung elastic recoil forces to some extent, and when the lungs are prepared by inflation with agarose, there is preload and after load to the contracting muscle that is not dissimilar to the forces acting on the airways *in vivo* (Adler *et al.*, 1998). At the same time the preparation retains the advantages of an *in vitro* system. Coarse slices have been used to evaluate the magnitude and velocity of contraction of airways on exposure to contractile agonists (Tao *et al.*, 1995; Wang *et al.*, 1997). Thin slices allow signalling events such as calcium transient to be assessed at the same time as contractions (Bergner and Sanderson, 2002).

The study of cultured ASM cells has also yielded useful information, although the contractility of these cells is not easily assessed. In addition, it is dependent on the conservation of contractile function in culture and is also influenced by the substratum on which the cells are cultured (Hirst *et al.*, 2000; Kelly and Tao, 1999). The substratum may affect the cell phenotype, but the density of adhesions between cell and substratum may also be a modulating factor. The contractions of cultured cells can be measured from changes in length or area of cells as well as by the assessment of the stiffness of the cells measured by the technique of magnetic twisting cytometry (Wang *et al.*, 1993a) or atomic force microscopy (Smith *et al.*, 2005). Magnetic twisting cytometry relies on the use of ferromagnetic beads attached to the cell via integrins that in turn are linked to the cytoskeleton. Contraction or relaxation of the cell is associated with changes in stiffness that can be followed in real-time. Atomic force microscopy can also be used to probe the viscoelastic properties of the ASM cells and the changes accompanying contraction of the cells.

9.3 Potential mechanisms by which ASM properties may contribute to airway responsiveness

Potential smooth muscle-related causes of AHR are illustrated in Figure 9.1. ASM *in vitro* is very responsive to contractile agonists and can contract to very short lengths (Lambert *et al.*, 1993; Stephens *et al.*, 1992). However, there are

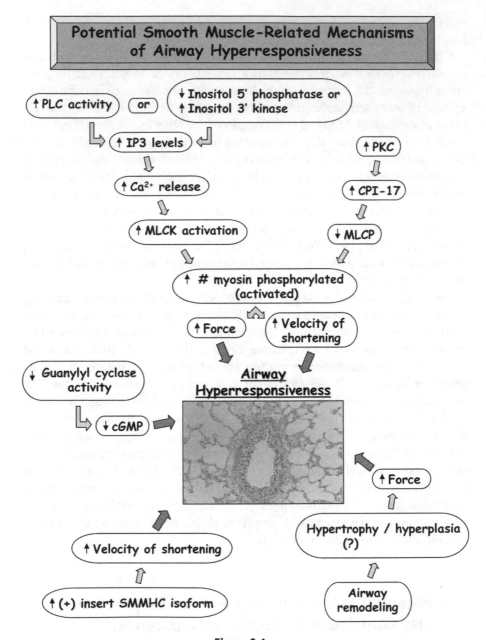

Figure 9.1

mechanical forces that limit shortening *in vivo*. Lung elastic recoil and the cyclical movement of the lungs act in concert to prevent significant airway narrowing. Whereas some forms of AHR may reflect changes in the mechanical load against ASM shortening, much of the time excessive airway narrowing likely reflects an increase in the capacity of ASM to overcome the mechanical impedances, since lung elastic recoil is not usually altered in asthma. An increase in the force of contraction by ASM is therefore a possible cause of AHR that could result from differences in the intrinsic contractile processes of the ASM cells or conceivably from an increase in ASM mass caused by remodelling (Lambert *et al.*, 1993). However, airway hyperesponsiveness needs to be considered from the standpoint of the dynamics of airway narrowing, and there is evidence that suggests alterations in contractile proteins leading to an altered velocity of contraction in asthma. Modulation of ASM contraction by neural pathways or by epithelial factors is also probable (Flavahan *et al.*, 1985). Previous experimentation has shown minor effects of interrupting neural pathways so it seems unlikely that alterations in neural control of the airways could act other than as modulators of airway responses (Shore *et al.*, 1985). Evaluating epithelial contributions to airway responsiveness is harder to do because there is no specific intervention to eliminate epithelial effects without significantly altering anything else. Many models have focused on the role of airway inflammation in AHR. Indeed, there is convincing evidence that inflammation of various sorts can augment airway responses to contractile agonists. The fact that bronchoconstriction reaches a plateau in normal

---

Figure 9.1   Potential mechanisms of smooth muscle-related AHR. Multiple factors are believed to contribute to AHR. An increase in phospholipase-C (PLC) activity leads to an increase in inositol triphosphate (IP$_3$), and thus an increase in calcium (Ca^{2+}) release from the sarcoplasmic reticulum. The greater intracellular Ca^{2+} leads to greater Ca^{2+}-calmodulin activation of myosin light chain kinase (MLCK). The greater MLCK activation increases the number of phosphorylated (activated) myosin molecules, potentially leading to an increase in smooth muscle force production and/or increase velocity of shortening (see text for details on how these mechanisms might lead to AHR). A decrease in inositol 5/-phosphatase or an increase in inositol 3/ kinase also leads to increases in IP$_3$ levels. Another mechanism that contributes to the increase in the number of phosphorylated myosin molecules is an increase in protein kinase (PKC), which increases the level of CPI-17, which itself inhibits myosin light chain phosphatase (MLCP). Cyclic guanosine monophosphate (cGMP) is a smooth muscle-relaxing second messenger, so its decreased production, due to a decrease in guanylyl cyclase activity, may also lead to AHR. Smooth muscle cell hyperplasia and potentially hypertrophy increase force production and may potentially participate in AHR. An increase expression of the (+)insert smooth muscle myosin heavy chain (SMMHC) isoform increases the velocity of shortening (see text for details).

human and other species at high doses of agonist and that the degree of constriction that can be induced is much less than seen in asthmatic subjects (Woolcock *et al.*, 1984) leads one to the conclusion that increasing constrictive agents alone is unlikely to cause AHR. Whereas changes in submaximal constriction as a result of the concurrent actions of several agonists are easily envisaged, changes in maximal responses are less easily attributable to such interactions. Inflammatory mediators presumably cause AHR by inducing functional changes in the ASM.

9.4 Experimental models

Innate AHR

Structure function studies relating airway responsiveness *in vivo* have been performed on human subjects and on a variety of animal species. Early studies performed on bronchial tissues harvested from subjects undergoing resectional lung surgery were unable to show clear relationships between responsiveness to histamine *in vivo* and responses of isolated tissue strips (Armour *et al.*, 1984; Taylor *et al.*, 1985). However, these studies preceded the recognition of the influence of airway epithelial cells on isolated tissue responses and were unable to take into consideration the orientation of muscle bundles within tissue preparations. Furthermore, there is always a possible influence of prior lung disorder on such outcomes. A comparison of airway responsiveness among different animal species with standardized challenge procedures demonstrated a good correlation with the mass of ASM measured morphometrically (Martin *et al.*, 1992). In guinea pigs, there was also a relationship between ASM mass and both the sensitivity of the ASM to the agonist and the maximal response, suggesting a possible link between factors determining the mass of muscle and the sensitivity to the agonist (Opazo-Saez *et al.*, 1996). This property will be referred to again below.

Screens of mice and rats have shown strain-dependent differences in AHR (Levitt and Mitzner, 1988). The genetic basis for the difference has been explored in mice (Desanctis *et al.*, 1995; Ewart *et al.*, 1996). There have been no reports of ASM-specific genes being linked to AHR, and the candidate genes postulated to account for AHR have in general been related to inflammation. A/J mice appear to have some similarities to Fisher (F344) rats in showing AHR to methacholine and serotonin compared to other strains (Martin *et al.*, 1992). These strains have been frequently contrasted with C57Bl/6 mice and Lewis rats, respectively, which have lower responses. A genetically determined AHR has also been described in the Basenji greyhound (Hirshman et al., 1980). Of all the species usually used as models of asthma, the guinea pig appears to be the one that has the most striking innate responsiveness to challenge with histamine or methacholine (Martin et al., 1992). Other species, such as rats and rabbits, that have been directly compared by similar methods require higher doses of agonists to evoke airway responses, and their level of airway constriction is limited compared to the responses of the guinea pig. The links between ASM structure and function are still tenuous in many of the models, and the muscle is implicated principally by its association with AHR rather than through experimentation specifically probing its function.

ASM mass and AHR

Of importance to the link between ASM and airway responsiveness is the ability to show such relationships ex vivo. The trachea and the intrapulmonary airways of the guinea pig have more muscle than the corresponding structures in the rat. A comparison of the responses of the tracheal smooth muscle from guinea pig to that of rat has demonstrated greater responsiveness comparable to that observed in vivo, and this is potentially caused by the difference in mass of ASM. However, sensitivity to agonists is also greater in guinea pigs, whereas differences in maximal contractions only would be expected if differences in mass of ASM entirely accounted for the greater responsiveness. Comparative studies of the biochemical characteristics of the ASM among different species are lacking.

A screen of different strains of rat has shown the F344 strain to be hyperresponsive ('susceptible') and the Lewis strain to be normoresponsive ('resistant'). The Fisher rat has more ASM than the Lewis strain (Eidelman et al., 1991). There is growing evidence that the difference in mass between different rat strains is not the only difference in the ASM properties. Tracheal preparations from these rats have demonstrated differences consistent with the in vivo findings of hyperresponsiveness (Florio et al., 1996). Explanted lung tissue in culture has been used to examine the responsiveness of airway tissues ex vivo (Tao et al., 1999; Wang et al., 1997). Agonists applied to the lung slices caused brisk airway narrowing that could be quantified by videomicroscopy. F344 rats were also hyperresponsive compared to the Lewis animals by this technique. Finally, single cells in culture from the F344 rats have been found to show greater responses to contractile agonists than Lewis rats (An et al., 2006; Tao et al., 1999; 2003). These experimental findings support the concept of an intrinsic difference in ASM properties.

Rate of shortening and AHR

A greater rate of ASM shortening has been observed at the muscle strip (Blanc et al., 2003; Fan et al., 1997) and airway explant (Duguet et al., 2000; Wang et al., 1997) levels in many models of AHR, and at the cellular (Ma et al., 2002) and muscle strip levels in human asthmatics (Mitchell et al., 1994). Why would a higher rate of contraction lead to the enhancement of airway responsiveness? Although this has not been completely elucidated, many theories have been postulated. Stephens and coworkers have shown in a dog model of AHR that the development of the maximum velocity of shortening occurs during the first 2 s of a 10-s contraction (Jiang et al., 1992). The remainder of the contraction was suggested to be handled by more slowly cycling cross-bridges. They also demonstrated in human asthmatic bronchial smooth muscle cells that 90 per cent of the shortening occurs during the first 1.5 s of the contraction (Ma et al., 2002). Thus, faster smooth muscle myosin isoforms in hyperresponsive airways would lead to a greater ASM shortening during the initial ~2-s phase of contraction, and thus a greater airway resistance. Another theory is that a greater rate of shortening keeps the muscle in a permanent state of contraction because it has time to constrict between each breath, thereby counteracting the relaxing effect of tidal volume (Solway and Fredberg, 1997). It is also possible that the combination of multiple mechanisms must be present to lead to AHR. For example, a greater rate of contraction combined with a decreased parenchymal load would lead to a greater extent of shortening, and thus a greater airway resistance.

Contractile protein expression and function

The expression and function of contractile proteins in innate AHR have been addressed in the F344 and Lewis rats. In particular, the expression of the various smooth muscle myosin heavy chain (SMMHC) isoforms has been investigated. There are four isoforms of the SMMHC that are generated by alternative splicing of a single gene (Eddinger and Murphy, 1988; White et al., 1993). Two of these isoforms differ by the presence ([+]insert) or absence ([−]insert) of a seven-amino-acid insert in the amino terminus region (White et al., 1993). The [+]insert isoform propels actin at twice the rate of the [−]insert isoform in the in vitro motility assay (Lauzon et al., 1998; Rovner et al., 1997). Two other isoforms differ in the carboxyl terminus by distinct sequences of 43 (SM1) or nine (SM2) amino acids (Babij and Periasamy, 1989; Nagai et al. 1989). No mechanical differences have been reported between these two isoforms. A greater proportion of the faster (+)insert SMMHC isoform is expressed in the hyperresponsive F344 than in the normoresponsive Lewis rat tracheae, whereas there is no difference in expression of other contractile proteins such as *h*-caldesmon, α-actin, or myosin light chain-17 (LC_{17}) isoforms (Gil *et al.*, 2006). The expression of these contractile proteins has yet to be addressed at the bronchial level. Bronchial expression would indeed be more relevant to the study of airway responsiveness, but its assessment is more difficult due to the contaminating contribution of blood vessel smooth muscle.

The fast [+]insert SMMHC isoform is an important determinant of the rate of smooth muscle contraction (Leguillette *et al.*, 2005). Its expression is greater in rapidly contracting phasic muscle such as the intestine and lower in slowly contracting tonic muscle such as the aorta (Leguillette *et al.*, 2005). The greater expression of the [+]insert isoform in the F344 rat trachea is consistent with the mechanical properties observed at the airway and whole muscle levels. In explant preparations, a greater rate of shortening in response to methacholine has also been observed for the F344 rats and this at any lung inflation volumes studied (Wang *et al.*, 1997). At the muscle strip level, the maximum unloaded shortening velocity and maximum extent of muscle shortening are higher in F344 than in the Lewis rats, but peak tension is not different (Blanc *et al.*, 2003). This is in agreement with the fact that the [+] and [−]insert myosin isoforms do not exhibit any difference in their molecular force production. A knockout mouse that lacks the [+]insert SMMHC isoform has also been engineered in a C57BL/6 background (Babu *et al.*, 2001). The time to reach maximum airway resistance after a challenge of methacholine is lengthened in these knockout mice (Tuck *et al.*, 2004). However, airway responsiveness is not decreased in these animals, potentially because of their hyporesponsive background (Tuck *et al.*, 2004).

Another factor that deserves attention in terms of muscle contractility is the level of activation. For an average dose of methacholine, the level of phosphorylation of the myosin regulatory light chain (LC_{20}) is greater in the F344 than in the Lewis rats (Gil *et al.*, 2006). However, the specific role of the greater phosphorylation level with respect to enhanced contractility remains controversial. It is clear that for an average dose of agonist, a greater level of phosphorylation leads to greater tension development (Gil *et al.*, 2006), but a link between phosphorylation level and rate of shortening remains to be established (Merkel *et al.*, 1990; Mitchell *et al.*, 2001).

Contractile signalling

Contraction of ASM is a complex process involving many steps. Any one of these steps is a potential site of difference in response. Several aspects of contractile signalling have been examined in the rats differing in responsiveness. Single-cell calcium imaging has demonstrated differences in responses of the F344 rat to serotonin and bradykinin compared to the normoresponsive Lewis strain (Tao *et al.* (1999; 2003). Both of these agonists cause greater contractile responses of explanted airways in F344 than in Lewis. The enhanced calcium signals are attributable to differences in inositol phosphate metabolism (Tao *et al.*, 2000). Inositol trisphosphate (IP-3), the intracellular mediator of calcium release from the sarcoplasmic reticulum, is present in greater concentrations in the cells of F344 rats after serotonin stimulation than in Lewis rats. The differences appear to be more a function of reduced degradation by an IP-3-specific phosphatase than differences in production of IP-3.

Changes in inositol phosphate metabolism are not the sole potential cause of enhanced airway responses. Part of the difference in responsiveness between F344 and Lewis rats *in vivo* may be explained by nitric oxide (NO) production in the airways (Jia *et al.*, 1996). Lewis rats *in vivo* become more responsive following the inhibition of nitric oxide synthase, whereas F344 rats are unaffected by the intervention. F344 airways *ex vivo* were relatively resistant to the effects of nitroprusside, a source of exogenous NO; tracheal tissues produced less 3,5-cyclic guanosine monophosphate when treated with nitroprusside (Jia *et al.*, 1995). It seems likely that the epithelium is a major source of NO in the rat and that it is the major epithelial mediator affecting airway responsiveness. Over-expression of NOS-2 in the airway epithelium of mice makes them hyporesponsive to methacholine (Hjoberg *et al.*, 2004). In conditional transgenic mice, the level of NO was increased in the airways and was demonstrated to affect airway responsiveness without any significant pro-inflammatory effects.

Calcium signalling and changes in airway calibre has been investigated in lung slices of inbred mice (C3H/HeJ, Balb/C, and A/J). Although their responsiveness to acetylcholine differed, their initial calcium transient and frequency of calcium oscillations were similar (Bergner and Sanderson, 2003). These results suggest that calcium sensitivity of the contractile apparatus, rather than calcium signalling, may be altered in AHR (Bergner and Sanderson, 2003).

Proliferative properties of ASM

As mentioned above, F344 rats have more ASM than Lewis rats. The relationship between the excess of ASM and hyperresponsiveness to contractile agonists is not clear. However, ASM cultured from these two rat strains also show differences in responsiveness to mitogenic stimuli (Zacour and Martin, 1996; 2000).

These differences are associated with protein kinase C isoforms (Zacour and Martin 2000) but by mechanisms that have not been elucidated. Analogous findings have been reported for spontaneously hypertensive rats that have vascular smooth muscle that is hyperresponsive not only to contractile agonists but also to growth-promoting agents (Hadrava *et al.*, 1992). Interestingly, there is also evidence that ASM from asthmatic subjects has enhanced proliferative responses to mitogens that may relate to the absence of an inhibitory transcription factor C/EBP-alpha, the influence of an altered extracellular matrix, and/or reduction in the synthesis of factors such as prostaglandin (PGE_2) (reviewed by Oliver and Black, 2006).

Animal models of allergic AHR

The interpretation of changes of airway responsiveness following allergen challenges is even more complex than that of innate hyperresponsiveness. The responses to allergen-derived mediators reflect cellular responsiveness (mast cells, T cells, etc.) to the challenge as well as measuring the responsiveness of the ASM to the mediators released (e.g., cysteinyl-leukotrienes, PGD_2, neurokinins). Innate AHR and increased responsiveness induced by stimuli such as allergen challenge may have different causes, and the extent to which responses to these challenges can be interpreted as reflecting smooth muscle function is somewhat uncertain.

Airway and muscle mechanics have been studied extensively in a canine model of allergic AHR. Ragweed pollen-sensitized dogs show marked increases in specific airway resistance when challenged with ragweed extracts or acetylcholine (Antonissen *et al.*, 1979; Becker *et al.*, 1989). The tracheal and bronchial smooth muscles of these sensitized dogs exhibit increased maximum shortening velocity and capacity, and prolonged relaxation time, but no differences in maximum force (Antonissen *et al.*, 1979; Jiang *et al.*, 1992). These enhanced contractile properties are associated with greater MLCK content and thus, greater LC_{20} phosphorylation levels, but not with greater Ca^{2+}-calmodulin activities (Jiang *et al.*, 1995). Significantly, Ma *et al.* (2002) demonstrated that human asthmatic bronchial smooth muscle cells exhibit similar mechanical properties. A greater velocity and capacity of shortening were reported as well as a greater MLCK mRNA content. These latter results attest to the relevance of this allergic model to the study of asthma.

More recently, a chronic model of allergic asthma has been developed in Balb/C mice exposed intranasally to house-dust mite. This model shows that the initial increase in airway responsiveness is immunologically driven by eosinophils, whereas sustained AHR post-allergen exposure is associated with airway remodelling (Leigh *et al.*, 2002). Furthermore, the changes in airway reactivity and sensitivity are observed early during the challenge period and are associated with increased smooth muscle area, whereas increases in maximal airway resistance are observed later, are sustained in the post-allergen phase, and are associated with collagen deposition (Leigh *et al.*, 2002). This study of the development and maintenance of AHR demonstrated the complexity of the interactions between the multiple factors involved as well as the importance of the timing of the measurements in relation to the allergen exposure.

Allergen-induced AHR: ASM growth

Allergen induced AHR involves a modest change in the dose-response curve to inhaled methacholine or other agonists. The basis for this form of AHR is still quite obscure. The links to inflammation have been solidly established, but the precise cause of the end-organ responsiveness remains unexplained. Although there is scant information concerning the alterations in ASM phenotype *in vivo* as a result of allergen-induced inflammation, there is now evidence of significant changes in contractile and cytoskeletal proteins (McVicker *et al.*, 2007). Whether there is any relationship between innate and acquired AHR remains to be determined. However, there are a number of potential mechanisms that may link the two forms of AHR.

There is abundant evidence of an increase in ASM mass in asthmatic airways (Carroll *et al.*, 1993; Dunnill *et al.*, 1969), and it has been postulated to be the basis for the AHR observed in asthma (Lambert *et al.*, 1993). Asthma frequently appears to have a slow onset, and it is an attractive hypothesis that this is attributable to the induction of changes in the airway structure. An increase in ASM mass has been demonstrated in a variety of allergen-driven animal models of asthma (Leigh *et al.*, 2002; Padrid *et al.*, 1995; Panettieri *et al.*, 1998; Sapienza *et al.*, 1991). The increase in mass results from hyperplasia, and several mediators have already been identified as causative. Cysteinyl-leukotrienes have been implicated in the growth of ASM *in vivo* by several investigators using different models (Henderson *et al.*, 2002; Salmon *et al.*, 1999; Wang *et al.*, 1993b). Endothelin antagonists have also been found to prevent ASM hyperplasia in a rodent model (Salmon *et al.*, 2000). Recent preliminary evidence suggests that the epidermal growth factor receptor (EGFR) may be involved in ASM hyperplasia (Vargaftig and Singer, 2003). Indeed, it is quite possible that cysteinyl-leukotrienes may act indirectly by releasing ligands of the EGFR or potentially by *trans*-activation of the EGFR (Ravasi *et al.*, 2006). G-protein-coupled receptors such as cysteinyl-leukotrienes are weaker mitogens than classical growth factors, so indirect mechanisms of the growth-promoting actions of the latter need to be considered.

Recently, the mechanical properties of the matrix have been shown to be key in the development of mesenchymal cell lineages (Engler *et al.*, 2006). Extreme stiffness favours the expression of osteogenic genes, whereas low stiffness favours the expression of neurons and intermediate-level muscle cells. Perhaps the alterations in airway wall compliance in asthma may have a place in the alterations in cell populations in the airway wall in disease.

Allergen-induced AHR: secretory phenotype of ASM

The contribution of an increase in ASM mass to AHR depends upon the new muscle having an appropriate orientation within the airway wall (Bates and Martin, 1990) as well as having normal or potentially enhanced contractile properties. Whether changes in ASM contractile properties occur *in vivo* requires further direct confirmation. However, there is evidence of changes in ASM in the allergen-sensitized canine and in a variety of cell model systems. There is considerable potential for both negative and positive effects on contractility. Phenotype changes in ASM have been best studied where muscle tends to manifest its secretory properties to a substantial extent. There are a few studies concerning alterations in phenotype of ASM *in vivo* (McVicker *et al.*, 2007; Moir *et al.*, 2003a). If ASM were to revert to a secretory phenotype following hyperplasia, then it would be expected to have a reduced capacity to generate force. In brown-Norway rats undergoing repeated allergen exposure, there is reportedly a fall in the content of sm-α in document -actin, SM1 sm-MHC, calponin, and smoothelin-A in bronchioles (Moir *et al.*, 2003b). Changes in airway responsiveness, proportional to the changes in ASM mass, occur after allergen challenge (Sapienza *et al.*, 1991). An increase in contractile responses of isolated airway preparations was found at 35 days after the challenge had ended, but the significance of this finding or its mechanism is uncertain because no changes in AHR *in vivo* were registered at that time point (Moir *et al.*, 2003b). An increase in fibronectin has been described in the airways of sensitized rats repeatedly challenged with ovalbumin (Palmans *et al.*, 2000), and this and other changes in the matrix following allergen challenge (Pini *et al.*, 2004) could possibly contribute to changes in the balance of contractile and secretory phenotypes.

Allergen-induced AHR: contractile phenotype of ASM

Retaining the contractile phenotype in culture so as to study the effects of various pro-inflammatory cytokines on ASM function has been an important issue. Indeed, total myosin content decreases when standard cell culture conditions are used (Arafat *et al.*, 2001; Babij *et al.*, 1992; Thyberg, 1996; Wong *et al.*, 1998). The expression of the [+]insert myosin isoform is also downregulated in standard cell culture conditions (e.g., the expression of the rabbit bladder [+]insert SMMHC isoform is down to 45 per cent during the third passage of primary culture (Arafat *et al.*, 2001)). In addition to laminin, collagen, and fibronectin, a homologous cell substrate has been reported to produce cells that are more contractile, although alterations in cell attachment have not been examined as an alternative explanation for the enhanced responses (Kelly and Tao, 1999). Despite the problems of cellular phenotype, investigators have used cultured cells to study the influence of potentially pertinent cytokines on both the contractile and secretory properties of ASM cells. *In vitro* studies have shown that cytokines modulate the contractility of ASM via both upregulation of CysLT1 receptors (IL-13) (Babij *et al.*, 1992) and altering calcium signalling in ASM (IL-13 and tumour necrosis factor-α (TNF-α)) (Amrani *et al.*, 1997; Eum *et al.*, 2005). The increase in calcium transients to contractile agonists after IL-13 pretreatment appears to be attributable to an upregulation of CD38, an enzyme responsible for the synthesis of cyclic-ADP ribose (Deshpande *et al.*, 2004).

Cells in the airways are subjected to periodic mechanical stress *in vivo*. Incorporating variations in mechanical strain into the design of cell-culture conditions has been attempted. Cultured ASM cells have been stressed via integrin-bound ferromagnetic beads on their surface by changing magnetic fields (magnetic twisting cytometry) (Fabry *et al.*, 2001; Wang *et al.*, 1993a) or by stretching of flexible membranes on which the cells are layered (Maksym *et al.*, 2005). Mechanical stress causes cytoskeletal rearrangement in ASM with increased contractility, stiffness and proliferation, demonstrating the plasticity of the cytoskeleton in ASM (Maksym *et al.*, 2005). Cyclical stress has been applied to simulate the effects of respiration by a commercial device (e.g., Flexcell™). Mechanical strain increases the contractility of ASM, but it also influences the synthetic functions of ASM, increasing IL-8 production by human ASM in an AP-1-dependent manner and also involving the MAPK pathway (Kumar *et al.*, 2003). Mechanical stress also interacts with the matrix in influencing ASM phenotype; collagen enhances growth more than laminin in cells subjected to cyclic mechanical strain, but strain reduces growth on both substrata (Bonacci *et al.*, 2003).

9.5 Conclusion

Despite the myriad of studies investigating the function of ASM, its precise role in health and disease remains obscure. Modest contractions of ASM are evoked in physiological circumstances, and excessive contractions are associated with disorder. The decrease in airway luminal area induced by ASM contraction is well known to be the final step in the cascade of events involved in asthma, the condition associated with the greatest degrees of airway responsiveness. The numerous tools available to study bronchoconstriction in animal and *in vitro* models have revealed that alterations at multiple sites can participate in inducing AHR. Other forms of hyperresponsiveness, such as enhanced proliferation in response to growth-promoting stimuli or enhanced secretion of pro-inflammatory mediators or matrix proteins may represent susceptibilities relevant to disease.

Acknowledgement

We thank Genevieve Bates for the artwork of Figure 9.1.

References

Adler, A., Cowley, E. A., Bates, J. H. *et al.* (1998). Airway-parenchymal interdependence after airway contraction in rat lung explants. *J Appl Physiol* **85**, 231–237.

Amrani, Y., Krymskaya, V., Maki, C. *et al.* (1997). Mechanisms underlying TNF-alpha effects on agonist-mediated calcium homeostasis in human airway smooth muscle cells. *Am J Physiol* **273** (pt 1), L1020–1028.

An, S. S., Fabry, B., Trepat, X. *et al.* (2006). Do biophysical properties of the airway smooth muscle in culture predict airway hyperresponsiveness? *Am J Respir Cell Mol Biol* **35**, 55–64.

Antonissen, L. A., Mitchell, R. W., Kroeger, E. A. *et al.* (1979). Mechanical alterations of airway smooth muscle in a canine asthmatic model. *J Appl Physiol Respir Environ Exerc Physiol* **46**, 681–687.

Arafat, H. A., Kim, G. S., DiSanto, M. E. *et al.* (2001). Heterogeneity of bladder myocytes in vitro: modulation of myosin isoform expression. *Tissue Cell* **33**, 219–232.

Armour, C. L., Lazar, N. M., Schellenberg, R. R. *et al.* 1984). A comparison of in vivo and in vitro human airway reactivity to histamine. *Am Rev Respir Dis* **129**, 907–910.

Babij, P., Kawamoto, S., White, S. *et al.* (1992). Differential expression of SM1 and SM2 myosin isoforms in cultured vascular smooth muscle. *Am J Physiol* **262**, C607–C613.

Babij, P. and Periasamy, M. (1989). Myosin heavy chain isoform diversity in smooth muscle is produced by differential RNA processing. *J Mol Biol* **210**, 673–679.

Babu, G. J., Loukianov, E., Loukianova, T. *et al.* (2001). Loss of SM-B myosin affects muscle shortening velocity and maximal force development. *Nat Cell Biol* **3**, 1025–1029.

Bates, J. H. and Martin, J. G. (1990). A theoretical study of the effect of airway smooth muscle orientation on bronchoconstriction. *J Appl Physiol* **69**, 995–1001.

Becker, A. B., Hershkovich, J., Simons, F. E. R. *et al.* (1989). Development of chronic airway hyperresponsiveness in ragweed-sensitized dogs. *J Appl Physiol* **66**, 2691–2697.

Bergner, A. and Sanderson, M. J. (2002). Acetylcholine-induced calcium signaling and contraction of airway smooth muscle cells in lung slices. *J Gen Physiol* **119**, 187–198.

Bergner, A. and Sanderson, M. J. (2003). Airway contractility and smooth muscle Ca(2+) signaling in lung slices from different mouse strains. *J Appl Physiol* **95**, 1325–1332.

Blanc, F. X., Coirault, C., Salmeron, S. *et al.* (2003). Mechanics and crossbridge kinetics of tracheal smooth muscle in two inbred rat strains. *Eur Respir J* **22**, 227–234.

Bonacci, J. V., Harris, T. and Stewart, A. G. (2003). Impact of extracellular matrix and strain on proliferation of bovine airway smooth muscle. *Clin Exp Pharmacol Physiol* **30**, 324–328.

Carroll, N., Morton, E. A. and James, A. (1993). The structure of large and small airways in non-fatal and fatal asthma. *Am Rev Respir Dis* **147**, 405–410.

Dandurand, R. T., Wang, C. G., Phillips, N. C. *et al.* (1993). Responsiveness of individual airways to methacholine in adult rat lung explants. *J Appl Physiol* **75**, 364–372.

Deshpande, D. A., Dogan, S., Walseth, T. F. *et al.* (2004). Modulation of calcium signaling by interleukin-13 in human airway smooth muscle: role of CD38/cyclic adenosine diphosphate ribose pathway. *Am J Respir Cell Mol Biol* **31**, 36–42.

Desanctis, G. T., Merchant, M., Beier, D. R. *et al.* (1995). Quantitative locus analysis of airway hyperresponsiveness in A/J and C57BL/6J mice. *Nat Genet* **11**, 150–154.

Ding, D. J., Martin, J. G. and Macklem, P. T. (1987). Effects of lung volume on maximal methacholine induced bronchoconstriction in normal human subjects. *J Appl Physiol* **62**, 1324–1330.

Duguet, A., Biyah, K., Minshall, E. *et al.* (2000). Bronchial responsiveness among inbred mouse strains. Role of airway smooth-muscle shortening velocity. *Am J Respir Crit Care Med* **161**, 839–848.

Dunnill, M. S., Masarrella, G. R. and Anderson, J. A. (1969). A comparison of the quantitative anatomy of the bronchi in normal subjects, in status asthmaticus, in chronic bronchitis and in emphysema. *Thorax* **24**, 176–179.

Eddinger, T. J. and Murphy, R. A. (1988). Two smooth muscle myosin heavy chains differ in their light meromyosin fragment. *Biochemistry* **27**, 3807–3811.

Eidelman, D. H., DiMaria, G. U., Bellofiore, S. *et al.* (1991). Strain-related differences in airway smooth muscle and airway responsiveness in the rat. *Am Rev Respir Dis* **144**, 792–796.

Engler, A. J., Sen, S., Sweeney, H. L. *et al.* (2006). Matrix elasticity directs stem cell lineage specification. *Cell* **126**, 677–689.

Espinosa, K., Bosse, Y., Stankova, J. *et al.* (2003). CysLT1 receptor upregulation by TGF-beta and IL-13 is associated with bronchial smooth muscle cell proliferation in response to LTD4. *J Allergy Clin Immunol* **111**, 1032–1040.

Eum, S. Y., Maghni, K., Tolloczko, B. *et al.* (2005). IL-13 may mediate allergen-induced hyperresponsiveness independently of IL-5 or eotaxin by effects on airway smooth muscle. *Am J Physiol Lung Cell Mol Physiol* **288**, L576–L584.

Ewart, S. L., Mitzner, W., DiSilvestre, D. A. *et al.* (1996). Airway hyperresponsiveness to acetylcholine: segregation analysis and evidence for linkage to murine chromosome 6. *Am J Respir Cell Mol Biol* **14**, 487–495.

Fabry, B., Maksym, G. N., Shore, S. A. *et al.* (2001). Selected contribution: time course and heterogeneity of contractile responses in cultured human airway smooth muscle cells. *J Appl Physiol* **91**, 986–994.

Fan, T., Yang, M., Halayko, A. *et al.* (1997). Airway responsiveness in two inbred strains of mouse disparate in IgE and IL-4 production. *Am J Respir Cell Mol Biol* **17**, 156–163.

Flavahan, N. A., Aarhus, L. L., Rimele, T. J. *et al.* (1985). Respiratory epithelium inhibits bronchial smooth muscle tone. *J Appl Physiol* **58**, 834–838.

Florio, C., Styhler, A., Heisler, S. *et al.* (1996). Mechanical responses of tracheal tissue in vitro: dependence on the tissue preparation employed and relationship to smooth muscle content. *Pulm Pharmacol* **9**, 157–166.

Fredberg, J. J., Inouye, D., Miller, B. *et al.* (1997). Airway smooth muscle, tidal stretches, and dynamically determined contractile states. *Am J Respir Crit Care Med* **156**, 1752–1759.

Gil, F. R., Zitouni, N. B., Azoulay, E. *et al.* (2006). Smooth muscle myosin isoform expression and LC20 phosphorylation in innate rat airway hyperresponsiveness. *Am J Physiol Lung Cell Mol Physiol* **291**, L932–L940.

Gunst, S. J. and Stropp, J. Q. (1995). Pressure–volume and length–stress relationships in canine bronchi in vitro. *J Appl Physiol* **64**, 2522–2531.

Hadrava, V., Tremblay, J., Sekaly, R. P. *et al.* (1992). Accelerated entry of aortic smooth muscle cells from spontaneously hypertensive rats into the S phase of the cell cycle. *Biochem Cell Biol* **70**, 599–604.

Henderson, W. R., Tang, L. O., Chu, S. J. *et al.* (2002). A role for cysteinyl leukotrienes in airway remodeling in a mouse asthma model. *Am J Respir Crit Care Med* **165**, 108–116.

Hirshman, C. A., Malley, A. and Downes, H. (1980). Basenji-greyhound dog model of asthma: reactivity to *Ascaris suum*, citric acid, and methacholine. *J Appl Physiol Respir Environ Exerc Physiol* **49**, 953–957.

Hirst, S. J., Twort, C. H. and Lee, T. H. (2000). Differential effects of extracellular matrix proteins on human airway smooth muscle cell proliferation and phenotype. *Am J Respir Cell Mol Biol* **23**, 335–344.

Hjoberg, J., Shore, S., Kobzik, L. *et al.* (2004). Expression of nitric oxide synthase-2 in the lungs decreases airway resistance and responsiveness. *J Appl Physiol* **97**, 249–259.

Jia, Y., Xu, L., Heisler, S. *et al.* (1995). Airways of a hyperresponsive rat strain show decreased relaxant responses to sodium nitroprusside. *Am J Physiol* **269**, L85–91.

Jia, Y., Xu, Turner, L., D. J. *et al.* (1996). Endogenous nitric oxide contributes to strain-related differences in airway responsiveness in rats. *J Appl Physiol* **80**, 404–410.

Jiang, H., Rao, K., Halayko, A. J. *et al.* (1992). Bronchial smooth muscle mechanics of a canine model of allergic airway hyperresponsiveness. *J Appl Physiol* **72**, 39–45.

Jiang, H., Rao, K., Liu, X. *et al.* (1995). Increased Ca^{2+} and myosin phosphorylation, but not calmodulin activity in sensitized airway smooth muscles. *Am J Physiol* **268**, L739–L746.

Kelly, S. M. and Tao, F. C. (1999). Modulation of contractile responses of airway smooth muscle cells by homologous cell substrate. *Am J Respir Crit Care Med* **159**, A259.

Kumar, A., Knox, A. J. and Boriek, A. M. (2003). CCAAT/enhancer-binding protein and activator protein-1 transcription factors regulate the expression of interleukin-8 through the mitogen-activated protein kinase pathways in response to mechanical stretch of human airway smooth muscle cells. *J Biol Chem* **278**, 18868–18876.

Lambert, R. K., Wiggs, B. R., Kuwano, K. *et al.* (1993). Functional significance of increased airway smooth muscle in asthma and COPD. *J Appl Physiol* **74**, 2771–2781.

Lauzon, A. M., Tyska, M. J., Rovner, A. S. *et al.* (1998). A 7-amino-acid insert in the heavy chain nucleotide binding loop alters the kinetics of smooth muscle myosin in the laser trap. *J Muscle Res Cell Motil* **19**, 825–837.

Leguillette, R., Gil, F. R., Zitouni, N. *et al.* (2005). (+)Insert smooth muscle myosin heavy chain (SM-B) isoform expression in human tissues. *Am J Physiol Cell Physiol* **289**, C1277–C1285.

Leigh, R., Ellis, R., Wattie, J. *et al.* (2002). Dysfunction and remodeling of the mouse airway persist after resolution of acute allergen-induced airway inflammation. *Am J Respir Cell Mol Biol* **27**, 526–535.

Levitt, R. C. and Mitzner, W. (1988). Expression of airway hyperreactivity to acetylcholine as a simple autosomal recessive trait in mice. *FASEB J* **2**, 2605–2608.

Ma, X., Cheng, Z., Kong, H. *et al.* (2002). Changes in biophysical and biochemical properties of single bronchial smooth muscle cells from asthmatic subjects. *Am J Physiol Lung Cell Mol Physiol* **283**, L1181–L1189.

Maksym, G. N., Deng, L., Fairbank, N. J. *et al.* (2005). Beneficial and harmful effects of oscillatory mechanical strain on airway smooth muscle. *Can J Physiol Pharmacol* **83**, 913–922.

Martin, J. G., Opazo-Saez, A., Du, T. *et al.* (1992). In vivo airway reactivity: predictive value of morphological estimates of airway smooth muscle [Review]. *Can J Physiol Pharmacol* **70**, 597–601.

McVicker, C. G., Leung, S. Y., Kanabar, V. *et al.* (2007). Repeated allergen inhalation induces cytoskeletal remodeling in smooth muscle from rat bronchioles. *Am J Respir Cell Mol Biol* **36**, 721–727.

Merkel, L., Gerthoffer, W. T. and Torphy, T. J. (1990). Dissociation between myosin phosphorylation and shortening velocity in canine trachea. *Am J Physiol* **258**, C524–C532.

Mitchell, R. W., Ruhlmann, E., Magnussen, H. *et al.* (1994). Passive sensitization of human bronchi augments smooth muscle shortening velocity and capacity. *Am J Physiol* **267**, L218–222.

Mitchell, R. W., Seow, C. Y., Burdyga, T. *et al.* (2001). Relationship between myosin phosphorylation and contractile capability of canine airway smooth muscle. *J Appl Physiol* **90**, 2460–2465.

Moir, L. M., Leung, S. Y., Eynott, P. R. *et al.* (2003a). Repeated allergen inhalation induces phenotypic modulation of smooth muscle in bronchioles of sensitized rats. *Am J Physiol Lung Cell Mol Physiol* **284**, L148–L159.

Moir, L. M., Leung, S. Y., Eynott, P. R. *et al.* (2003b). Repeated allergen inhalation induces phenotypic modulation of smooth muscle in bronchioles of sensitized rats. *Am J Physiol Lung Cell Mol Physiol* **284**, L148–L159.

Moreno, R. H., Hogg, J. C. and Pare, P. D. (1986). Mechanics of airway narrowing. *Am Rev Respir Dis* **133**, 1171–1180.

Moreno, R. H., Lisboa, C., Hogg, J. C. *et al.* (1993). Limitation of airway smooth muscle shortening by cartilage stiffness and lung elastic recoil in rabbits. *J Appl Physiol* **75**, 738–744.

Nagai, R., Kuro-o, M., Babij, P. *et al.* (1989). Identification of two types of smooth muscle myosin heavy chain isoforms by cDNA cloning and immunoblot analysis. *J Biol Chem* **264**, 9734–9737.

Oliver, B. G. and Black, J. L. (2006). Airway smooth muscle and asthma. *Allergol Int* **55**, 215–223.

Opazo-Saez, A., Du, T.,Wang, N. S. *et al.* (1996). Methacholine induced bronchoconstriction and airway smooth muscle in the guinea pig. *J Appl Physiol* **80**, 437–444.

Padrid, P., Snook, S., Finucane, T. *et al.* (1995). Persistent airway hyperresponsiveness and histologic alterations after chronic antigen challenge in cats. *Am J Respir Crit Care Med* **151**, 184–193.

Palmans, E., Kips, J. C. and Pauwels, R. A. (2000). Prolonged allergen exposure induces structural airway changes in sensitized rats. *Am J Respir Crit Care Med* **161**, 627–635.

Panettieri, R. A., Kurray, R. K., Eszterhas, A. J. (1998). Repeated allergen inhalations induce DNA synthesis in airway smooth muscle and epithelial cells in vivo. *Am J Physiol Lung Cell Mol Physiol* **274**, L417–L424.

Pini, L., Martin, J. G., Hamid, Q. *et al.* (2004). Extracellular matrix (ECM) remodeling is more prominent in the peripheral airways of OA-challenged BN rats. *Am J Respir Crit Care Med* **169** (A699), 5–1–2004.

Ravasi, S., Citro, S., Viviani, B. *et al.* (2006). CysLT1 receptor-induced human airway smooth muscle cells proliferation requires ROS generation, EGF receptor transactivation and ERK1/2 phosphorylation. *Respir Res* **7**, 42.

Rovner, A. S., Freyzon, Y. and Trybus, K. M. (1997). An insert in the motor domain determines the functional properties of expressed smooth muscle myosin isoforms. *J Muscle Res Cell Motil* **18**, 103–110.

Salmon, M., Liu, Y. C., Mak, J. C. *et al.* (2000). Contribution of upregulated airway endothelin-1 expression to airway smooth muscle and epithelial cell DNA synthesis after repeated allergen exposure of sensitized brown-Norway rats. *Am J Respir Cell Mol Biol* **23**, 618–625.

Salmon, M., Walsh, D. A., Huang, T. J. *et al.* (1999). Involvement of cysteinyl leukotrienes in airway smooth muscle cell DNA synthesis after repeated allergen exposure in sensitized brown Norway rats. *Br J Pharmacol* **127**, 1151–1158.

Sapienza, S., Du, T., Eidelman, D. H. *et al.* (1991). Structural changes in the airways of sensitized brown Norway rats after antigen challenge. *Am Rev Respir Dis* **144**, 423–427.

Shore, S. A., Bai, T. R., Wang, C. G., *et al.* (1985). Central and local cholinergic components of histamine-induced bronchoconstriction in dogs. *J Appl Physiol* **58**, 443–451.

Smith, B. A., Tolloczko, B., Martin, J. G. *et al.* (2005). Probing the viscoelastic behavior of cultured airway smooth muscle cells with atomic force microscopy: stiffening induced by contractile agonist. *Biophys J* **88**, 2994–3007.

Solway, J. and Fredberg, J. J. (1997). Perhaps airway smooth muscle dysfunction contributes to asthmatic bronchial hyperresponsiveness after all. *Am J Respir Cell Mol Biol* **17**, 144–146.

Stephens, N. L., Halayko, A. and Jiang, H. (1992). Normalization of contractile parameters in canine airway smooth muscle: morphological and biochemical [Review]. *Can J Physiol Pharmacol* **70**, 635–644 [Erratum 1436].

Tao, F. C., Michoud, M. C., Tolloczko, B. *et al.* (1995). Differences in intracellular Ca^{2+} between airway smooth muscle of two inbred strains of rats differing in airway responsiveness. *Am J Respir Crit Care Med* **151**, A285.

Tao, F. C., Shah, S., Pradhan, A. A. *et al.* (2003). Enhanced calcium signaling to bradykinin in airway smooth muscle from hyperresponsive inbred rats. *Am J Physiol Lung Cell Mol Physiol* **284**, L90–L99.

Tao, F. C., Tolloczko, B., Eidelman, D. H. *et al.* (1999). Enhanced Ca(2+) mobilization in airway smooth muscle contributes to airway hyperresponsiveness in an inbred strain of rat. *Am J Respir Crit Care Med* **160**, 446–453.

Tao, F. C., Tolloczko, B., Mitchell, C. A. *et al.* (2000). Inositol (1,4,5)trisphosphate metabolism and enhanced calcium mobilization in airway smooth muscle of hyperresponsive rats. *Am J Respir Cell Mol Biol* **23**, 514–520.

Taylor, S. M., Pare, P. D., Armour, C. L. *et al.* (1985). Airway reactivity in chronic obstructive pulmonary disease. Failure of in vivo methacholine responsiveness to correlate with cholinergic, adrenergic, or nonadrenergic responses in vitro. *Am Rev Respir Dis* **132**, 30–35.

Thyberg, J. (1996). Differentiated properties and proliferation of arterial smooth muscle cells in culture. *Int Rev Cytol* **169**, 183–265.

Tuck, S. A., Maghni, K., Poirier, A. *et al.* (2004). Time course of airway mechanics of the (+)insert myosin isoform knockout mouse. *Am J Respir Cell Mol Biol* **30**, 326–332.

Vargaftig, B. B. and Singer, M. (2003). Leukotrienes mediate part of Ova-induced lung effects in mice via EGFR. *Am J Physiol Lung Cell Mol Physiol* **285**, L808–L818.

Wagers, S., Lundblad, L. K. A., Ekman, M. *et al.* (2004). The allergic mouse model of asthma: normal smooth muscle in an abnormal lung? *J Appl Physiol* **96**, 2019–2027.

Wang, C. G., Almirall, J. J., Dolman, C. S. *et al.* (1997). In vitro bronchial responsiveness in two highly inbred rat strains. *J Appl Physiol* **82**, 1445–1452.

Wang, N., Butler, J. P. and Ingber, D. E. (1993a). Mechanotransduction across the cell surface and through the cytoskeleton. *Science* **260**, 1124–1127.

Wang, C. G., Du, T., Xu, L. J. *et al.* (1993b). Role of leukotriene D4 in allergen-induced increases in airway smooth muscle in the rat. *Am Rev Respir Dis* **148**, 413–417.

White, S., Martin, A. F. and Periasamy, M. (1993). Identification of a novel smooth muscle myosin heavy chain cDNA: isoform diversity in the S1 head region. *Am J Physiol* **264**, C1252–C1258.

Woolcock, A. J., Salome, C. M. and Yan, K. (1984). The shape of the dose-response curve to histamine in asthmatic and normal subjects. *Am Rev Respir Dis* **130**, 71–75.

Wong, J. Z., Woodcock-Mitchell, J., Mitchell, J. *et al.* (1998). Smooth muscle actin and myosin expression in cultured airway smooth muscle cells. *Am J Physiol* **274**, L786–L792.

Yager, D., Kamm, R. D. and Drazen, J. M. (1995). Airway wall liquid. Sources and role as an amplifier of bronchoconstriction [Review]. *Chest* **107**, 105S-110S.

Zacour, M. E. and Martin, J. G. (1996). Enhanced growth response of airway smooth muscle in inbred rats with airway hyperresponsiveness. *Am J Respir Cell Mol Biol* **15**, 590–599.

Zacour, M. E. and Martin, J. G. (2000). Protein kinase C is involved in enhanced airway smooth muscle cell growth in hyperresponsive rats. *Am J Physiol Lung Cell Mol Physiol* **278**, C59–L67.

10

Altered properties of airway smooth muscle in asthma

Judith Black, Janette Burgess, Brian Oliver and Lyn Moir

Discipline of Pharmacology, University of Sydney, Woolcock Institute of Medical Research, Sydney, Australia

10.1 Introduction

Our understanding of the biology and pharmacology of the airway smooth muscle (ASM) cell, in both health and disease, has increased exponentially over the last two decades. This is due to a number of factors, including the availability of techniques to culture ASM (Panettieri *et al.*, 1989), as well as access to newer technologies, such as laser capture microdissection (Burgess *et al.*, 2003a) and gene microarrays (Woodruff *et al.*, 2004). Coincident with these technological advances has been the realization that the importance of the ASM is not limited to its ability to contract and relax in response to provoking stimuli and bronchodilators. The ASM cell's properties of proliferation, cytokine and growth factor production, expression of cell-surface molecules, and communication with the extracellular matrix (ECM) and inflammatory cells have redefined its role in asthma. Whereas the contraction of ASM has been regarded as an event secondary to the presence of inflammatory cells and mediators, it is now accepted that the cell itself can

Airway Smooth Muscle   Edited by Kian Fan Chung
© 2008 John Wiley & Sons, Ltd

produce all these factors (Hirst, 2003), raising the possibility of abnormalities issuing from the muscle cell itself.

Recently, the ability to culture ASM from biopsies derived from asthmatic volunteers has become a reality (Chan *et al.*, 2006; Johnson *et al.*, 2001), and these studies have revealed important differences in the *in vitro* behaviour of ASM cells derived from asthmatic and non-asthmatic volunteers. The first of these was the demonstration that asthmatic cells proliferate more rapidly than their non-asthmatic counterparts (Johnson *et al.*, 2001). Subsequently, it was found that, although glucocorticoids can inhibit proliferation of non-asthmatic cells, they fail to do so in asthmatic cell cultures (Roth *et al.*, 2004). The mechanism underlying this finding proved to be an absence of the transcription factor CCAAT enhancer binding protein alpha (C/EBP-α). This observation was consistent with the fact that C/EBP-α is an essential component of the complex formed with the activated glucocorticoid receptor, which subsequently inhibits proliferation via upregulation of p21[waf1cip1] (Roth *et al.*, 2002). Several other differences in *in vitro* properties of the two cell types have since emerged: release of more connective tissue growth factor (CTGF) (Burgess *et al.*, 2003a); alterations in the profile and signalling pathways of ECM proteins (Burgess *et al.*, 2006b); decreased release of prostaglandin E_2 (PGE $_2$) (Chambers *et al.*, 2003); greater release of CXCL10 (IP10) (Brightling *et al.*, 2005), which then attracts more mast cells to the muscle (Brightling *et al.*, 2002); and differences in response to rhinovirus exposure (Oliver and Black, 2006). Details of these differences are expanded upon below. A unifying explanation for these differences that could relate them to the deficiency in C/EBP-α is as yet unavailable, but they suggest that asthma is intrinsically and essentially an abnormality of ASM. What remains to be seen is whether ablation of the smooth muscle, which is currently under investigation in trials of bronchial thermoplasty (Cox *et al.*, 2006), will add credence to this possibility.

10.2 The extracellular matrix (ECM) and the airway smooth muscle (ASM)

Within the structure of the airway, the ASM is associated with a complex network of interlacing macromolecules – the ECM. The ECM forms a supporting structure for the airway wall, but also has the potential to influence cellular functions, including proliferation, differentiation and migration. The ECM proteins are mainly interstitial collagens, glycoproteins such as fibronectin, laminin and tenascin, elastin, and proteoglycans. The profile of ECM proteins is altered in the airways of asthmatic individuals compared to non-asthmatic individuals. Histological studies have demonstrated increased deposition of collagens I, III, and V; fibronectin; tenascin; hyaluronan; versican; laminin $\alpha 2/\beta 2$; lumican and biglycan (Laitinen *et al.*, 1996; 1997; Roberts and Burke, 1998; Roche *et al.*, 1989), while the levels of collagen IV and elastin are decreased (Bousquet *et al.*, 1992) in asthmatic airways.

A variety of cells, including those from the fibroblast/myofibroblast lineage and ASM cells, have the capacity to produce ECM proteins in the airways. Only a small number of histological studies have examined the alterations in ECM proteins specifically in the area of the ASM in asthmatic airways. Bai and colleagues observed an increase in the amount of total ECM around individual ASM cells in fatal asthma cases (Bai *et al.*, 2000), while others have reported an increase in collagen (Thompson and Schellenberg, 1998), hyaluronan and versican (Roberts and Burke, 1998) in asthma. A disintegrin and metalloproteinase 33 (ADAM33) has also recently been reported to be expressed to a greater extent in the muscle layer in biopsies from asthmatic individuals than from non-asthmatics (Ito *et al.*, 2007). The production and release of ECM proteins from ASM cells has been examined *in vitro*. We have recently reported that ASM cells from asthmatic individuals release a profile of ECM proteins different from those of non-asthmatic individuals (Johnson *et al.*, 2004). The production of perlecan and collagen I by asthmatic ASM cells was significantly increased, while laminin $\alpha 1$ and collagen IV were decreased, and chondroitin sulphate was detected only in cells from non-asthmatic individuals. Chan and colleagues reported that the production of fibronectin was also increased in asthmatic ASM cells (Chan *et al.*, 2006).

There is a unique relationship between ASM cells and the ECM, with each able to influence the characteristics and behaviour of the other. The ECM released from asthmatic ASM cells has the capacity to alter the behaviour of non-asthmatic ASM cells. When non-asthmatic ASM cells were grown on ECM isolated from asthmatic ASM cells, they exhibited a greater rate of proliferation (Figure 10.1). Asthmatic cells also had a greater rate of proliferation on asthmatic ECM than non-asthmatic ECM (Burgess *et al.*, 2005). Similarly, when both asthmatic and non-asthmatic ASM cells were plated on asthmatic ECM, the release of eotaxin induced by interleukin (IL)-13 was enhanced compared to plating on non-asthmatic ECM (Chan *et al.*, 2006) (Figure 10.2). In all circumstances tested, the release of eotaxin was greater from the asthmatic ASM cells. These studies provide evidence that the alteration in the ECM protein profile in asthmatic airways contributes to the behaviour and characteristics of ASM cells, which in turn contribute to the altered pathophysiology observed in the asthmatic airway.

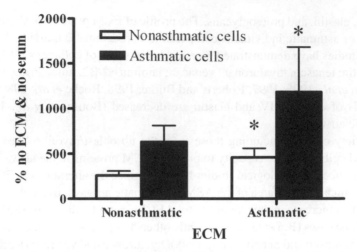

Figure 10.1 Extracellular matrix (ECM) derived from asthmatic ASM cells induces greater proliferation. Non-asthmatic and asthmatic ASM cells were grown on ECM derived from non-asthmatic or asthmatic ASM cells in the presence of 5 per cent fetal bovine serum (FBS) for 3 days ($n = 9$ and 9, respectively). Data are expressed as the percentage of proliferation compared to cells grown in the absence of ECM or FBS $\pm$ SEM. *Significantly greater than non-asthmatic ECM

Transforming growth factor β

The production of ECM proteins from ASM cells is regulated by growth factors present in the airways. Transforming growth factor (TGF) β is increasingly implicated in profibrotic events in the airway, and has been implicated in airway remodelling in asthma (Coutts *et al.*, 2001; McKay *et al.*, 1998; Redington *et al.*, 1997). TGFβ is produced by a range of cells, including platelets, macrophages, epithelial cells and smooth muscle cells. In asthma it has been reported that the concentration of TGFβ is increased in bronchial lavage fluid (Redington *et al.*, 1997) and that TGFβ gene expression is increased in bronchial tissue (Minshall *et al.*, 1997; Ohno *et al.*, 1996; Vignola *et al.*, 1997). TGFβ immunoreactivity is also increased in bronchial biopsies and submucosal eosinophils from asthmatic subjects (Minshall *et al.*, 1997; Vignola *et al.*, 1997). ASM-derived TGFβ localizes in the vicinity of the ASM cells through an association with the TGFβ latency binding protein, which is also expressed by ASM cells. Plasmin cleaves this complex to release the biologically active TGFβ, which in turn can act in an autocrine manner to induce ECM protein production by ASM cells (Coutts *et al.*, 2001). *In vitro*, TGFβ induces the production and release of collagen I, fibronectin and

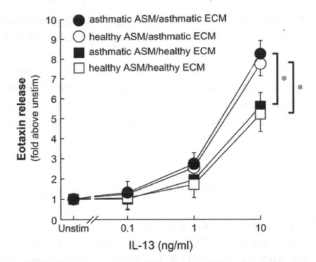

Figure 10.2 Native extracellular matrix (ECM) derived from asthmatic ASM cells enhances IL-13-dependent eotaxin release at 24 h from ASM cells cultured from either healthy control subjects or subjects with asthma. Data are shown as the fold increase above unstimulated and are mean ± SEM of duplicate values from independent experiments using cells cultured from six subjects, which were seeded on ECM substrates from at least three subjects. Baseline values for eotaxin (ng/ml/million cells) were 16.35 ± 12.24 (*open squares*), 38.07 ± 18.22 (*closed squares*), 20.89 ± 14.27 (*open circles*), and 38.35 ± 20.89 (*closed circles*). *$p < 0.05$ compared with ECM from healthy subjects. (Reprinted from the American Journal of Respiratory and Critical Care Medicine, Vol 174. pp. 379–385, (2006). © American Thoracic Society)

versican from both asthmatic and non-asthmatic ASM cells (Burgess *et al.*, 2006c; Johnson *et al.*, 2006).

Many of the effects of TGFβ in the airways, including induction of cell proliferation, ECM synthesis, and regulation of migration of cells to a wound site are, at least in part, mediated via connective tissue growth factor (CTGF) (Duncan *et al.*, 1999; Igarashi *et al.*, 1993; Oemar and Luscher, 1997).

Connective tissue growth factor (CTGF)

CTGF is thought to have a role in normal wound healing processes (Blom *et al.*, 2001; Igarashi *et al.*, 1993; Pawar *et al.*, 1995), and its over-expression has been associated with many fibrotic disease states (Clarkson *et al.*, 1999; Grotendorst, 1997; Igarashi *et al.*, 1996).

We, and others, have recently reported that CTGF is produced by and released from ASM cells (Burgess *et al.*, 2003b; Xie *et al.*, 2005). The levels of CTGF released following stimulation with TGFβ are significantly greater in cells isolated from asthmatic individuals than in those from non-asthmatic individuals (Burgess et al., 2003b).

The presence of both CTGF mRNA and protein has been demonstrated in ASM cells in vitro, and the presence of CTGF mRNA has been confirmed in ASM cells obtained from tissue sections by laser capture microdissection (Burgess et al., 2003b). TGFβ also induces CTGF expression in bronchial rings from non-asthmatic individuals, but whether the same effect occurs in asthmatic bronchial rings, or even whether a larger induction occurs, is not known due to difficulty in accessing tissue for these types of experiments.

CTGF induces the production and release of the ECM proteins collagen I and fibronectin from both asthmatic and non-asthmatic ASM cells in culture, and also in bronchial rings from non-asthmatic individuals (Johnson et al., 2006). This induction mediates, in part, the TGFβ-induced collagen I and fibronectin release from ASM cells, although other CTGF-independent mechanisms also contribute to this release. PGE_2 regulates the TGFβ-induced expression of CTGF, collagen I and fibronectin in ASM cells. The presence of PGE_2 inhibits the induction of the mRNA for these genes in ASM cells in culture and the deposition of the proteins in bronchial rings following TGFβ stimulation. (Burgess et al., 2006a). These studies link the TGFβ-mediated upregulation of CTGF with a corresponding upregulation in fibronectin and collagen I production in both asthmatic and non-asthmatic ASM cells.

Signalling pathways

The signalling events leading to the induction of CTGF, collagen I and fibronectin by TGFβ are different in ASM cells. The involvement of the phosphoinositol-3 kinase (PI3K), extracellular regulated kinase (ERK) or the p38 MAPK (p38) pathways differ between the proteins, but there seems to be an increased dependence of the PI3K pathway in asthmatic ASM cells (Table 10.1) (Johnson et al., 2006). It is possible that the SMAD pathway is also involved in the induction of these genes, as TGFβ induces SMAD2 phosphorylation in both asthmatic and the non-asthmatic ASM cells; however, the involvement of this signalling pathway has not been directly examined. Asthmatic ASM cells may potentially have a greater reliance on the SMAD signalling pathway, as the phosphorylation of SMAD2 following TGFβ stimulation was significantly increased in these cells compared to the non-asthmatic ASM cells (Johnson et al., 2006). It is interesting to note the importance of the PI3K pathway in asthmatic ASM cells. In this

Table 10.1 Signalling events leading to the induction of CTGF, collagen I and fibronectin in ASM cells

	Asthmatic			Non-asthmatic		
	MAPK	P38	PI3K	MAPK	P38	PI3K
CTGF	✓	×	✓	×	✓	×
Collagen I	×	×	✓	×	✓	✓
Fibronectin	×	✓	✓	×	×	✓

× not involved; ✓ involved in induction of mRNA.

study, two inhibitors, LY294002 and wortmannin, were used to block the PI3K pathway. Differing results were obtained with the two inhibitors, suggesting that there may be specific roles for the different subclasses of PI3K in the asthmatic ASM cells. Further studies are being conducted to elucidate the importance of the PI3K subclass isoforms in the regulation of events in the asthmatic ASM cells.

Vascular endothelial growth factor

TGFβ also induces the release of vascular endothelial growth factor (VEGF) from ASM cells and a variety of other cells (Burgess *et al.*, 2006a; Kazi *et al.*, 2004; Knox *et al.*, 2001; Wen *et al.*, 2003). VEGF is an important regulator of endothelial cell growth and is thought to play a role in controlling the angiogenesis that is observed in the asthmatic airway. VEGF levels in bronchial lavage fluid and induced sputum are higher in asthmatics than non-asthmatics (Asai *et al.*, 2003; Lee and Lee, 2001), and VEGF gene expression is also increased in bronchial tissue in asthmatics (Hoshino *et al.*, 2001). We have recently reported that CTGF can also induce VEGF release from asthmatic and non-asthmatic ASM cells, although the levels do not differ between the cell types (Burgess *et al.*, 2006a).

Recombinant CTGF and $VEGF_{165}$ are known to form a complex which alters the functionality of VEGF (Inoki *et al.*, 2002). CTGF and VEGF released from ASM cells also associate and co-localize in the ECM surrounding ASM cells in culture or in bronchial rings after stimulation with TGFβ (Burgess *et al.*, 2006a). The localization of this complex is dependent on the presence of CTGF, as PGE_2 increases the release of VEGF into the cell culture medium in which the ASM cells are grown but inhibits the localization of CTGF and VEGF to the ECM

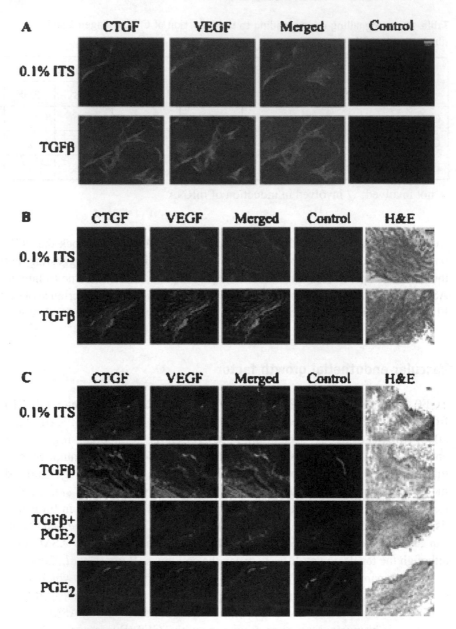

Figure 10.3   Reproduced with permission from Am J Physiol Lung Cell Mol Physiol, Jan 2006; 290: L153–L161. (For a colour reproduction of this figure, please see the colour section, located towards the centre of the book).

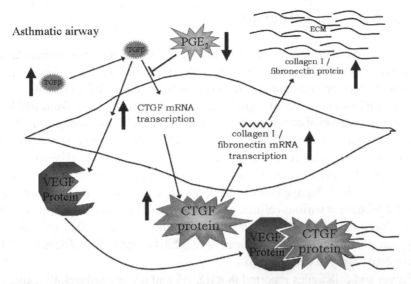

Figure 10.4   Schematic representing the relationship between the ASM cell, the extracellular matrix (ECM), and growth factors in the asthmatic airway

(Figure 10.3). Asthmatic and non-asthmatic ASM cells release the same amount of VEGF following TGFβ stimulation; however, it is not known whether the increased release of CTGF and the decreased release of PGE$_2$ from asthmatic ASM results in a greater amount of VEGF localized in the vicinity of ASM. Figure 10.4 summarizes the relationship of the factors described above in the asthmatic airway.

Figure 10.3   Colocalization of CTGF and VEGF and modulation by TGF-β and PGE$_2$ in ASM cells and tissue sections. ASM cells (*A*) were incubated in 0.1 per cent ITS DMEM or 0.1 per cent ITS DMEM plus TGF-β (1 ng/ml) or human bronchial rings from the same patient, or were incubated in 0.1 per cent ITS DMEM or 0.1 per cent ITS DMEM plus TGF-β (1 ng/ml) (*B*) or 0.1 per cent ITS DMEM, 0.1 per cent ITS DMEM plus TGF-β (1 ng/ml), 0.1 per cent ITS DMEM plus TGF-β plus PGE$_2$ (10 μM) or 0.1 per cent ITS DMEM plus PGE$_2$ (*C*) for 24 h. Cells and tissue sections were simultaneously stained with rabbit anti-CTGF coupled with donkey antirabbit TRITC (red staining) and mouse antihuman VEGF coupled with goat antimouse FITC (green staining). The images were merged by imaging software, and regions of colocalization were identified (yellow staining). Controls are representative of the staining seen with both isotype control antibodies. The haematoxylin and eosin (H&E)-stained sequential sections were used to identify airway morphology. The results are representative of tissue from three non-asthmatic patients.(Reproduced with permission from Am J Physiol Lung Cell Mol Physiol, Jan 2006; 290: L153–L161)

10.3 ASM and integrins

Cell–ECM interactions are mediated via a family of cell-surface receptors known as integrins (Giancotti and Ruoslahti, 1999). Integrins are heterodimeric transmembrane glycoproteins consisting of two non-covalently linked subunits, α and β, which allow bidirectional signalling; that is, they can signal from inside the cell (e.g., regulation of attachment of the cell to the ECM) or from outside the cell (e.g., signals are transmitted from the extracellular environment to the cell), and hence the extracellular environment can modulate cell function (Hynes, 2002).

Human ASM cells in culture express the integrin subunits $\alpha 1$-$\alpha 6$, αV and $\beta 1$ (Freyer *et al.*, 2001; Nguyen *et al.*, 2005), and these receptors have been implicated in the regulation of various cellular processes, including cytokine secretion (Chan *et al.*, 2006; Peng *et al.*, 2005), cell survival (Freyer *et al.*, 2001), cell adhesion (Nguyen *et al.*, 2005; Parameswaran *et al.*, 2004), and cell proliferation (Moir *et al.*, 2005; Nguyen *et al.*, 2005).

Freyer and colleagues reported that $\alpha 5$, $\beta 1$ and αV are universally expressed on human ASM cells but that fewer than one-third of cells express $\alpha 1$, $\alpha 3$ and $\alpha 4$ (Freyer *et al.*, 2001), whereas Nguyen *et al.* (2005) reported universal expression of $\alpha 5$ and $\beta 1$, but that $\alpha 1$, $\alpha 2$, $\alpha 3$ and αV are expressed on 50-70 per cent of cells. Although the reason for the differences in expression levels between these two studies is not clear, exposure of ASM cells to mediators and ECM proteins known to be increased in the asthmatic airway can modulate integrin expression; platelet derived growth factor (PDGF), fibronectin and collagen I increase $\alpha 3$, TNFα increases $\beta 1$, and TGFβ_1 increases $\alpha 5 \beta 1$ expression (Moir *et al.*, 2006; Nguyen *et al.*, 2005; Parameswaran *et al.*, 2004), suggesting that differences in expression may reflect altered environmental factors. Expression of individual subunits may therefore be modulated by the inflammatory and profibrotic environment within the airway; hence, transient expression of integrin subunits may allow altered cellular function.

We have preliminary data suggesting that integrin expression on proliferating asthmatic ASM cells in culture is different from that of non-asthmatic ASM cells (unpublished observations) and that different integrins contribute to the proliferation of ASM cells from asthmatic subjects (Moir *et al.*, 2005). Nguyen and colleagues reported that the $\beta 1$-family integrins $\alpha 2 \beta 1$, $\alpha 4 \beta 1$ and $\alpha 5 \beta 1$ mediate the ECM-enhanced mitogen-induced proliferation of non-asthmatic ASM cells (Nguyen *et al.*, 2005). However, we found that blocking antibodies to $\beta 1$ partially attenuated serum-induced proliferation of both non-asthmatic and asthmatic ASM cells, and that blocking antibodies to $\alpha 4$ and $\beta 1$ in combination further reduced proliferation of non-asthmatic ASM cells, but had no additional effect on asthmatic cells (Moir *et al.*, 2005). Since integrin receptors have been shown to have a functional role in many different cellular processes, altered expression of these receptors in the asthmatic airway may have a dramatic effect on cellular function, and hence perpetuate both the inflammatory and structural remodelling processes associated with asthma.

Role of integrins in current asthma therapies

Current asthma therapies have been shown to have little or no effect on the structural changes observed in the asthmatic airway. Glucocorticoids, which are effective at reducing proliferation of non-asthmatic ASM cells in culture, are ineffective at reducing the enhanced proliferation of cells from asthmatic subjects (Roth *et al.*, 2004). A recent study by Bonacci and colleagues reported that the antiproliferative action of glucocorticoids on non-asthmatic ASM cells is lost when cells are cultured on collagen I, and that blocking cell-collagen I interaction via the integrin $\alpha 2\beta 1$ restores this antimitogenic response (Bonacci *et al.*, 2006). Therefore, the altered matrix expression in asthma may modulate the actions of asthma medications.

10.4 The ASM cell and inflammation

GM-CSF and RANTES released from human smooth muscle cells in culture was first documented in 1997 (John *et al.*, 1997; Saunders *et al.*, 1997), and subsequently these cells have been shown to produce a plethora of cytokines, chemokines and other inflammatory mediators, such as IL-1β, IL-5, IL-6, IL-8, IL-11, MCP-1, MCP-2, MCP-3, IFN-β, eotaxin, leukaemia inhibitory factor, and prostanoids such as PGE$_2$ (Pascual *et al.*, 2003). The induction of many of these factors is the result of stimulation with pro-inflammatory cytokines, and as such has helped to establish the role of the smooth muscle cell in the airway inflammatory milieu. The ability to respond to inflammatory mediators is dependent upon the presence of the cytokine and chemokine receptors on the surface of these cells, suggesting that both autocrine and paracrine signalling control ASM function *in vivo*. Furthermore, it is likely that the smooth muscle cell functions as a cell of the innate immune system, since the presence of a number of innate pattern recognition receptors (PRRs) has been demonstrated (Sukkar *et al.*, 2006), and synergistic cooperation between ASM cells and monocytes has been shown to occur following PRRs activation (Morris *et al.*, 2005). ASM cells from non-asthmatics cannot function as antigen-presenting cells (APC), a surprising finding given that they can express MHC class II (Lazaar *et al.*, 1997), and the co-stimulatory molecules CD80 and CD86 (Hakonarson *et al.*, 2001). However, alveolar macrophages from asthmatic volunteers can function as APC (Balbo *et al.*, 2001), but it remains to be determined whether the same is true of asthmatic muscle cells. Smooth muscle cells can bind many immune cells such as eosinophils (Hughes *et al.*, 2000), T cells (Lazaar *et al.*, 1994) and mast cells (Liu *et al.*, 2005).

The interaction between the smooth muscle cell and other immune cells is mediated by constitutively expressed cell-surface receptors, such as intercellular adhesion molecule (ICAM)-1 and vascular cell adhesion molecule (VCAM)-1, to which the activated T cells bind (Lazaar *et al.*, 1994). It is also possible that smooth muscle–T cell interactions are mediated via CD44, which is present upon the surface of both cells. Hyaluronan is thought to act as a cellular bridge binding to its cell-surface receptor CD44 on both the smooth muscle cell and the T cell (Lazaar *et al.*, 1994), inducing the proliferation of the smooth muscle cell. Other potential agents regarding the interaction between T cells and smooth muscle cells in orchestrating the immune response include CD40 (Lazaar *et al.*, 1998) and OX40 ligand (Burgess *et al.*, 2004), both of which have been shown to be present and functional upon the surface of the smooth muscle cell. In the presence of inflammatory stimuli, the expression of both CD40 and OX40 ligand is increased on the surface of asthmatic ASM cells compared with non-asthmatic cells *in vitro* (Burgess *et al.*, 2005).

Given the recent observation that mast cell localization within smooth muscle bundles is increased in asthma *in vivo* (Brightling *et al.*, 2002), considerable effort has been made to detect the mechanism by which this occurs. Increased mast cell numbers may be partly due to the elevated chemokine and cytokine production observed in asthma, in combination with smooth muscle-derived stem cell factor (SCF) (Kassel *et al.*, 1999). It is also likely that binding of the chemokine CXCL10 (IP-10), which is preferentially expressed in the muscle of asthmatic patients and *ex vivo* asthma-derived smooth muscle cells, to its receptor CXCR3, which has also been shown to be preferentially expressed on mast cells located in the muscle, contributes to the increased accumulation of the mast cells observed in the asthmatic ASM (Brightling *et al.*, 2005). The adhesion of the mast cell to the smooth muscle cells is mediated to some extent by tumour suppressor in lung cancer-1 (TSLC1), but not by ICAM-1, VCAM-1, CD18, α4 or β1 integrins

(Yang *et al.*, 2006). p38 MAPK-dependent eotaxin release occurs upon binding of the mast cell to smooth muscle cells (Liu *et al.*, 2005). Furthermore, mast cells situated within the ASM bundles of asthmatic subjects produce the TH_2 cytokines IL-4 and IL-13 (Brightling *et al.*, 2003). It is thought that IL-4 and IL-13 contribute to the pathogenesis of asthma; in addition, they have been shown to have direct effects upon ASM cells, such as the induction of VEGF (Faffe *et al.*, 2006) and induction of bronchial hyperreactivity (Kellner *et al.*, 2007). Furthermore, asthma severity correlates with the *in vivo* degranulation of mast cells, and the increased degranulation observed in cartilaginous versus membranous bronchioles suggests that an inhaled stimulus is activating these cells (Carroll *et al.*, 2002).

10.5 The ASM cell and infection

There is increasing evidence that pathogens can induce asthma exacerbations, of which at least 70 per cent are associated with viral infection. While viruses are accepted as pathogens in the lungs, disagreement surrounds the issue of whether ASM can become infected *in vivo* with virus. Previous studies have shown that both submucosal (Papadopoulos *et al.*, 2000) and smooth muscle cells (Morbini and Arbustini, 2001) can be infected with viruses *in vivo*. Of the many different virus types which have been isolated from individuals experiencing exacerbations, rhinovirus is by far the most common type found; furthermore, infection of ASM cells has been demonstrated *in vitro* (Grunstein *et al.*, 2000). We, as well as others, have shown that rhinovirus induces the release of pro-inflammatory cytokines from ASM cells; however, we were the first to demonstrate greater rhinovirus-induced cytokine release in smooth muscle cells from asthmatics, thereby providing a potential mechanism by which rhinovirus-induced exacerbations occur (Hakonarson *et al.*, 1999).

During a virus-induced asthma exacerbation, the clinical utility of beta adrenoceptor agonists is reduced (Reddel *et al.*, 1999), leading to the hypothesis that viral infection induces dysfunctional smooth muscle beta adrenoceptor expression. Some preliminary evidence exists in support of this, derived from research with airway tissue segments (Hakonarson *et al.*, 1998), and recently beta adrenoceptor downregulation has been shown to occur in ASM (Moore *et al.*, 2006), while increased expression was shown to occur in epithelial cells (Tsutsumi *et al.*, 1999).

References

Asai, K., Kanazawa, H., Kamoi, H. *et al.* (2003). Increased levels of vascular endothelial growth factor in induced sputum in asthmatic patients. *Clin Exp Allergy* **33**, 595–599.

Bai, T. R., Cooper, J., Koelmeyer, T. *et al.* (2000). The effect of age and duration of disease on airway structure in fatal asthma. *Am J Respir Crit Care Med* **162**, 663–669.

Balbo, P., Silvestri, M., Rossi, G. A. *et al.* (2001). Differential role of CD80 and CD86 on alveolar macrophages in the presentation of allergen to T lymphocytes in asthma. *Clin Exp Allergy* **31**, 625–636.

Blom, I. E., van Dijk, A. J., Wieten, L. *et al.* (2001). In vitro evidence for differential involvement of CTGF, TGF beta, and PDGF-BB in mesangial response to injury. *Nephrol Dial Transplant* **16**, 1139–1148.

Bonacci, J. V., Schuliga, M., Harris, T. *et al.* (2006). Collagen impairs glucocorticoid actions in airway smooth muscle through integrin signalling. *Br J Pharmacol* **149**, 365–373.

Bousquet, J., Chanez, P., Lacoste, J. Y. *et al.* (1992). Asthma: a disease remodeling the airways. *Allergy* **47**, 3–11.

Brightling, C. E., Ammit, A. J., Kaur, D. *et al.* (2005). The CXCL10/CXCR3 axis mediates human lung mast cell migration to asthmatic airway smooth muscle. *Am J Respir Crit Care Med* **171**, 1103–1108.

Brightling, C. E., Bradding, P., Symon, F. A. *et al.* (2002). Mast-cell infiltration of airway smooth muscle in asthma. *N Engl J Med* **346**, 1699–1705.

Brightling, C. E., Symon, F. A., Holgate, S. T. *et al.* (2003). Interleukin-4 and -13 expression is co-localized to mast cells within the airway smooth muscle in asthma. *Clin Exp Allergy* **33**, 1711–1716.

Burgess, J. K., Blake, A. E., Boustany, S. *et al.* (2005). CD40 and OX40 ligand are increased on stimulated asthmatic airway smooth muscle. *J Allergy Clin Immunol* **115**, 302–308.

Burgess, J. K., Carlin, S., Pack, R. A. *et al.* (2004). Detection and characterization of OX40 ligand expression in human airway smooth muscle cells: a possible role in asthma? *J Allergy Clin Immunol* **113**, 683–689.

Burgess, J. K., Ge, Q., Poniris, M. H. *et al.* (2006a). Connective tissue growth factor and vascular endothelial growth factor from airway smooth muscle interact with the extracellular matrix. *Am J Physiol Lung Cell Mol Physiol* **290**, L153–161.

Burgess, J. K., Johnson, P. R., Ge, Q. *et al.* (2003a). Expression of connective tissue growth factor in asthmatic airway smooth muscle cells. *Am J Respir Crit Care Med* **167**, 71–77.

Burgess, J. K., Johnson, P. R., Ge, Q. *et al.* (2003b). Expression of connective tissue growth factor in asthmatic airway smooth muscle cells. *Am J Respir Crit Care Med* **167**, 71–77.

Burgess, J. K., Oliver, B. G., Poniris, M. H. *et al.* (2006b). A phosphodiesterase 4 inhibitor inhibits matrix protein deposition in airways in vitro. *J Allergy Clin Immunol* **118**, 649–657.

Burgess, J. K., Oliver, B. G. G., Poniris, M. H. *et al.* (2006c). A phosphodiesterase 4 inhibitor inhibits matrix protein deposition in airways in vitro. *J Allergy Clin Immunol* **118**, 649–657.

Carroll, N. G., Mutavdzic, S. and James, A. L. (2002). Distribution and degranulation of airway mast cells in normal and asthmatic subjects. *Eur Respir J* **19**, 879–885.

Chambers, L. S., Black, J. L., Ge, Q. *et al.* (2003). PAR-2 activation, PGE_2, and COX-2 in human asthmatic and nonasthmatic airway smooth muscle cells. *Am J Physiol Lung Cell Mol Physiol* **285**, L619–627.

Chan, V., Burgess, J. K., Ratoff, J. C. *et al.* (2006). Extracellular matrix regulates enhanced eotaxin expression in asthmatic airway smooth muscle cells. *Am J Respir Crit Care Med* **174**, 379–385.

Clarkson, M. R., Gupta, S., Murphy, M. *et al.* (1999). Connective tissue growth factor: a potential stimulus for glomerulosclerosis and tubulointerstitial fibrosis in progressive renal disease. *Curr Opin Nephrol Hypertens* **8**, 543–548.

Coutts, A., Chen, G., Stephens, N. *et al.* (2001). Release of biologically active TGF-beta from airway smooth muscle cells induces autocrine synthesis of collagen. *Am J Physiol Lung Cell Mol Physiol* **280**, L999–1008.

Cox, G., Miller, J. D., McWilliams, A. *et al.* (2006). Bronchial thermoplasty for asthma. *Am J Respir Crit Care Med* **173**, 965–969.

Duncan, M. R., Frazier, K. S., Abramson, S. *et al.* (1999). Connective tissue growth factor mediates transforming growth factor beta-induced collagen synthesis: down-regulation by cAMP. *FASEB J* **13**, 1774–1786.

Faffe, D. S., Flynt, L., Bourgeois, K. *et al.* (2006). Interleukin-13 and interleukin-4 induce vascular endothelial growth factor release from airway smooth muscle cells: role of vascular endothelial growth factor genotype. *Am J Respir Cell Mol Biol* **34**, 213–218.

Freyer, A. M., Johnson, S. R. and Hall, I. P. (2001). Effects of growth factors and extracellular matrix on survival of human airway smooth muscle cells. *Am J Respir Cell Mol Biol* **25**, 569–576.

Giancotti, F. G. and Ruoslahti, E. (1999). Integrin signaling. *Science* **285**, 1028–1032.

Grotendorst, G. R. (1997). Connective tissue growth factor: a mediator of TGF-beta action on fibroblasts. *Cytokine Growth Factor Rev* **8**, 171–179.

Grunstein, M. M., Hakonarson, H., Maskeri, N. *et al.* (2000). Autocrine cytokine signaling mediates effects of rhinovirus on airway responsiveness. *Am J Physiol Lung Cell Mol Physiol* **278**, L1146–L1153.

Hakonarson, H., Carter, C., Maskeri, N. *et al.* (1999). Rhinovirus-mediated changes in airway smooth muscle responsiveness: induced autocrine role of interleukin-1beta. *Am J Physiol* **277**, L13–L21.

Hakonarson, H., Kim, C., Whelan, R. *et al.* (2001). Bi-directional activation between human airway smooth muscle cells and T lymphocytes: role in induction of altered airway responsiveness. *J Immunol* **166**, 293–303.

Hakonarson, H., Maskeri, N., Carter, C. *et al.* (1998). Mechanism of rhinovirus-induced changes in airway smooth muscle responsiveness. *J Clin Invest* **102**, 1732–1741.

Hirst, S. J. (2003). Regulation of airway smooth muscle cell immunomodulatory function: role in asthma. *Respir Physiol Neurobiol* **137**, 309–326.

Hoshino, M., Nakamura, Y. and Hamid, Q. A. (2001). Gene expression of vascular endothelial growth factor and its receptors and angiogenesis in bronchial asthma. *J Allergy
Clin Immunol* **107**, 1034–1038.

Hughes, J. M., Arthur, C. A., Baracho, S. *et al.* (2000). Human eosinophil–airway smooth
muscle cell interactions. *Mediators Inflamm* **9**, 93–99.

Hynes, R. O. (2002). Integrins: bidirectional, allosteric signaling machines. *Cell* **110**,
673–687.

Igarashi, A., Nashiro, K., Kikuchi, K. *et al.* (1996). Connective tissue growth factor gene
expression in tissue sections from localized scleroderma, keloid, and other fibrotic skin
disorders. *J Invest Dermatol* **106**, 729–733.

Igarashi, A., Okochi, H., Bradham, D. M. *et al.* (1993). Regulation of connective tissue
growth factor gene expression in human skin fibroblasts and during wound repair.
Mol Cell Biol **4**, 637–645.

Inoki, I., Shiomi, T., Hashimoto, G. *et al.* (2002). Connective tissue growth factor binds
vascular endothelial growth factor (VEGF) and inhibits VEGF-induced angiogenesis.
FASEB J **16**, 219–221.

Ito, I., Laporte, J. D., Fiset, P. O. *et al.* (2007). Downregulation of a disintegrin and
metalloproteinase 33 by IFN-gamma in human airway smooth muscle cells. *J Allergy
Clin Immunol* **119**, 89–97.

John, M., Hirst, S. J., Jose, P. J. *et al.* (1997). Human airway smooth muscle cells express
and release RANTES in response to T helper 1 cytokines: regulation by T helper 2
cytokines and corticosteroids. *J Immunol* **158**, 1841–1847.

Johnson, P. R., Burgess, J. K., Ge, Q. *et al.* (2006). Connective tissue growth factor induces
extracellular matrix in asthmatic airway smooth muscle. *Am J Respir Crit Care Med*
173, 32–41.

Johnson, P. R., Burgess, J. K., Underwood, P. A. *et al.* (2004). Extracellular matrix proteins modulate asthmatic airway smooth muscle cell proliferation via an autocrine
mechanism. *J Allergy Clin Immunol* **113**, 690–696.

Johnson, P. R., Roth, M., Tamm, M. *et al.* (2001). Airway smooth muscle cell proliferation
is increased in asthma. *Am J Respir Crit Care Med* **164**, 474–477.

Kassel, O., Schmidlin, F., Duvernelle, C. *et al.* (1999). Human bronchial smooth muscle
cells in culture produce stem cell factor. *Eur Respir J* **13**, 951–954.

Kazi, A. S., Lotfi, S., Goncharova, E. A. *et al.* (2004). Vascular endothelial growth factor-
induced secretion of fibronectin is ERK dependent. *Am J Physiol Lung Cell Mol Physiol*
286, L539–545.

Kellner, J., Gamarra, F., Welsch, U. *et al.* (2007). IL-13Ralpha2 reverses the effects
of IL-13 and IL-4 on bronchial reactivity and acetylcholine-induced Ca^+ signaling.
Int Arch Allergy Immunol **142**, 199–210.

Knox, A. J., Corbett, L., Stocks, J. *et al.* (2001). Human airway smooth muscle cells
secrete vascular endothelial growth factor: up-regulation by bradykinin via a protein
kinase C and prostanoid-dependent mechanism. *FASEB J* **15**, 2480–2488.

Laitinen, A., Altraja, A., Kampe, M. *et al.* (1997). Tenascin is increased in airway basement
membrane of asthmatics and decreased by an inhaled steroid. *Am J Respir Crit Care
Med* **156**, 951–958.

Laitinen, L. A., Laitinen, A., Altraja, A. *et al.* (1996). Bronchial biopsy findings in intermittent or 'early' asthma. *J Allergy Clin Immunol* **98**, S3–6; discussion S33–40.

Lazaar, A. L., Albelda, S. M., Pilewski, J. M. *et al.* (1994). T lymphocytes adhere to airway smooth muscle cells via integrins and CD44 and induce smooth muscle cell DNA synthesis. *J Exp Med* **180**, 807–816.

Lazaar, A. L., Amrani, Y., Hsu, J. *et al.* (1998). CD40-mediated signal transduction in human airway smooth muscle. *J Immunol* **161**, 3120–3127.

Lazaar, A. L., Reitz, H. E., Panettieri, R. A. *et al.* (1997). Antigen receptor-stimulated peripheral blood and bronchoalveolar lavage-derived T cells induce MHC class II and ICAM-1 expression on human airway smooth muscle. *Am J Respir Cell Mol Biol* **16**, 38–45.

Lee, Y. C. and Lee, H. K. (2001). Vascular endothelial growth factor in patients with acute asthma. *J Allergy Clin Immunol* **107**, 1106.

Liu, L., Yang, J. and Huang, Y. (2006). Human airway smooth muscle Cells express eotaxin in response to signaling following mast cell contact. *Respiration* **73**, 227–235.

McKay, S., De Jongste, J. C., Saxena, P. R. *et al.* (1998). Angiotensin II induces hypertrophy of human airway smooth muscle cells: expression of transcription factors and transforming growth factor-beta1. *Am J Respir Cell Mol Biol* **18**, 823–833.

Minshall, E. M., Leung, D. Y., Martin, R. J. *et al.* (1997). Eosinophil-associated TGF-beta1 mRNA expression and airways fibrosis in bronchial asthma. *Am J Respir Cell Mol Biol* **17**, 326–333.

Moir, L. M., Burgess, J. K., Johnson, P. R. A. *et al.* (2005). Asthmatic and non-asthmatic airway smooth muscle cell proliferation is mediated via different integrins. *Proc Am Thorac Soc* **2**, A337.

Moir, L. M., Poniris, M. H., Burgess, J. K. *et al.* (2006). Transforming growth factor-beta increases integrin expression on human airway smooth muscle cells in culture. *Proc Am Thorac Soc* **3**, A261.

Moore, P. E., Cunningham, G., Calder, M. M. *et al.* (2006). Respiratory syncytial virus infection reduces beta2-adrenergic responses in human airway smooth muscle. *Am J Respir Cell Mol Biol* **35**, 559–564.

Morbini, P. and Arbustini, E. (2001). In situ characterization of human cytomegalovirus infection of bronchiolar cells in human transplanted lung. *Virchows Arch* **438**, 558–566.

Morris, G. E., Whyte, M. K., Martin, G. F. *et al.* (2005). Agonists of Toll-like receptors 2 and 4 activate airway smooth muscle via mononuclear leukocytes. *Am J Respir Crit Care Med* **171**, 814–822.

Nguyen, T. T., Ward, J. P. and Hirst, S. J. (2005). beta1-integrins mediate enhancement of airway smooth muscle proliferation by collagen and fibronectin. *Am J Respir Crit Care Med* **171**, 217–223.

Oemar, B. S. and Luscher, T. F. (1997). Connective tissue growth factor. Friend or foe? *Arterioscler Thromb Vasc Biol* **17**, 1483–1489.

Ohno, I., Nitta, Y., Yamauchi, K. *et al.* (1996). Transforming growth factor beta 1 (TGF beta 1) gene expression by eosinophils in asthmatic airway inflammation. *Am J Respir Cell Mol Biol* **15**, 404–409.

Oliver, B. G. and Black, J. L. (2006). Airway smooth muscle and asthma. *Allergol Int* **55**, 215–223.

Panettieri, R. A., Murray, R. K., Depalo, L. R. *et al.* (1989). A human airway smooth muscle cell line that retains physiological responsiveness. *Am J Physiol* **256**, C329–335.

Papadopoulos, N. G., Bates, P. J., Bardin, P. G. *et al.* (2000). Rhinoviruses infect the lower airways. *J Infect Dis* **181**, 1875–1884.

Parameswaran, K., Radford, K., Zuo, J. *et al.* (2004). Extracellular matrix regulates human airway smooth muscle cell migration. *Eur Respir J* **24**, 545–551.

Pascual, R. M., Awsare, B. K., Farber, S. A. *et al.* (2003). Regulation of phospholipase A_2 by interleulin-1 in human airway smooth muscle. *Chest* **123**, 433S–434S.

Pawar, S., Kartha, S. and Toback, F. G. (1995). Differential gene expression in migrating renal epithelial cells after wounding. *J Cell Physiol* **165**, 556–565.

Peng, Q., Lai, D., Nguyen, T. T. *et al.* (2005). Multiple beta 1 integrins mediate enhancement of human airway smooth muscle cytokine secretion by fibronectin and type I collagen. *J Immunol* **174**, 2258–2264.

Reddel, H., Ware, S., Marks, G. *et al.* (1999). Differences between asthma exacerbations and poor asthma control. *Lancet* **353**, 364–369.

Redington, A. E., Madden, J., Frew, A. J. *et al.* (1997). Transforming growth factor-beta 1 in asthma. Measurement in bronchoalveolar lavage fluid. *Am J Respir Crit Care Med* **156**, 642–647.

Roberts, C. R. and Burke, A. K. (1998). Remodelling of the extracellular matrix in asthma: proteoglycan synthesis and degradation. *Can Respir J* **5**, 48–50.

Roche, W. R., Beasley, R., Williams, J. H. *et al.* (1989). Subepithelial fibrosis in the bronchi of asthmatics. *Lancet* **1**, 520–524.

Roth, M., Johnson, P. R., Borger, P. *et al.* (2004). Dysfunctional interaction of C/EBPalpha and the glucocorticoid receptor in asthmatic bronchial smooth-muscle cells. *N Engl J Med* **351**, 560–574.

Roth, M., Johnson, P. R., Rudiger, J. J. *et al.* (2002). Interaction between glucocorticoids and beta2 agonists on bronchial airway smooth muscle cells through synchronised cellular signalling. *Lancet* **360**, 1293–1299.

Saunders, M. A., Mitchell, J. A., Seldon, P. M. *et al.* (1997). Release of granulocyte-macrophage colony stimulating factor by human cultured airway smooth muscle cells: suppression by dexamethasone. *Br J Pharmacol* **120**, 545–546.

Sukkar, M. B., Xie, S., Khorasani, N. M. *et al.* (2006). Toll-like receptor 2, 3, and 4 expression and function in human airway smooth muscle. *J Allergy Clin Immunol* **118**, 641–648.

Thompson, R. J. and Schellenberg, R. R. (1998). Increased amount of airway smooth muscle does not account for excessive bronchoconstriction in asthma. *Can Respir J* **5**, 61–62.

Tsutsumi, H., Ohsaki, M., Seki, K. *et al.* (1999). Respiratory syncytial virus infection of human respiratory epithelial cells enhances both muscarinic and beta2-adrenergic receptor gene expression. *Acta Virol* **43**, 267–270.

Vignola, A. M., Chanez, P., Chiappara, G. *et al.* (1997). Transforming growth factor-beta expression in mucosal biopsies in asthma and chronic bronchitis. *Am J Respir Crit Care Med* **156**, 591–599.

Wen, F. Q., Liu, X., Manda, W. *et al.* (2003). TH$_2$ cytokine-enhanced and TGF-beta-enhanced vascular endothelial growth factor production by cultured human airway smooth muscle cells is attenuated by IFN-gamma and corticosteroids. *J Allergy Clin Immunol* **111**, 1307–1318.

Woodruff, P. G., Dolganov, G. M., Ferrando, R. E. *et al.* (2004). Hyperplasia of smooth muscle in mild to moderate asthma without changes in cell size or gene expression. *Am J Respir Crit Care Med* **169**, 1001–1006.

Xie, S., Sukkar, M. B., Issa, R. *et al.* (2005). Regulation of TGF-β1-induced connective tissue growth factor expression in airway smooth muscle cells. *Am J Physiol Lung Cell Mol Physiol* **288**, L68–L76.

Yang, W., Kaur, D., Okayama, Y. *et al.* (2006). Human lung mast cells adhere to human airway smooth muscle, in part, via tumor suppressor in lung cancer-1. *J Immunol* **176**, 1238–1243.

11

The airway smooth muscle in chronic obstructive pulmonary disease (COPD)

Maria B. Sukkar and Kian Fan Chung

Section of Airways Disease, National Heart and Lung Institute, Imperial College, London, UK

11.1 Introduction

Chronic obstructive pulmonary disease (COPD) is an important disease causing significant morbidity and mortality. The prevalence of COPD varies between 4 and 10 per cent of the adult population and in adults older than 40 years, it is estimated to be 3 million people in the UK and 17.1 million in the USA. COPD was ranked sixth among causes of death globally in 1990 but is expected to become the third common cause of death by 2020. In this chapter, the pathophysiological abnormalities in COPD will be reviewed in relation to the basis for the progressive airflow obstruction and airway hyperresponsiveness in this disease, while emphasizing the potential contribution of the airway smooth muscle (ASM) to airflow obstruction, airway inflammation and airway remodelling. Much more emphasis has been given so far to the role of ASM in the airflow obstruction of asthma, but increasing attention is now being given to its potential role in COPD.

Airway Smooth Muscle   Edited by Kian Fan Chung
© 2008 John Wiley & Sons, Ltd

11.2 Definition and assessment of COPD

The Global Initiative on Obstructive Lung Disease (GOLD) (*http://www.goldcopd .com*) defines COPD as a preventable and treatable disease with some significant extrapulmonary effects that may contribute to the severity in individual patients. Its pulmonary component is characterized by airflow limitation that is not fully reversible. The airflow limitation is usually progressive and associated with an abnormal inflammatory response of the lung to noxious particles or gases.

COPD encompasses the disease labels of chronic bronchitis and emphysema. Chronic bronchitis is a clinical description of the presence of a chronic increase of bronchial secretions characterized by a productive cough on most days for a minimum of 3 months per year, for at least 2 successive years. Emphysema is an anatomical diagnosis defined by permanent destructive enlargement of air spaces distal to the terminal bronchioles. COPD is caused primarily by cigarette smoking, but other factors include outdoor and indoor air pollution derived from combustion of biomass, occupational exposures, perinatal events and childhood respiratory events. Contributory factors include genetic factors, chronic mucus hypersecretion, airway hyperresponsiveness and asthma.

In the early stages of the disease, there may be few symptoms. Usually, there is a cough with sputum production, to which the patient may become accustomed. In later stages of the disease, symptoms of breathlessness, usually on exertion, settle in insidiously, and the patient may not be bothered until the disease is far advanced. An accurate assessment of airflow limitation is obtained by using forced expiratory tests, in particular the forced expiratory volume in 1 s (FEV_1), and its ratio to the forced vital capacity (FVC), that is, FEV_1/FVC ratio, which may lead to earlier diagnosis.

COPD is now largely classified according to the degree of impairment of the FEV_1 associated with an FEV_1/FVC ratio of less than 0.7:

- stage I   >80 per cent predicted
- stage II   50–80 per cent predicted
- stage III   30–50 per cent predicted
- stage IV   <30 per cent predicted.

In those patients who develop COPD, the FEV_1 declines more rapidly than in non-smoking volunteers. However, FEV_1 only provides one indicator of severity of COPD, particularly regarding advanced disease. A composite staging system that comprises the body mass index, dyspnoea index and exercise capacity index, in addition to FEV_1 as a measure of airflow limitation (Celli *et al.*, 2004), provides a better predictor of mortality than FEV_1 alone. This staging system takes into account the potential systemic contribution of COPD. Arterial blood gases, which may show hypoxaemia and/or hypercapnia in advanced disease, are another useful index of the severity of COPD. High-resolution computed tomography of the lungs will reveal the presence and extent of emphysema.

11.3 Pathological features of COPD

The characteristic pathology of stable COPD includes chronic inflammation of the small bronchi and bronchioles less than 2 mm in diameter, together with structural changes to the airways and lung emphysema. Chronic bronchitis, characterized by glandular hypertrophy and hyperplasia with dilated ducts of 2–4 mm internal diameter, may be present with mucus plugging, goblet cell hyperplasia, cellular inflammation, increased ASM mass, and distortion due to fibrosis of bronchioles. The chronic inflammation of COPD is characterized by an accumulation of neutrophils, macrophages, B cells, lymphoid aggregates and $CD8^+$ T cells, particularly in the small airways (Hogg, 2004), and the degree of inflammation increases with the severity of disease as classified by GOLD (Hogg *et al.*, 2004). Neutrophils are localized particularly to the bronchial epithelium and the bronchial glands (Saetta *et al.*, 1997), and also in close apposition to ASM bundles (Baraldo *et al.*, 2004). $CD8^+$ T cells are increased throughout the airways and in lung parenchyma (O'Shaughnessy *et al.*, 1997) and have also been localized to the ASM bundles (Baraldo *et al.*, 2004). Because these cells express IFN-γ and CXCR3, they are likely to be type 1 T-helper cells, which could be a prime driver of the inflammatory response in COPD (Saetta *et al.*, 2002). The observation of increased numbers of B cells in more advanced COPD (Gosman *et al.*, 2006; Hogg *et al.*, 2004) and the presence of lymphoid follicles may represent an adaptive immune response to chronic infection.

Although inflammatory cells contribute to the volume of the tissue wall in the small airways in COPD, structural changes such as epithelial metaplasia, increase in ASM, goblet cell hyperplasia, and submucosal gland hypertrophy are other constituents of this thickening, which also increase with severity (Hogg *et al.*, 2004). The degree of airflow limitation as measured by FEV_1 is also correlated with the degree of airway wall thickness, providing indirect evidence for a role for airway wall remodelling in airflow obstruction of COPD. Emphysema may also contribute to airflow limitation resulting from loss of lung elastic recoil (Kim *et al.*, 1991).

Epithelial changes

The epithelium can be damaged by cigarette smoke, and its subsequent response may represent attempts by the airway epithelium to protect itself and repair the injury caused by cigarette smoke (Puchelle *et al.*, 2006). Significant changes in oxidant responsive genes have been observed in the epithelium of healthy smokers as well as COPD patients (Pierrou *et al.*, 2007). A small increase in the proliferative rate of the epithelium of small airways of COPD patients has been reported (Pilette *et al.*, 2007), together with the increased expression of galectin-1 in the epithelium, which could be involved in epithelial proliferation and apoptosis. Cigarette smoke induces the release of IL-1, IL-8 and G-CSF from bronchial epithelial cells through oxidative pathways (Mio *et al.*, 1997), accounting for potential neutrophil and monocytic chemotactic activities released from the epithelium (Masubuchi *et al.*, 1998). A higher expression of MCP-1, TGF-β1, IL-8 mRNA and protein has been observed in bronchiolar epithelium of smokers with COPD compared with those smokers without COPD (de Boer *et al.*, 2000; Takizawa *et al.*, 2001; Vignola *et al.*, 1997). Cultured epithelial cells from smokers and COPD release more TGF-β *in vitro* than those from normals (Takizawa *et al.*, 2001). In addition, the expression of FGF-1 and FGF-2 is increased in the bronchial epithelium of COPD patients (Kranenburg *et al.*, 2005). These growth factors could regulate the proliferation and inflammatory activity of ASM cells in COPD.

Goblet cells, submucosal glands and mucus production

While an earlier study indicated no predictive value of mucus hypersecretion for mortality in COPD (Peto *et al.*, 1983), other studies have associated chronic sputum production to the risk of hospitalization, excessive yearly decline in FEV_1, and COPD development (de Marco *et al.*, 2007). Post-mortem study of lungs from patients who died of COPD showed increased amount of intraluminal mucus in the bronchioles compared to controls without respiratory disease (Aikawa *et al.*, 1989). In surgically resected lung tissues, increasing accumulation of inflammatory exudates with mucus in the small airways has been noted with increasing severity of disease (Hogg *et al.*, 2004). Submucosal gland hypertrophy is also seen in the large airways (Dunnill *et al.*, 1969; Reid and Melb, 1954). A disproportionate increase in mucous acini and reduction in serous acini have been reported in chronic bronchitis (Glynn and Michaels, 1960), but there is no correlation between the mucus gland enlargement and sputum production (Mullen *et al.*, 1985; Nagai *et al.*, 1985).

Goblet cell hyperplasia is a feature of both large and small airways in chronic bronchitis (Reid and Melb, 1954). Goblet cells are usually sparse in the small airways, but are increased in small airways (diameter < 1 mm) of patients with COPD (Saetta *et al.*, 2000). This has been associated with a neutrophilic inflammation, supporting the concept that neutrophils, through the release of neutrophil elastase and cathepsin G, may directly cause degranulation of goblet cells (Sommerhoff *et al.*, 1990). The mechanism of goblet cell hyperplasia itself may involve the activation of the epidermal growth factor receptor (EGFR), which may be upregulated by oxidants in cigarette smoke and by release of cytokines such as TNFα, IL-8 or IL-13 (Shim *et al.*, 2001; Takeyama *et al.*, 2000). Increased expression of the mucin, MUC5B, in the bronchiolar lumen and of MUC5AC in the bronchiolar epithelium has been reported (Caramori *et al.*, 2004; Innes *et al.*, 2006).

Extracellular matrix changes

Subepithelial basement membrane thickness in COPD is not usually increased (Benayoun *et al.*, 2003; Chanez *et al.*, 1997), except in a subset of patients with reversible airways obstruction, in whom an increase was observed together with tissue eosinophilia. An increase in total collagens I and III in the surface epithelial basement membrane, bronchial lamina propria and adventitia (Kranenburg *et al.*, 2006), together with an increase in laminin-β_2 in ASM cells, has been reported, indicating the presence of fibrotic changes in the airway matrix. Other studies also support the increase in matrix deposition in the adventitial compartments of the small airways (Adesina *et al.*, 1991; Cosio *et al.*, 1978; Hogg *et al.*, 2004). This raises the possibility that this could contribute to fixed airflow limitation by preventing the ASM from relaxing completely either during hyperinflation or during pharmacologically induced relaxation. On the other hand, reduced expression of interstitial proteoglycans, such as decorin and biglycan, in the peribronchial area of small airways without any change in the expression of types I, II, and IV collagen, laminin or fibronectin (van Straaten *et al.*, 1999) has also been reported. These conflicting data indicate the need for further studies on changes of the extracellular matrix in COPD and their relationship to airflow obstruction.

ASM changes

ASM changes are more prominent in the small airways than in the large airways. In the large airways, alterations in smooth muscle mass are not observed, and the amount of ASM does not correlate with airflow limitation (Tiddens *et al.*, 1995), although the wall area internal to the muscle is thickened and is associated with a reduction in FEV_1/FVC ratio. In biopsies of large airways, ASM area and size were not increased (Benayoun *et al.*, 2003). Furthermore, the smooth muscle protein isoforms were not increased, and, although myosin light-chain kinase was slightly increased, there was no increase in phosphorylated myosin light chain (Benayoun *et al.*, 2003). In patients with severe persistent asthma, hyperplasia and hypertrophy of the ASM were observed in large proximal airways (Benayoun *et al.*, 2003; Ebina *et al.*, 1993). In addition, culture of ASM cells obtained from patients with emphysema showed a normal proliferation (Roth *et al.*, 2004) while from asthmatics, there was an increased proliferative response to serum persisting over several passages (Johnson *et al.*, 2001). Thus, there are important differences between asthma and COPD in terms of ASM abnormalities.

The ASM cells in the small airways have been less studied, but their properties may be different from those in the proximal airways. An increase in ASM in the small airways of COPD patients has been reported in several studies (Bosken *et al.*, 1990; Cosio *et al.*, 1980; Saetta *et al.*, 1998; 2000). The number of ASM has been inversely correlated with FEV_1 (per cent predicted) (Saetta *et al.*, 2000). The number of ASM was also increased by nearly 50 per cent in patients with more severe COPD at GOLD stages 3 and 4 (Hogg *et al.*, 2004). The ASM mass of the small airways was the only differentiating feature in comparing non-obstructed COPD patients with asthmatics (Kuwano *et al.*, 1993). It is not known whether this process is due to hyperplasia or hypertrophy, or to both.

11.4 COPD mechanisms (Figure 11.1)

The abnormalities found in the airways and lungs in COPD allows one to postulate the mechanisms underlying them (Barnes *et al.*, 2003). The basic abnormality is the response of airway cells, particularly epithelial cells and macrophages, to toxic gases and particulates in cigarette smoke that generates an inflammatory and immune response with the release of cytokines, chemokines, growth factors and proteases. In COPD, as contrasted to healthy smokers, these mechanisms may be amplified by the effect of oxidative stress, perhaps by viruses and by genetic factors. The inflammatory response includes an excessive recruitment of

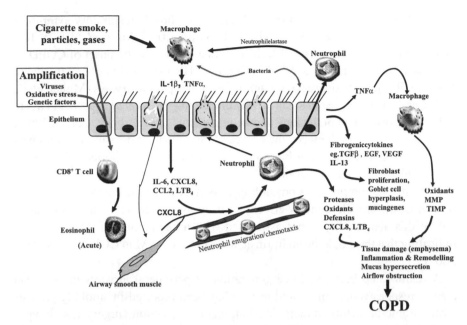

Figure 11.1 Inflammatory pathways in COPD leading to emphysema and small airway inflammation and remodelling. Constituents of cigarette smoke interact with epithelial cells and macrophages to cause inflammation; this process may be amplified by oxidant stress or viruses. CD8 T cells and neutrophils may participate in the chronic inflammatory process. Release of proteases contributes to emphysema. Small airways inflammation and remodelling are characteristic features of COPD, being the cause of the chronic airflow obstruction. ASM cells in the small airways contribute to and are also responsive to the chronic inflammatory process. (For a colour reproduction of this figure, please see the colour section, located towards the centre of the book).

neutrophils and immune cells, including CD8 cytotoxic T cells. There is increasing elastolysis, with the involvement of serine proteases, cathepsins and matrix metalloproteinase leading to alveolar wall destruction (emphysema). The chronic inflammatory process, together with airway wall remodelling, affects particularly the small airways. Within these mechanisms, ASM cells are involved and may contribute particularly to the processes occurring in the small airways.

Bronchial hyperresponsiveness (BHR) in COPD

BHR is present in most patients with mild to moderate COPD (Tashkin *et al.*, 1992; 1996) and is a risk factor for accelerated decline in lung function, the presence of BHR preceding the onset of FEV_1 decline. Irrespective of smoking status, BHR usually predicts the development of chronic respiratory symptoms and accelerated decline in FEV_1 (Brutsche *et al.*, 2006; Xu *et al.*, 1997). It is therefore likely that the onset of BHR occurs early before the onset of COPD.

What is the mechanism of BHR in COPD? BHR in COPD has been related to baseline FEV_1, to the degree of inflammation in peripheral airways (Mullen *et al.*, 1985), and , to the degree of small airway thickening (Riess *et al.*, 1996), but not to indices of impairment of the lung parenchymal structure (Verhoeven *et al.*, 2000). A positive correlation between the degree of inflammation measured by numbers of $CD8^+$ T cells in mucosal biopsies and BHR has been reported (Finkelstein *et al.*, 1995; Rutgers *et al.*, 2000).

Deep inspiration provides physiological protection against airway narrowing. In COPD, the bronchodilatory effect of a deep inspiration is reduced (Fairshter, 1985); this abnormality is not associated with inflammation of ASM bundles, indicating that this could be an intrinsic defect of the ASM in COPD (Slats *et al.*, 2007).

A relationship between force generation of peripheral airways *in vitro* and airway responsiveness measured *in vivo* has been reported in smoking patients with chronic bronchitis undergoing lung cancer resection surgery (De Jongste *et al.*, 1987; Schmidt *et al.*, 2001), but other studies reported no such relationship (Armour *et al.*, 1984; Cerrina *et al.*, 1986; Vincenc *et al.*, 1983). Small airways from COPD patients with airflow obstruction generate greater isometric force and isotonic shortening than those of non-obstructed patients (Opazo Saez *et al.*, 2000), but maximal isotonic shortening is not significantly different between obstructed and non-obstructed patients. Force and stress generated correlated well with lung function measurements. Therefore, the increased ability of ASM from small airways to generate force may contribute to BHR of COPD, along with loss of lung recoil and fibrosis of the small airways.

Effects of cytokines on BHR and ASM contractility

The increased contractility of ASM in COPD may be caused by many mechanisms. It has become evident that cytokines, which are important in driving inflammatory and remodelling abnormalities in COPD, may also directly induce changes in the contractile properties of ASM by increasing force generation, calcium sensitivity and mobilization. TNF-α potentiates contractile responses of isolated human bronchial tissue *in vitro* (Anticevich *et al.*, 1995; Sukkar *et al.*, 2001). Because Ca^{2+} plays a central role in regulating ASM contraction, alterations in intracellular Ca^{2+} signalling have been proposed as a key mechanism underlying cytokine-induced BHR (Amrani *et al.*, 1997; Deshpande *et al.*, 2005).

TNF-α, and the Th2-type cytokine IL-13 may induce BHR by modulating CD38/cyclic ADP ribose (cADPr)-mediated calcium signalling in ASM cells (Deshpande *et al.*, 2003; 2004). CD38 is a membrane-bound protein that is responsible for the synthesis and degradation of cADPr, a β-nicotinamide adenine dinucleotide (NAD) metabolite that mediates agonist-induced elevation of intracellular calcium concentration in ASM cells (Deshpande *et al.*, 2005). In CD38-deficient mice, reduced airway responsiveness is associated with reduced agonist-induced increases in intracellular Ca^{2+} responses of isolated ASM cells, indicating the importance of CD38/cADPr-mediated calcium signalling in the regulation of airway responsiveness (Deshpande *et al.*, 2005). In cultured human ASM cells, TNF-α, IL-13, and, to a lesser extent, IL-1β and IFN-γ, increase CD38 expression, ADP-ribosyl activity, and intracellular calcium responses to G-protein-coupled receptor (GPCR) agonists, whereas cADPr antagonists attenuate cytokine-mediated increases of agonist-induced calcium responses (Deshpande *et al.*, 2003; 2004). Interestingly, while TNF-α-induced CD38 expression and cADPr activity in human ASM cells are attenuated by corticosteroids (Kang *et al.*, 2006), the synergistic induction of CD38 expression by TNF-α and IFN-γ is insensitive to corticosteroid inhibition (Tliba *et al.*, 2006), perhaps suggesting that CD38/cADPr signalling is an important mechanism of BHR in COPD, where corticosteroids have limited efficacy.

Cytokine-induced changes in intracellular calcium signalling have also been attributed to specific modulation of the transient receptor potential C3 (TRPC3), a member of the transient receptor potential channel subfamily which regulates receptor- and store-operated Ca^{2+} influx. Among the TRPC proteins (TRPC1, 3, 4, 5 and 6), TNF-α specifically induces expression of TRPC3 in ASM cells, and specific knockdown of TRPC3 by small interfering RNA attenuates TNF-α-mediated increases in store-operated calcium entry and TNF-α-mediated enhancement of GPCR-induced increases in intracellular Ca^{2+} responses (White *et al.*, 2006).

IL-13 and BHR in COPD

IL-13 has been directly linked to the development of BHR in mouse models of allergic asthma because IL-13-deficient mice do not develop BHR after challenge with specific allergen (Taube *et al.*, 2004), and allergen-induced BHR is significantly attenuated in the presence of blocking anti-IL-13 antibody (Eum *et al.*, 2005). The direct administration of IL-13 to naive mice induces BHR (Wills-Karp *et al.*, 1998), and genetically mutated mice lacking T-bet (T-box expressed in T cells), a Th1-specific transcription factor, spontaneously develop BHR alongside airway inflammatory and remodelling changes characteristic of human asthma (Finotto *et al.*, 2002; 2005). Neutralization of IL-13, but not of IL-4 or of TGF-β, in these mice ameliorates spontaneous induction of BHR (Finotto *et al.*, 2005). BHR in T-bet-deficient mice is associated with enhanced calcium responses to acetylcholine in ASM cells (Bergner *et al.*, 2006), consistent with evidence that IL-13 increases airway responsiveness by inducing alterations in intracellular calcium signalling (Guedes *et al.*, 2006; Tliba *et al.*, 2003). However, IL-13-mediated BHR is not only a feature of the allergic/Th2 phenotype, because over-expression of IL-13 in the lung induces BHR alongside airway and lung abnormalities consistent with a COPD phenotype (Fulkerson *et al.*, 2006; Zheng *et al.*, 2000; Zhu *et al.*, 1999), and increased frequency of an IL-13 gene promoter polymorphism (−1055 C to T) has been demonstrated in COPD patients (van der Pouw Kraan *et al.*, 2002). Moreover, increased numbers of IL-13-expressing cells have been detected in the bronchial submucosa of smokers with chronic bronchitis (Miotto *et al.*, 2003). CD4[+] and CD8[+] T cells isolated from the bronchoalveolar lavage fluid of COPD patients also produce significantly more IL-13 than CD4[+] and CD8[+] T cells from smokers with normal lung function and non-smokers, and the level of IL-13 production is related to the degree of airway obstruction (Barcelo *et al.*, 2006). Thus, the fact that IL-13 is linked to development of BHR in animal models and directly modulates ASM contractile and calcium responses *in vitro,* together with evidence that IL-13 is over-expressed in patients with COPD, indicates a likely role for IL-13 in mediating BHR in COPD.

Arginase I is an enzyme that converts arginine to ornithine, the precursor of proline and polyamines. Proline is used in collagen and mucus production, whereas polyamines increase cell proliferation. Increased arginase activity has been demonstrated in asthmatics (Zimmermann *et al.*, 2003; Zimmermann and Rothenberg, 2006), and the development, persistence and resolution of IL-13-induced BHR correlates with the level of arginase I expression. Furthermore, use of small interfering RNA to inhibit specifically arginase I expression in the lung abrogates the development of IL-13-induced BHR (Yang *et al.*, 2006). Since smoking is an important risk factor for the development of COPD, evidence of increased arginase I expression in the epithelium and smooth muscle bundles of bronchial biopsies obtained from smoking compared to non-smoking asthmatic patients (Bergeron *et al.*, 2007) suggests that IL-13–arginase I interactions may be relevant to the mechanism of BHR in COPD.

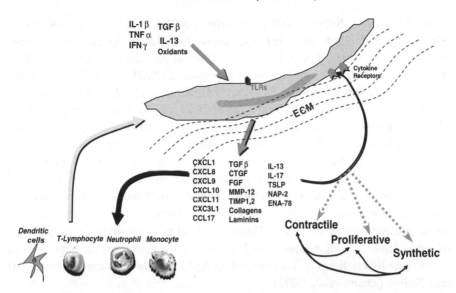

Figure 11.2 Potential role of airway smooth muscle cells in COPD. Pro-inflammatory cytokines and oxidant stress cause cytokine and chemokines, proteases and components of the extracellular matrix to be released from ASM cells. These in turn interact in an autocrine fashion with the ASM cell, and that could change its phenotype, contributing to contractile, proliferative and synthetic processes. In addition, this may attract inflammatory and immune cells to the smooth muscle cell for direct cell–cell interactions. (For a colour reproduction of this figure, please see the colour section, located towards the centre of the book).

ASM component of inflammation and remodelling in COPD (Figure 11.2)

Cytokines/chemokines

There is a potential for the ASM to contribute to the inflammatory and remodelling processes in the small airways. The ASM cell is not only a cell with contractile properties but it can also express and release cytokines, chemokines, growth factors and proteases (Howarth *et al.*, 2004; Jarai *et al.*, 2004), thereby participating in the inflammatory and remodelling process (Chung, 2000). ASM cells also produce matrix proteins, and their behaviour may depend on interactions with their own matrix (Johnson, 2001).

Potential cytokines and chemokines of interest in COPD that may be released from ASM include GM-CSF (Saunders *et al.*, 1997; Sukkar *et al.*, 2000), IL-6 (McKay *et al.*, 2000), TGF-β (Berger *et al.*, 2003; Coutts *et al.*, 2001; Lee *et al.*,

2006; Xie *et al.*, 2007), CXCL10 (IP-10), CXCL11 (ITAC), CXCL9 (MIG) (Brightling *et al.*, 2005; Hardaker *et al.*, 2004), CXCL8 (IL-8) (John *et al.*, 1998; Watson *et al.*, 1998), CXCL1 (GRO-α) (Issa *et al.*, 2006), CCL-2 (MCP-1) (Pype *et al.*, 1999), neutrophil-activating protein-2 (NAP-2), CXCL5 (epithelial neutrophil-activating peptide 78, ENA-78) (Catley *et al.*, 2006), CX3CL1 (fractalkine) (Sukkar *et al.*, 2004) and CCL17 (TARC) (Faffe *et al.*, 2003). Many cytokines and chemokines can be induced in ASM cells upon activation with IL-1β and/or TNF-α, both of which have been associated with induction of emphysema in transgenic mouse models (Churg *et al.*, 2004; Fujita *et al.*, 2001; Lappalainen *et al.*, 2005). Cigarette smoke directly activates ASM cells to synthesize IL-8 and augments TNF-α-mediated IL-8 release (Oltmanns *et al.*, 2005), suggesting direct involvement of ASM-derived mediators in response to a prime causative factor of COPD. Thymic stromal lymphopoietin (TSLP), which triggers dendritic cell-mediated Th2 inflammatory responses, is expressed in ASM of COPD patients and is released by ASM cells *in vitro* exposed to IL-1β and TNF-α (Zhang *et al.*, 2007).

The TH₁-type cytokine IFN-γ induces pulmonary emphysema in mice (Wang *et al.*, 2000) and is also a potent regulator of ASM cell chemokine production. IFN-γ synergizes with TNF-α in the induction of CXCL10 in ASM cells, a chemokine detectable in the lung tissue of patients who have died of emphysema, but not in lung tissue of patients who were 'healthy' at the time of death. Moreover, CXCL10 is detectable within the smooth muscle layer in the lung tissues of emphysema patients, suggesting that ASM cells are likely to be a source of CXCL10 *in vivo* (Hardaker *et al.*, 2004). IFN-γ also synergizes with TNF-α to induce CX3CL1 (fractalkine) production in ASM cells (Sukkar *et al.*, 2004). CX3CL1 may be linked to the pathology of COPD, as there is increased expression of this chemokine in lung tissue samples from COPD (GOLD-2) patients compared with healthy smokers (Ning *et al.*, 2004), and it mediates mononuclear cell adherence to airway epithelial cell cultures *in vitro* (Fujimoto *et al.*, 2001).

The Th2-type cytokines IL-4 and IL-13 are also important regulators of ASM chemokine production (Hirst *et al.*, 2002; John *et al.*, 1997; 1998; Jarai *et al.*, 2004; Moore *et al.*, 2002). Studies in transgenic mice demonstrate a critical role for IL-13 in the induction of emphysema (Hoshino *et al.*, 2007; Zheng *et al.*, 2000). Furthermore, IL-4 and IL-13 cooperate with TNF-α to augment CCL17 (TARC) release from ASM cells (Faffe *et al.*, 2003). This is of interest because TARC and its receptor, CCR4, are upregulated in the bronchoalveolar lavage fluid after acute and chronic cigarette smoke exposure in animal models (Lee *et al.*, 2007; Ritter *et al.*, 2005). CCL17 induces selective chemotaxis of Th2 cells, and its over-expression in asthma is associated with increased expression of CCR4 and infiltration of Th2 cells. However, in cigarette-induced airway inflammation in rats, CCL17 over-expression is not associated with increased CCR4 expression nor infiltration of Th2 cells (Ritter *et al.*, 2005), but, rather, with neutrophils and macrophages in bronchoalveolar lavage fluid.

IL-17A is a member of the IL-17 family of cytokines, which consists of six members, IL-17A–F, secreted by a recently identified lineage of TH cells (TH17) (Weaver *et al.*, 2007). Expression of IL-17 receptors (IL-17R) has been demonstrated in cultured ASM cells and within ASM bundles in airway sections obtained from COPD patients (Rahman *et al.*, 2005). Furthermore, IL-17A induces IL-8 secretion in ASM cells (Rahman *et al.*, 2005; Vanaudenaerde *et al.*, 2003; Wuyts *et al.*, 2005). Activation of ASM cells by IL-1β or TNF-α does not induce expression of IL-17A itself (Henness *et al.*, 2004), but IL-17A augments the effects of these cytokines on IL-6 (Henness *et al.*, 2004) and IL-8 (Dragon *et al.*, 2007; Henness *et al.*, 2006) release.

Toll-like receptors (TLRs)

ASM cells have been shown to express Toll-like receptors, a family of 10 (TLR 1-10) pattern-recognition receptors that mediate innate and adaptive immune responses to infectious pathogens (Sukkar *et al.*, 2006). TLRs recognize microorganism-associated molecular patterns (MAMPs), which include a diverse group of lipid, protein and nucleic acid structures. Expression of TLRs in ASM is regulated by inflammatory cytokines, including IL-1β, TNF-α and IFN-γ (Morris *et al.*, 2005; 2006; Sukkar *et al.*, 2006), and activation of TLRs in ASM cells may lead to production of pro-inflammatory mediators, facilitate the interaction between ASM cells and infiltrating inflammatory cells, and alter the contractile properties of the muscle (Bachar *et al.*, 2004; Morris *et al.*, 2005; 2006; Lee *et al.*, 2004; Lin *et al.*, 2007; Niimi *et al.*, 2007; Shan *et al.*, 2006; Sukkar *et al.*, 2006). Since infections of the respiratory tract by various pathogens, including bacteria, viruses and fungi, are the major cause of disease exacerbation in COPD, activation of TLRs in ASM by microbial products may be a possible mechanism of disease exacerbation in COPD.

Exposure of ASM cells to the TLR4 ligand LPS, the major cell-wall component of Gram-negative bacteria, induces synthesis of IL-6, IL-8 and eotaxin (Morris *et al.*, 2005; Shan *et al.*, 2006; Sukkar *et al.*, 2006). The viral replicative intermediate dsRNA, which is a ligand for TLR3, however, is a more potent inducer of ASM mediator release and induces the expression of several cytokines and chemokines, including IL-6, CXCL8 (IL-8), CCL11 (eotaxin), CCL5 (RANTES) and CXCL10 (IP-10) (Morris *et al.*, 2006; Niimi *et al.*, 2007; Sukkar *et al.*, 2006). dsRNA itself is also a potent regulator of TLR expression in ASM cells. It induces gene expression of its own receptor (TLR3), as well as other TLRs (TLR2 and TLR4), with most potent effects on TLR2 (Sukkar *et al.*, 2006). Exposure of ASM cells to the TLR2 ligand, lipoteichoic acid (LTA), induces calcium mobilization and activates ERK1/2 signalling pathways. Antibodies directed against TLR2 or transfection with the dominant negative mutant TLR2 inhibits LTA-induced ERK1/2 phosphorylation, providing functional evidence of TLR2 activation in ASM cells (Lee *et al.*, 2004).

Cellular interactions between infiltrating immune cells and ASM cells may have a profound impact on airway inflammatory and remodelling processes in COPD. Stimulation of ASM cell and peripheral blood mononuclear cell (PBMC) co-cultures with TLR2, TLR4, TLR7, or TLR8 ligands has been associated with enhanced production of IL-6, CXCL-8 and CCL2 compared to TLR stimulation of either of these cell types alone. Amplification of inflammatory responses by the TLR4 agonist LPS was mediated by monocytes rather than T cells present in ASM cell and PBMC co-cultures. In addition, IL-1β produced by LPS-activated monocytes was responsible, to some extent, for amplification of inflammatory interactions between ASM cells and PBMCs (Morris *et al.*, 2005; 2006). LPS and dsRNA may also promote interactions between ASM cells and inflammatory cells via induction of cell adhesion molecules. dsRNA has been shown to induce expression of ICAM-1 (Morris *et al.*, 2006), while LPS induces expression of VCAM-1 and mediates VCAM-1-induced neutrophil adhesion in ASM cells (Lin *et al.*, 2007).

Extracellular matrix components

Airway wall remodelling in COPD patients is associated with increased bronchial deposition of ECM proteins, such as collagen subtypes I, III and IV; fibronectin; and laminin. Of these ECM proteins, only laminin was found to be over-expressed in the ASM compartment; moreover, its expression in ASM was found to be inversely correlated with the degree of airway obstruction (Kranenburg *et al.*, 2006). The extracellular matrix surrounding the ASM can influence ASM cell function, such as cytokine production, proliferation and apoptosis, and can also determine phenotypic modulation of ASM cells. Endogenous expression of laminin by ASM cells was shown to be important for their maturation towards a contractile phenotype (Tran *et al.*, 2006). Furthermore, ASM cells grown on laminin express more contractile proteins and proliferate less in comparison to cells grown on collagen I and fibronectin (Hirst *et al.*, 2000). The migration of ASM cells grown on laminin is also reduced compared with cells grown on collagens III and IV and fibronectin (Parameswaran *et al.*, 2004). Over-expression of laminin in the ASM of COPD patients may represent a protective mechanism to prevent modulation of ASM cells towards a proliferative/synthetic phenotype that promotes thickening and remodelling of the airway wall.

Expression of surface integrin-α_5 and integrin-β_2 is involved in survival signalling from collagen IV, collagen V, laminin and fibronectin which provide antiapoptotic signals to the ASM (Freyer *et al.*, 2001). Disruption of the integrins leads to apoptosis of the ASM cell, and this may occur when neutrophils interact with ASM. Release of serine proteases from neutrophils, including neutrophil elastase and cathepsin G, induces fibronectin degradation with disruption of integrin binding, leading to apoptosis by 'detachment' of ASM cells (Oltmanns *et al.*, 2004). Neutrophils are prominent cells within ASM bundles in COPD (Baraldo *et al.*, 2004), and this may be a mechanism by which the homeostasis of the ASM mass may be maintained in the presence of proliferative signals. T cells are also seen in ASM bundles in COPD (Baraldo *et al.*, 2004), and adhesion of activated T cells can induce ASM proliferation through CD44 and VCAM (Lazaar *et al.*, 1994). Thus, proliferative and apoptotic pathways are simultaneously present, and the balance of these factors may determine the ASM mass.

Matrix metalloproteinases

Matrix metalloproteinases (MMP) are zinc-dependent proteolytic enzymes that play a major role in matrix turnover, remodelling and angiogenesis in COPD (Shapiro, 2002), and ASM cells may be an important source. Human ASM cells constitutively express pro-MMP-2, MMP-3, and MT1-MMP (Elshaw *et al.*, 2004). MMP-3 is usually bound to ASM-derived matrix, consistent with the staining for MMP-3 in the submucosal matrix of patients with chronic asthma (Dahlen *et al.*, 1999). Over-expression of IL-1β in mouse lung induces inflammatory and morphological changes consistent with a COPD phenotype. Of note, the inflammatory response is characterized by increased production of MMP-9 and MMP-12 (Lappalainen *et al.*, 2005). This is of interest because IL-1β activates human ASM cells to secrete MMP-9 (Liang *et al.*, 2007) and MMP-12 (Xie *et al.*, 2005), possibly indicating a role for ASM-derived MMP-9/-12 in the development of COPD. The importance of MMP-9 and MMP-12 in the development of emphysema has been demonstrated in MMP-12 knockout mice, which are completely protected against the development of cigarette smoke-induced emphysema (Hautamaki *et al.*, 1997), and in a more recent study in which a chemical MMP-9 and MMP-12 inhibitor prevented the development of smoke-induced emphysema and small airways remodelling in guinea pigs (Churg *et al.*, 2007). Furthermore, elevated expression of MMP-9, with a concomitant decrease in the expression of its inhibitor, TIMP-1, is observed during exacerbations of COPD (Mercer *et al.*, 2005), while active MMP-12 is present in ASM cells of small airways of smokers and COPD patients (Xie *et al.*, 2005). In addition to acting as a gelatinase in which the extracellular matrix may be degraded, and contribute to emphysema (close proximity of ASM bundle in small airways to emphysema), MMP-12 also releases TNF-α from pro-TNF-α (Chandler *et al.*, 1996).

ASM cells also produce tissue inhibitor of metalloproteinase (TIMP), particularly TIMP-1 and TIMP-2, which counteract the proteolytic activity of secreted MMPs (Elshaw *et al.*, 2004). Whether there is a shift in the MMP/TIMP balance in ASM cells in airway disease remains to be determined. However, there is evidence that MMPs may contribute to airway wall remodelling by modulating ASM proliferation and migration. MMPs regulate ASM hyperplasia by causing the release of immobilized growth factors, such as the release of TGFβ when the ECM proteoglycan, decorin, is degraded by MMPs (Imai *et al.*, 1997). In addition, MMPs may degrade insulin-like growth factor-binding proteins, causing the release of insulin-growth factor (Fowlkes *et al.*, 1994). IGF-II is released from ASM cells and induces ASM proliferation (Noveral *et al.*, 1994). The cells in the lung are constantly exposed to forces of stretch and relaxation, and excessive stretch in airway disease due to excessive airway narrowing may potentially contribute to airway wall remodelling by promoting ASM cell proliferation and migration. Mechanical strain applied to human ASM cells in culture increases their proliferation and migration by inducing expression of extracellular MMP inducer (EMMPRIN), leading to the subsequent release and activation of MMP-1, MMP-2, MMP-3 and MT1-MMP (Hasaneen *et al.*, 2005).

Growth factors

Abnormalities of several growth factor/receptor pathways have been implicated in the pathogenesis of COPD. Dysregulated expression and/or function of growth factor/receptor systems, including TGF-β and TGF-β type I and II receptors, VEGF, CTGF, PDGF, FGF/FGFR, EGF/EGFR, has been documented in COPD patients, and is associated with airway wall remodelling and emphysema.

Role of TGF-β Airway wall remodelling in COPD may be due to direct profibrotic effects of cigarette smoke on airway structural cells. Thus, exposure of rat tracheal explants to cigarette smoke was associated with rapid activation of TGF-β mediated signalling and TGF-β mediated fibrotic responses, such as induction of CTGF and pro-collagen (Wang *et al.*, 2003; 2005). Furthermore, in an *in vivo* mouse model, significant increases in mRNA expression of TGF-β, CTGF and PDGF (A and B) and pro-collagen were observed in the small airways after a single cigarette smoke exposure, indicating that pro-fibrotic effects are a primary response rather than a secondary process due to persistent inflammation. Repeated smoke exposure for up to 6 months in these animals was associated with sustained expression of pro-collagen, CTGF and PDGF (Churg *et al.*, 2006).

Evidence of neutrophilic infiltration within the ASM in COPD (Baraldo *et al.*, 2004)[5], together with evidence that neutrophil-derived elastase, a serine protease over-expressed in the airways of patients with COPD (Lapperre *et al.*, 2007), activates ASM cells to release TGF-β (Lee *et al.*, 2006), suggests that ASM-cell-derived TGF-β may contribute to inflammatory and remodelling changes in the airways of COPD patients. Other serine proteases, such as tryptase and plasmin, also activate ASM cells to synthesize TGF-β (Berger *et al.*, 2003; Coutts *et al.*, 2001). Aside from serine proteases, mechanical injury to ASM is the only other mechanism known to activate TGF-β release in ASM cells (Chen and Khalil, 2002), although cytokines such as GM-CSF have been shown to upregulate expression of TGF-β receptors and initiate TGF-β-mediated signalling in ASM cells (Chen *et al.*, 2003). TGF-β released by ASM cells may act in an autocrine manner to induce expression of ECM proteins, such as collagen and fibronectin and growth factors, particularly CTGF and VEGF (Burgess *et al.*, 2003; 2006; Coutts *et al.*, 2001; Johnson *et al.*, 2006; Xie *et al.*, 2005). Indeed, concomitant upregulation of both TGF-β and CTGF was identified in a comprehensive gene expression profile study of COPD patients (Ning *et al.*, 2004), and TGF-β-mediated expression of ECM proteins was shown to be partly mediated by induction of CTGF (Johnson *et al.*, 2006).

While the predominant effects of TGF-β in ASM cells concern the synthesis of ECM proteins and growth factors (Jarai *et al.*, 2004), TGF-β also regulates inflammatory functions of ASM. It has been shown to induce expression of cytokines and chemokines, such as IL-8 (Fong *et al.*, 2000), IL-6 and the IL-6-type cytokines IL-11 and leukaemia inhibitory factor (LIF) (Elias *et al.*, 1997). It synergizes with IFN-γ/TNF-α to augment IL-8 production but inhibits IFN-γ/TNF-α-dependent CX3CL1 (fractalkine) production (Sukkar *et al.*, 2004). It also differentially regulates IL-4/IL-13-mediated CCL11 (eotaxin) and CCL26 (eotaxin-3) expression, enhancing release of the former and inhibiting release of the latter (Zuyderduyn *et al.*, 2004).

Several studies have demonstrated hyperplastic effects of TGF-β in human ASM cells (Cohen *et al.*, 2004; Sturrock *et al.*, 2007; Xie *et al.*, 2007), but an equal number of studies have not been able to demonstrate such effects (Bosse *et al.*, 2006; Cohen *et al.*, 1997; Espinosa *et al.*, 2003). Although different culture and experimental conditions were used in the different studies, there does not appear to be a consistent methodological reason for the discrepancy in findings. More complicating is the fact that TGF-β appears to have differential effects on growth responses to other ASM cell mitogens, causing inhibition of thrombin- or EGF-induced ASM proliferation (Cohen *et al.*, 1997) while interacting in synergy with FGF-2 to augment ASM proliferation (Bosse *et al.*, 2006). Where demonstrated, the growth-stimulatory effects of TGF-β have been shown to occur secondary to the release and activity of other growth factors, such as insulin-like growth factor-binding protein-3 (IGFBP-3) (Cohen *et al.*, 2004) and PDGF (Bosse *et al.*, 2006). In addition, TGF-β has also been shown to promote ASM proliferation in response to LTD_4 via a mechanism involving upregulation of CysLT1 receptors on ASM cells (Espinosa *et al.*, 2003). While the mitogenic potential of TGF-β in human ASM cells remains a matter of controversy, there is more consistent evidence demonstrating hypertrophic effects of TGF-β on ASM cells (Black *et al.*, 1996; Goldsmith *et al.*, 2007; Sturrock *et al.*, 2007). Hypertrophic responses to TGF-β include increases in cell size, protein synthesis, protein abundance of α-smooth muscle actin and smooth muscle myosin heavy chain (smMHC), formation of actomyosin filaments, and shortening to acetylcholine, and they require phosphorylation of the eukaryotic translation initiation factor-4E-binding protein, a signalling event specifically involved in translational control (Goldsmith *et al.*, 2006). In a synergistic manner with IL-4, TGF-β2 induces cardiotrophin-1 expression in ASM cells (Zhou *et al.*, 2003). Cardiotrophin-1 is an IL-6 family cytokine that induces hypertrophic responses in ASM cells and may be a possible mechanism by which growth factors may lead to ASM hypertrophy in COPD.

Role of vascular endothelial growth factor (VEGF)   VEGF is a prototypic growth factor for endothelial cells but also regulates many aspects of endothelial cell function pertinent to lung function (Voelkel *et al.*, 2006). VEGF may be a critical mediator of a 'lung structure maintenance programme' (LSMP) that protects the integrity of the adult lung; failure of the LSMP, due to aberrant expression and function of the VEGF/VEGFR system, may lead to development of emphysema (Voelkel *et al.*, 2006). This hypothesis is based on reduced VEGF/VEGFR expression in lung tissue of patients with severe emphysema (Kasahara *et al.*, 2000; 2001; Santos *et al.*, 2003), as well as the evidence that experimental emphysema may be induced following blockade of VEGF receptors in rats (Kasahara *et al.*, 2000). However, in patients with mild to moderate COPD, there is increased expression of VEGF and VEGF receptors in the bronchial and bronchiolar airways and the alveolar parenchyma (Kranenburg *et al.*, 2005; Santos *et al.*, 2003). It appears, therefore, that over-experession of VEGF in COPD has paradoxical effects; where it may contribute to airway remodelling protect against the development of emphysema. Increased VEGF expression in the bronchial and bronchiolar airways in COPD patients has, among other cell types (particularly epithelial and vascular smooth muscle cells), been attributed to increased expression levels in the ASM layer (Kranenburg *et al.*, 2005), suggesting that ASM cells may contribute to aberrant regulation of VEGF in COPD. Indeed, ASM cells in culture have been shown to synthesize VEGF following stimulation with a wide range of cytokines, including pro-inflammatory cytokines (IL-1β and TNF-α), Th2-type cytokines (IL-4, IL-5 and IL-13), the IL-6 family cytokine oncostatin M, and, as mentioned above, TGF-β (TGF-β1, -β2, -β3) (Alagappan *et al.*, 2005; Faffe *et al.*, 2006; Stocks *et al.*, 2005; Wen *et al.*, 2003). In addition, ASM contractile agents, including bradykinin, angiotensin II and endothelin-1, also induce ASM cells to synthesize VEGF (Alagappan *et al.*, 2007; Knox *et al.*, 2001). Stimulation of VEGF by IL-1β and bradykinin involves COX-2-mediated PGE$_2$ production, whereas TGF-β-induced VEGF production occurs independently of endogenous prostanoids (Bradbury *et al.*, 2005; Knox *et al.*, 2001; Stocks *et al.*, 2005). Synthesis of VEGF by ASM cells may possibly contribute to increased ASM bulk, as ASM cells proliferate in response to VEGF *in vitro* (Zou *et al.*, 2005). In contrast to most inflammatory stimuli, IFN-γ is a negative regulator of VEGF production in ASM cells, inhibiting spontaneous VEGF release and Th2 cytokine- or TGF-β-induced VEGF release (Wen *et al.*, 2003). Interestingly, VEGF polymorphisms in ASM cells appear to determine the extent of VEGF release in response to Th2 cytokines (Faffe *et al.*, 2006). Whether or not genetic factors may be associated with VEGF production by ASM cells from COPD patients remains to be determined.

Role of fibroblast growth factors (FGFs)   Increased expression of FGF-1 (acidic FGF), FGF-2 (basic FGF) and their receptor FGFR-1 was demonstrated in COPD patients. Increased expression of FGF-2 and FGFR-1 was demonstrated in the epithelial, airway and vascular smooth muscle compartments, whereas increases in FGF-1 expression were restricted to the epithelial compartment of the airway wall (Kranenburg *et al.*, 2005). In another study where patient selection was based on a diagnosis of chronic bronchitis, increased expression of FGF-2, but not FGFR-1, was demonstrated in the glandular, but not smooth muscle or vessel, compartments of the airway wall (Guddo *et al.*, 2006). Some studies, however, have not been able to confirm increased expression of FGF-2 in the small or large airways of COPD patients (Hashimoto *et al.*, 2005), although the reasons for this discrepancy are not clear. Expression of FGF-2 and FGFR1 in ASM was shown to be inversely correlated with FEV_1 and FEV_1/FVC in COPD patients (Kranenburg *et al.*, 2005). FGF-2 is a well documented and potent mitogen of human ASM cells (Fernandes *et al.*, 2004; Kranenburg *et al.*, 2005; Ravenhall *et al.*, 2000) and also augments ASM proliferative responses induced by other ASM mitogens, such as PDGF (Bonner *et al.*, 1996) and TGF-β (Bosse *et al.*, 2006). FGF-2 also induces migration of ASM cells (Goncharova *et al.*, 2003) and activates ASM cells to synthesize inflammatory mediators such as GM-CSF (Bonacci *et al.*, 2003).

11.5 Conclusion

Although the ASM has not been thoroughly studied in COPD, it is clear that it is likely to contribute to the inflammatory and remodelling processes that have been seen in the small airways. In addition, it is likely that the ASM phenotype is altered in COPD in a different way from that in asthma. The fixed airflow obstruction seen in COPD could be the result of excessive extracellular matrix production by the ASM and by an abnormal interaction between components of the extracellular matrix and the muscle itself. Not excluded is the possibility of a pharmacological defect in β_2-receptor signalling in ASM that could explain the poor bronchodilator response of β_2-adrenergic agonists (Koto *et al.*, 1996). Further research may lead to treatments that could reverse this poor response to β_2-agonists.

References

Adesina, A. M., Vallyathan, V., McQuillen, E. N. *et al.* (1991). Bronchiolar inflammation and fibrosis associated with smoking. A morphologic cross-sectional population analysis. *Am Rev Respir Dis* **143**, 144–149.

Aikawa, T., Shimura, S., Sasaki, H. *et al.* (1989). Morphometric analysis of intraluminal mucus in airways in chronic obstructive pulmonary disease. *Am Rev Respir Dis* **140**, 477–482.

Alagappan, V. K., McKay, S., Widyastuti, A. *et al.* (2005). Proinflammatory cytokines upregulate mRNA expression and secretion of vascular endothelial growth factor in cultured human airway smooth muscle cells. *Cell Biochem Biophys* **43**, 119–129.

Alagappan, V. K., Willems-Widyastuti, A., Seynhaeve, A. L. *et al.* (2007). Vasoactive peptides upregulate mRNA expression and secretion of vascular endothelial growth factor in human airway smooth muscle cells. *Cell Biochem Biophys* **47**, 109–118.

Amrani, Y., Krymskaya, V., Maki, C. *et al.* (1997). Mechanisms underlying TNF-alpha effects on agonist-mediated calcium homeostasis in human airway smooth muscle cells. *Am J Physiol* **273**, L1020–1028.

Anticevich, S. Z., Hughes, J. M., Black, J. L. *et al.* (1995). Induction of human airway hyperresponsiveness by tumour necrosis factor-alpha. *Eur J Pharmacol* **284**(1–2), 221–225.

Armour, C. L., Lazar, N. M., Schellenberg, R. R. *et al.* (1984). A comparison of in vivo and in vitro human airway reactivity to histamine. *Am Rev Respir Dis* **129**, 907–910.

Bachar, O., Adner, M., Uddman, R. *et al.* (2004). Toll-like receptor stimulation induces airway hyper-responsiveness to bradykinin, an effect mediated by JNK and NF-kappa B signaling pathways. *Eur J Immunol* **34**, 1196–1207.

Baraldo, S., Turato, G., Badin, C. *et al.* (2004). Neutrophilic infiltration within the airway smooth muscle in patients with COPD. *Thorax* **59**, 308–312.

Barcelo, B., Pons, J., Fuster, A. *et al.* (2006). Intracellular cytokine profile of T lymphocytes in patients with chronic obstructive pulmonary disease. *Clin Exp Immunol* **145**, 474–479.

Barnes, P. J., Shapiro, S. D. and Pauwels, R. A. (2003). Chronic obstructive pulmonary disease: molecular and cellular mechanisms. *Eur Respir J* **22**, 672–688.

Benayoun, L., Druilhe, A., Dombret, M. C. *et al.* (2003). Airway structural alterations selectively associated with severe asthma. *Am J Respir Crit Care Med* **167**, 1360–1368.

Berger, P., Girodet, P. O., Begueret, H. *et al.* (2003). Tryptase-stimulated human airway smooth muscle cells induce cytokine synthesis and mast cell chemotaxis. *FASEB J* **17**, 2139–2141.

Bergeron, C., Boulet, L. P., Page, N. *et al.* (2007). Influence of cigarette smoke on the arginine pathway in asthmatic airways: increased expression of arginase I. *J Allergy Clin Immunol* **119**, 391–397.

Bergner, A., Kellner, J., Silva, A. K. *et al.* (2006). Ca^{2+}-signaling in airway smooth muscle cells is altered in T-bet knock-out mice. *Respir Res* **7**, 33.

Black, P. N., Young, P. G. and Skinner, S. J. (1996). Response of airway smooth muscle cells to TGF-beta 1: effects on growth and synthesis of glycosaminoglycans. *Am J Physiol* **271**, L910–917.

Bradbury, D., Clarke, D., Seedhouse, C. *et al.* (2005). Vascular endothelial growth factor induction by prostaglandin E_2 in human airway smooth muscle cells is mediated by E prostanoid EP2/EP4 receptors and SP-1 transcription factor binding sites. *J Biol Chem* **280**, 29993–30000.

Bonacci, J. V., Harris, T., Wilson, J. W. *et al.* (2003). Collagen-induced resistance to glucocorticoid anti-mitogenic actions: a potential explanation of smooth muscle hyperplasia in the asthmatic remodelled airway. *Br J Pharmacol* **138**, 1203–1206.

Bonner, J. C., Badgett, A., Lindroos, P. M. *et al.* (1996). Basic fibroblast growth factor induces expression of the PDGF receptor- alpha on human bronchial smooth muscle cells. *Am J Physiol* **271**, L880–888.

Bosken, C. H., Wiggs, B. R., Pare, P. D. *et al.* (1990). Small airway dimensions in smokers with obstruction to airflow. *Am Rev Respir Dis* **142**, 563–570.

Bosse, Y., Thompson, C., Stankova, J. *et al.* (2006). Fibroblast growth factor 2 and transforming growth factor beta1 synergism in human bronchial smooth muscle cell proliferation. *Am J Respir Cell Mol Biol* **34**, 746–753.

Brightling, C. E., Ammit, A. J., Kaur, D. *et al.* (2005). The CXCL10/CXCR3 axis mediates human lung mast cell migration to asthmatic airway smooth muscle. *Am.J Respir Crit Care Med* **171**, 1103–1108.

Brutsche, M. H., Downs, S. H., Schindler, C. *et al.* (2006). Bronchial hyperresponsiveness and the development of asthma and COPD in asymptomatic individuals: SAPALDIA cohort study. *Thorax* **61**, 671–677.

Burgess, J. K., Ge, Q., Poniris, M. H. *et al.* (2006). Connective tissue growth factor and vascular endothelial growth factor from airway smooth muscle interact with the extracellular matrix. *Am J Physiol Lung Cell Mol Physiol* **290**, L153–L161.

Burgess, J. K., Johnson, P. R., Ge, Q. *et al.* (2003). Expression of connective tissue growth factor in asthmatic airway smooth muscle cells. *Am J Respir Crit Care Med* **167**, 71–77.

Caramori, G., Di Gregorio, C., Carlstedt, I. *et al.* (2004). Mucin expression in peripheral airways of patients with chronic obstructive pulmonary disease. *Histopathology* **45**, 477–484.

Catley, M. C., Sukkar, M. B., Chung, K. F. *et al.* (2006). Validation of the anti-inflammatory properties of small-molecule IkappaB kinase (IKK)-2 inhibitors by comparison with adenoviral-mediated delivery of dominant-negative IKK1 and IKK2 in human airways smooth muscle.*Mol Pharmacol* **70**, 697–705.

Celli, B. R., Cote, C. G., Marin, J. M. *et al.* (2004). The body-mass index, airflow obstruction, dyspnea, and exercise capacity index in chronic obstructive pulmonary disease. *N Engl J Med* **350**, 1005–1012.

Cerrina, J., Ladurie, M. L., Lebat, G. *et al.* (1986). Comparison of human bronchial muscle response to histamine in vivo with histamine and isoproterenol agonists in vitro. *Am Rev Respir Dis* **134**, 57–61.

Chandler, S., Cossins, J., Lury, J. *et al.* (1996). Macrophage metalloelastase degrades matrix and myelin proteins and processes a tumour necrosis factor-alpha fusion protein. *Biochem Biophys Res Commun* **228**, 421–429.

Chanez, P., Vignola, A. M., O'Shaugnessy, T. *et al.* (1997). Corticosteroid reversibility in COPD is related to features of asthma. *Am J Respir Crit Care Med* **155**, 1529–1534.

Chen, G., Grotendorst, G., Eichholtz, T. *et al.* (2003). GM-CSF increases airway smooth muscle cell connective tissue expression by inducing TGF-beta receptors. *Am J Physiol Lung Cell Mol Physiol* **284**, L548–L556.

Chen, G. and Khalil, N. (2002). In vitro wounding of airway smooth muscle cell monolayers increases expression of TGF-beta receptors. *Respir Physiol Neurobiol* **132**, 341–346.

Chung, K. F. (2000). Airway smooth muscle cells: contributing to and regulating airway mucosal inflammation? *Eur Respir J* **15**, 961–968.

Churg, A., Tai, H., Coulthard, T. *et al.* (2006). Cigarette smoke drives small airway remodeling by induction of growth factors in the airway wall. *Am J Respir Crit Care Med* **174**, 1327–1334.

Churg, A., Wang, R. D., Tai, H. *et al.* (2004). Tumor necrosis factor-alpha drives 70% of cigarette smoke-induced emphysema in the mouse. *Am J Respir Crit Care Med* **170**, 492–498.

Churg, A., Wang, R., Wang, X. *et al.* (2007). An MMP-9/-12 inhibitor prevents smoke-induced emphysema and small airway remodeling in guinea pigs. *Thorax* (in press).

Cohen, M D, Ciocca, V, Panettieri, R A, Jr. (1997). TGF-beta 1 modulates human airway smooth-muscle cell proliferation induced by mitogens. *Am J Respir Cell Mol Biol* **16**, 85–90.

Cohen, P., Rajah, R., Rosenbloom, J. *et al.* (2004). IGFBP-3 mediates TGF-beta1-induced cell growth in human airway smooth muscle cells. *Am J Physiol Lung Cell Mol Physiol* **278**, L545–L551.

Cosio, M., Ghezzo, H., Hogg, J. C. *et al.* (1978). The relations between structural changes in small airways and pulmonary-function tests. *N Engl J Med* **298**, 1277–1281.

Cosio, M. G., Hale, K. A. and Niewoehner, D. E. (1980). Morphologic and morphometric effects of prolonged cigarette smoking on the small airways. *Am Rev Respir Dis* **122**, 265–271.

Coutts, A., Chen, G., Stephens, N. *et al.* (2001). Release of biologically active TGF-beta from airway smooth muscle cells induces autocrine synthesis of collagen. *Am J Physiol Lung Cell Mol Physiol* **280**, L999–1008.

Dahlen, B., Shute, J. and Howarth, P. (1999). Immunohistochemical localisation of the matrix metalloproteinases MMP-3 and MMP-9 within the airways in asthma. *Thorax* **54**, 590–596.

de Boer, W. I., Sont, J. K., van Schadewijk, A. *et al.* (2000). Monocyte chemoattractant protein 1, interleukin 8, and chronic airways inflammation in COPD. *J Pathol* **190**, 619–626.

De Jongste, J. C., Mons, H., Van Strik, R. *et al.* (1987). Comparison of human bronchiolar smooth muscle responsiveness in vitro with histological signs of inflammation. *Thorax* **42**, 870–876.

de Marco, R., Accordini, S., Cerveri, I. *et al.* (2007). Incidence of chronic obstructive pulmonary disease in a cohort of young adults according to the presence of chronic cough and phlegm. *Am J Respir Crit Care Med* **175**, 32–39.

Deshpande, D. A., Dogan, S., Walseth, T. F. *et al.* (2004). Modulation of calcium signaling by interleukin-13 in human airway smooth muscle: role of CD38/cyclic adenosine diphosphate ribose pathway. *Am.J Respir Cell Mol Biol* **31**, 36–42.

Deshpande, D. A., Walseth, T. F., Panettieri, R. A. *et al.* (2003). CD38/cyclic ADP-ribose-mediated Ca^{2+} signaling contributes to airway smooth muscle hyper-responsiveness. *FASEB J* **17**, 452–454.

Deshpande, D. A., White, T. A., Dogan, S. *et al.* (2005). CD38/cyclic ADP-ribose signaling: role in the regulation of calcium homeostasis in airway smooth muscle. *Am J Physiol Lung Cell Mol Physiol* **288**, L773–L788.

Deshpande, D. A., White, T. A., Guedes, A. G. *et al.* (2005). Altered airway responsiveness in CD38-deficient mice. *Am J Respir Cell Mol Biol* **32**, 149–156.

Dragon, S., Rahman, M. S., Yang, J. *et al.* (2007). IL-17 enhances IL-1beta-mediated CXCL-8 release from human airway smooth muscle cells. *Am J Physiol Lung Cell Mol Physiol* **292**, L1023–L1029.

Dunnill, M. S., Massarella, G. R. and Anderson, J. A. A. (1969). Comparison of the quantitative anatomy of the bronchi in normal subjects, in status asthmaticus, in chronic bronchitis, and in emphysema. *Thorax* **24**, 176–179.

Ebina, M., Takahashi, T., Chiba, T. *et al.* (1993). Cellular hypertrophy and hyperplasia of airway smooth muscle underlying bronchial asthma. *Am Rev Respir Dis* **148**, 720–726.

Elias, J. A., Wu, Y., Zheng, T. *et al.* (1997). Cytokine- and virus-stimulated airway smooth muscle cells produce IL-11 and other IL-6-type cytokines. *Am J Physiol* **273**(3 pt 1), L648–655.

Elshaw, S. R., Henderson, N., Knox, A. J. *et al.* (2004). Matrix metalloproteinase expression and activity in human airway smooth muscle cells. *Br J Pharmacol* **142**, 1318–1324.

Espinosa, K., Bosse, Y., Stankova, J. *et al.* (2003). CysLT1 receptor upregulation by TGF-beta and IL-13 is associated with bronchial smooth muscle cell proliferation in response to LTD4. *J Allergy Clin Immunol* **111**, 1032–1040.

Eum, S. Y., Maghni, K., Tolloczko, B. *et al.* (2005). IL-13 may mediate allergen-induced hyperresponsiveness independently of IL-5 or eotaxin by effects on airway smooth muscle. *Am J Physiol Lung Cell Mol Physiol* **288**, L576–L584.

Faffe, D. S., Flynt, L., Bourgeois, K. *et al.* (2006). Interleukin-13 and interleukin-4 induce vascular endothelial growth factor release from airway smooth muscle cells: role of vascular endothelial growth factor genotype. *Am J Respir Cell Mol Biol* **34**, 213–218.

Faffe, D. S., Whitehead, T., Moore, P. E. *et al.* (2003). IL-13 and IL-4 promote TARC release in human airway smooth muscle cells: role of IL-4 receptor genotype. *Am J Physiol Lung Cell Mol Physiol* **285**, L907–L914.

Fairshter, R. D. (1985). Airway hysteresis in normal subjects and individuals with chronic airflow obstruction. *J Appl Physiol* **58**, 1505–1510.

Fernandes, D. J., Ravenhall, C. E., Harris, T. *et al.* (2004). Contribution of the p38MAPK signalling pathway to proliferation in human cultured airway smooth muscle cells is mitogen-specific. *Br J Pharmacol* **142**, 1182–1190.

Finkelstein, R., Ma, H. D., Ghezzo, H. *et al.* (1995). Morphometry of small airways in smokers and its relationship to emphysema type and hyperresponsiveness. *Am J Respir Crit Care Med* **152**, 267–276.

Finotto, S., Hausding, M., Doganci, A. *et al.* (2005). Asthmatic changes in mice lacking T-bet are mediated by IL-13. *Int Immunol* **17**, 993–1007.

Finotto, S., Neurath, M. F., Glickman, J. N. *et al.* (2002). Development of spontaneous airway changes consistent with human asthma in mice lacking T-bet. *Science* **295**(5553), 336–338.

Fong, C. Y., Pang, L., Holland, E. *et al.* (2000). TGF-beta1 stimulates IL-8 release, COX-2 expression, and PGE(2) release in human airway smooth muscle cells. *Am J Physiol Lung Cell Mol Physiol* **279**, L201–L207.

Fowlkes, J. L., Enghild, J. J., Suzuki, K. *et al.* (1994). Matrix metalloproteinases degrade insulin-like growth factor-binding protein-3 in dermal fibroblast cultures. *J Biol Chem* **269**, 25742–25746.

Freyer, A. M., Johnson, S. R. and Hall, I. P. (2001). Effects of growth factors and extracellular matrix on survival of human airway smooth muscle cells. *Am J Respir Cell Mol Biol* **25**, 569–576.

Fujimoto, K., Imaizumi, T., Yoshida, H. *et al.* (2001). Interferon-gamma stimulates fractalkine expression in human bronchial epithelial cells and regulates mononuclear cell adherence. *Am J Respir Cell Mol Biol* **25**, 233–238.

Fujita, M., Shannon, J. M., Irvin, C. G. *et al.* (2001). Overexpression of tumor necrosis factor-alpha produces an increase in lung volumes and pulmonary hypertension. *Am J Physiol Lung Cell Mol Physiol* **280**, L39–L49.

Fulkerson, P. C., Fischetti, C. A., Hassman, L. M. *et al.* (2006). Persistent effects induced by IL-13 in the lung. *Am J Respir Cell Mol Biol* **35**, 337–346.

Glynn, A. A. and Michaels, L. (1960). Bronchial biopsy in chronic bronchitis and asthma. *Thorax* **15**, 142–153.

Goldsmith, A. M., Bentley, J. K., Zhou, L. *et al.* (2006). Transforming growth factor-beta induces airway smooth muscle hypertrophy. *Am J Respir Cell Mol Biol* **34**, 247–254.

Goldsmith, A. M., Hershenson, M. B., Wolbert, M. P. *et al.* (2007). Regulation of airway smooth muscle alpha-actin expression by glucocorticoids. *Am J Physiol Lung Cell Mol Physiol* **292**, L99–L106.

Goncharova, E. A., Billington, C. K., Irani, C. *et al.* (2003). Cyclic AMP-mobilizing agents and glucocorticoids modulate human smooth muscle cell migration. *Am J Respir Cell Mol Biol* **29**, 19–27.

Gosman, M. M., Willemse, B. W., Jansen, D. F. *et al.* (2006). Increased number of B-cells in bronchial biopsies in COPD. *Eur Respir J* **27**, 60–64.

Guddo, F., Vignola, A. M., Saetta, M. *et al.* (2006). Upregulation of basic fibroblast growth factor in smokers with chronic bronchitis. *Eur Respir J* **27**, 957–963.

Guedes, A. G., Paulin, J., Rivero-Nava, L. *et al.* (2006). CD38-deficient mice have reduced airway hyperresponsiveness following IL-13 challenge. *Am J Physiol Lung Cell Mol Physiol* **291**, L1286–L1293.

Hardaker, E. L., Bacon, A. M., Carlson, K. *et al.* (2004). Regulation of TNF-alpha- and IFN-gamma-induced CXCL10 expression: participation of the airway smooth muscle in the pulmonary inflammatory response in chronic obstructive pulmonary disease. *FASEB J* **18**, 191–193.

Hasaneen, N. A., Zucker, S., Cao, J. *et al.* (2005). Cyclic mechanical strain-induced proliferation and migration of human airway smooth muscle cells: role of EMMPRIN and MMPs. *FASEB J* **19**, 1507–1509.

Hashimoto, M., Tanaka, H. and Abe, S. (2005). Quantitative analysis of bronchial wall vascularity in the medium and small airways of patients with asthma and COPD. *Chest* **127**, 965–972.

Hautamaki, R. D., Kobayashi, D. K., Senior, R. M. *et al.* (1997). Requirement for macrophage elastase for cigarette smoke-induced emphysema in mice. *Science* **277**(5334), 2002–2004.

Henness, S., Johnson, C. K., Ge, Q. *et al.* (2004). IL-17A augments TNF-alpha-induced IL-6 expression in airway smooth muscle by enhancing mRNA stability. *J Allergy Clin Immunol* **114**, 958–964.

Henness, S., van Thoor, E., Ge, Q. *et al.* (2006). IL-17A acts via p38 MAPK to increase stability of TNF-alpha-induced IL-8 mRNA in human ASM. *Am J Physiol Lung Cell Mol Physiol* **290**, L1283–L1290.

Hirst, S. J., Hallsworth, M. P., Peng, Q. *et al.* (2002). Selective induction of eotaxin release by interleukin-13 or interleukin-4 in human airway smooth muscle cells is synergistic with interleukin-1beta and is mediated by the interleukin-4 receptor alpha-chain. *Am J Respir Crit Care Med* **165**, 1161–1171.

Hirst, S. J., Twort, C. H. and Lee, T. H. (2000). Differential effects of extracellular matrix proteins on human airway smooth muscle cell proliferation and phenotype. *Am J Respir Cell Mol Biol* **23**, 335–344.

Hogg, J. C. (2004). Pathophysiology of airflow limitation in chronic obstructive pulmonary disease. *Lancet* **364**(9435), 709–721.

Hogg, J. C., Chu, F., Utokaparch, S. *et al.* (2004). The nature of small-airway obstruction in chronic obstructive pulmonary disease. *N Engl J Med* **350**, 2645–2653.

Hoshino, T., Kato, S., Oka, N. *et al.* (2007). Pulmonary inflammation and emphysema: role of the cytokines IL-18 and IL-13. *Am J Respir Crit Care Med* **176**, 49–62.

Howarth, P. H., Knox, A. J., Amrani, Y. *et al.* (2004). Synthetic responses in airway smooth muscle. *J Allergy Clin.Immunol* **114**(2 Suppl), S32–S50.

Imai, K., Hiramatsu, A., Fukushima, D. *et al.* (1997). Degradation of decorin by matrix metalloproteinases: identification of the cleavage sites, kinetic analyses and transforming growth factor-beta1 release. *Biochem J* **322**, 809–814.

Innes, A.L., Woodruff, P. G., Ferrando, R. E. *et al.* (2006). Epithelial mucin stores are increased in the large airways of smokers with airflow obstruction. *Chest* **130**, 1102–1108.

Issa, R., Xie, S., Lee, K. Y. *et al.* (2006). GRO-alpha regulation in airway smooth muscle by IL-1beta and TNF-alpha: role of NF-kappaB and MAP kinases. *Am J Physiol Lung Cell Mol Physiol* **291**, L66–L74.

Jarai, G., Sukkar, M., Garrett, S. *et al.* (2004). Effects of IL-1b, IL-13 & TGFβ on gene expression in human airway smooth muscle using gene microarrays. *Eur J Pharmacol* **497**, 255–265.

John, M., Au, B. T., Jose, P. J. *et al.* (1998). Expression and release of interleukin-8 by human airway smooth muscle cells: inhibition by Th-2 cytokines and corticosteroids. *Am J Respir Cell Mol Biol* **18**, 84–90.

John, M., Hirst, S. J., Jose, P. J. *et al.* (1997). Human airway smooth muscle cells express and release RANTES in response to Th-1 cytokines: regulation by Th-2 cytokines and corticosteroids. *J Immunol* **158**, 1841–1847.

Johnson, P. R. (2001). Role of human airway smooth muscle in altered extracellular matrix production in asthma. *Clin Exp Pharmacol Physiol* **28**, 233–236.

Johnson, P. R., Burgess, J. K., Ge, Q. *et al.* (2006). Connective tissue growth factor induces extracellular matrix in asthmatic airway smooth muscle. *Am J Respir Crit Care Med* **173**, 32–41.

Johnson, P. R., Roth, M., Tamm, M. *et al.* (2001). Airway smooth muscle cell proliferation is increased in asthma. *Am J Respir Crit Care Med* **164**, 474–477.

Lazaar, A. L., Albelda, S. M., Pilewski, J. M. *et al.* (1994). T lymphocytes adhere to airway smooth muscle cells via integrins and CD44 and induce smooth muscle cell DNA synthesis. *J Exp Med* **180**, 807–816.

Kang, B. N., Tirumurugaan, K. G., Deshpande, D. A. *et al.* (2006). Transcriptional regulation of CD38 expression by tumor necrosis factor-alpha in human airway smooth muscle cells: role of NF-kappaB and sensitivity to glucocorticoids. *FASEB J* **20**, 1000–1002.

Kasahara, Y., Tuder, R. M., Cool, C. D. *et al.* (2001). Endothelial cell death and decreased expression of vascular endothelial growth factor and vascular endothelial growth factor receptor 2 in emphysema. *Am J Respir Crit Care Med* **163**(3 pt 1), 737–744.

Kasahara, Y., Tuder, R. M., Taraseviciene-Stewart, L. *et al.* (2000). Inhibition of VEGF receptors causes lung cell apoptosis and emphysema. *J Clin Invest* **106**, 1311–1319.

Kim, W. D., Eidelman, D. H., Izquierdo, J. L. *et al.* (1991). Centrilobular and panlobular emphysema in smokers. Two distinct morphologic and functional entities. *Am Rev Respir Dis* **144**, 1385–1390.

Knox, A. J., Corbett, L., Stocks, J. *et al.* (2001). Human airway smooth muscle cells secrete vascular endothelial growth factor: up-regulation by bradykinin via a protein kinase C and prostanoid-dependent mechanism. *FASEB J* **15**, 2480–2488.

Koto, H., Mak, J. C., Haddad, E. B. *et al.* (1996). Mechanisms of impaired beta-adrenoceptor-induced airway relaxation by interleukin-1beta in vivo in the rat. *J Clin Invest* **98**, 1780–1787.

Kranenburg, A. R., de Boer, W. I., Alagappan, V. K. *et al.* (2005). Enhanced bronchial expression of vascular endothelial growth factor and receptors (Flk-1 and Flt-1) in patients with chronic obstructive pulmonary disease. *Thorax* **60**, 106–113.

Kranenburg, A. R., Willems-Widyastuti, A., Moori, W. J. *et al.* (2005). Chronic obstructive pulmonary disease is associated with enhanced bronchial expression of FGF-1, FGF-2, and FGFR-1. *J Pathol* **206**, 28–38.

Kranenburg, A. R., Willems-Widyastuti, A., Moori, W. J. *et al.* (2006). Enhanced bronchial expression of extracellular matrix proteins in chronic obstructive pulmonary disease. *Am J Clin Pathol* **126**, 725–735.

Kuwano, K., Bosken, C. H., Pare, P. D. *et al.* (1993). Small airways dimensions in asthma and in chronic obstructive pulmonary disease. *Am Rev Respir Dis* **148**, 1220–1225.

Lappalainen, U., Whitsett, J. A., Wert, S. E. *et al.* (2005). Interleukin-1beta causes pulmonary inflammation, emphysema, and airway remodeling in the adult murine lung. *Am J Respir Cell Mol Biol* **32**, 311–318.

Lapperre, T. S., Willems, L. N., Timens, W. *et al.* (2007). Small airways dysfunction and neutrophilic inflammation in bronchial biopsies and BAL in COPD. *Chest* **131**, 53–59.

Lee, C. W., Chien, C. S. and Yang, C. M. (2004). Lipoteichoic acid-stimulated p42/p44 MAPK activation via Toll-like receptor 2 in tracheal smooth muscle cells. *Am J Physiol Lung Cell Mol Physiol* **286**, L921–L930.

Lee, K. M., Renne, R. A., Harbo, S. J. *et al.* (2007). 3-week inhalation exposure to cigarette smoke and/or lipopolysaccharide in AKR/J mice. *Inhal Toxicol* **19**, 23–35.

Lee, K. Y., Ho, S. C., Lin, H. C. *et al.* (2006). Neutrophil-derived elastase induces TGF-beta1 secretion in human airway smooth muscle via NF-kappaB pathway. *Am J Respir Cell Mol Biol* **35**, 407–414.

Liang, K. C., Lee, C. W., Lin, W. N. *et al.* (2007). Interleukin-1beta induces MMP-9 expression via p42/p44 MAPK, p38 MAPK, JNK, and nuclear factor-kappaB signaling pathways in human tracheal smooth muscle cells. *J Cell Physiol* **211**, 759–770.

Lin, W. N., Luo, S. F., Lee, C. W. *et al.* (2007). Involvement of MAPKs and NF-kappaB in LPS-induced VCAM-1 expression in human tracheal smooth muscle cells. *Cell Signal* **19**, 1258–1267.

Masubuchi, T., Koyama, S., Sato, E. *et al.* (1998). Smoke extract stimulates lung epithelial cells to release neutrophil and monocyte chemotactic activity. *Am J Pathol* **153**, 1903–1912.

McKay, S., Hirst, S. J., Haas, M. B. *et al.* (2000). Tumor necrosis factor-alpha enhances mRNA expression and secretion of interleukin-6 in cultured human airway smooth muscle cells. *Am J Respir Cell Mol Biol* **23**, 103–111.

Mercer, P. F., Shute, J. K., Bhowmik, A. *et al.* (2005). MMP-9, TIMP-1 and inflammatory cells in sputum from COPD patients during exacerbation. *Respir Res* **6**, 151.

Mio, T., Romberger, D. J., Thompson, A. B. *et al.* (1997). Cigarette smoke induces interleukin-8 release from human bronchial epithelial cells. *Am J Respir Crit Care Med* **155**, 1770–1776.

Miotto, D., Ruggieri, M. P., Boschetto, P. *et al.* (2003). Interleukin-13 and -4 expression in the central airways of smokers with chronic bronchitis. *Eur Respir J* **22**, 602–608.

Moore, P. E., Church, T. L., Chism, D. D. *et al.* (2002). IL-13 and IL-4 cause eotaxin release in human airway smooth muscle cells: a role for ERK. *Am J Physiol Lung Cell Mol Physiol* **282**, L847–L853.

Morris, G. E., Parker, L. C., Ward, J. R. *et al.* (2006). Cooperative molecular and cellular networks regulate Toll-like receptor-dependent inflammatory responses. *FASEB J* **20**, 2153–2155.

Morris, G. E., Whyte, M. K., Martin, G. F. *et al.* (2005). Agonists of Toll-like receptors 2 and 4 activate airway smooth muscle via mononuclear leukocytes. *Am J Respir Crit Care Med* **171**, 814–822.

Mullen, J. B., Wright, J. L., Wiggs, B. R. *et al.* (1985). Reassessment of inflammation of airways in chronic bronchitis. *Br Med J (Clin Res Ed)* **291**(6504), 1235–1239.

Nagai, A., West, W. W. and Thurlbeck, W. M. (1985). The National Institutes of Health Intermittent Positive-Pressure Breathing trial: pathology studies. II. Correlation between morphologic findings, clinical findings, and evidence of expiratory air-flow obstruction. *Am Rev Respir Dis* **132**, 946–953.

Niimi, K., Asano, K., Shiraishi, Y. *et al.* (2007). TLR3-mediated synthesis and release of eotaxin-1/CCL11 from human bronchial smooth muscle cells stimulated with double-stranded RNA. *J Immunol* **178**, 489–495.

Ning, W., Li, C. J., Kaminski, N. *et al.* (2004). Comprehensive gene expression profiles reveal pathways related to the pathogenesis of chronic obstructive pulmonary disease. *Proc Natl Acad Sci U S A* **101**, 14895–14900.

Noveral, J. P., Bhala, A., Hintz, R. L. *et al.* (1994). Insulin-like growth factor axis in airway smooth muscle cells. *Am J Physiol* **267**(6 pt 1), L761–L765.

Oltmanns, U., Chung, K. F., Walters, M. *et al.* (2005). Cigarette smoke induces IL-8, but inhibits eotaxin and RANTES release from airway smooth muscle. *Respir Res* **6**, 74.

Oltmanns, U., Sukkar, M. B., Issa, R. *et al.* (2004). Neutrophils induce apoptosis of human airway smooth muscle cells via neutrophil elastase. *Am J Respir Crit Care Med* **169**, A195.

Opazo Saez, A. M., Seow, C. Y. and Pare, P. D. (2000). Peripheral airway smooth muscle mechanics in obstructive airways disease. *Am J Respir Crit Care Med* **161**(3 pt 1), 910–917.

O'Shaughnessy, T. C., Ansari, T. W., Barnes, N. C. *et al.* (1997). Inflammation in bronchial biopsies of subjects with chronic bronchitis: inverse relationship of CD8+ T lymphocytes with FEV_1. *Am J Respir Crit Care Med* **155**, 852–857.

Parameswaran, K., Radford, K., Zuo, J. *et al.* (2004). Extracellular matrix regulates human airway smooth muscle cell migration. *Eur Respir J* **24**, 545–551.

Peto, R., Speizer, F. E., Cochrane, A. L. *et al.* (1983). The relevance in adults of air-flow obstruction, but not of mucus hypersecretion, to mortality from chronic lung disease. Results from 20 years of prospective observation. *Am Rev Respir Dis* **128**, 491–500.

Pierrou, S., Broberg, P., O'Donnell, R. A. *et al.* (2007). Expression of genes involved in oxidative stress responses in airway epithelial cells of smokers with chronic obstructive pulmonary disease. *Am J Respir Crit Care Med* **175**, 577–586.

Pilette, C., Colinet, B., Kiss, R. *et al.* (2007). Increased galectin-3 expression and intraepithelial neutrophils in small airways in severe chronic obstructive pulmonary disease. *Eur Respir J* **29**, 914–922.

Puchelle, E., Zahm, J. M., Tournier, J. M. *et al.* (2006). Airway epithelial repair, regeneration, and remodeling after injury in chronic obstructive pulmonary disease. *Proc Am Thorac Soc* **3**, 726–733.

Pype, J. L., Dupont, L. J., Menten, P. *et al.* (1999). Expression of monocyte chemotactic protein (MCP)-1, MCP-2, and MCP-3 by human airway smooth-muscle cells. Modulation by corticosteroids and T-helper 2 cytokines. *Am J Respir Cell Mol Biol* **21**, 528–536.

Rahman, M. S., Yang, J., Shan, L. Y. *et al.* (2005). IL-17R activation of human airway smooth muscle cells induces CXCL-8 production via a transcriptional-dependent mechanism. *Clin Immunol* **115**, 268–276.

Ravenhall, C., Guida, E., Harris, T. *et al.* (2000). The importance of ERK activity in the regulation of cyclin D1 levels and DNA synthesis in human cultured airway smooth muscle. *Br J Pharmacol* **131**, 17–28.

Reid, L. M. and Melb, M. B. (1954). Pathology of chronic bronchitis. *Lancet* **i**, 275–278.

Riess, A., Wiggs, B., Verburgt, L. *et al.* (1996). Morphologic determinants of airway responsiveness in chronic smokers. *Am J Respir Crit Care Med* **154**, 1444–1449.

Ritter, M., Goggel, R., Chaudhary, N. *et al.* (2005). Elevated expression of TARC (CCL17) and MDC (CCL22) in models of cigarette smoke-induced pulmonary inflammation. *Biochem Biophys Res Commun* **334**, 254–262.

Roth, M., Johnson, P. R., Borger, P. *et al.* (2004). Dysfunctional interaction of C/EBPalpha and the glucocorticoid receptor in asthmatic bronchial smooth-muscle cells. *N Engl J Med* **351**, 560–574.

Rutgers, S. R., Timens, W., Tzanakis, N. *et al.* (2000). Airway inflammation and hyperresponsiveness to adenosine 5'-monophosphate in chronic obstructive pulmonary disease. *Clin Exp Allergy* **30**, 657–662.

Saetta, M., Di, S. A., Turato, G. *et al.* (1998). CD8$^+$ T-lymphocytes in peripheral airways of smokers with chronic obstructive pulmonary disease. *Am J Respir Crit Care Med* **157**(3 pt 1), 822–826.

Saetta, M., Mariani, M., Panina-Bordignon, P. *et al.* (2002). Increased expression of the chemokine receptor CXCR3 and its ligand CXCL10 in peripheral airways of smokers with chronic obstructive pulmonary disease. *Am J Respir Crit Care Med* **165**, 1404–1409.

Saetta, M., Turato, G., Baraldo, S. *et al.* (2000). Goblet cell hyperplasia and epithelial inflammation in peripheral airways of smokers with both symptoms of chronic bronchitis and chronic airflow limitation. *Am J Respir Crit Care Med* **161**(3 pt 1), 1016–1021.

Saetta, M., Turato, G., Facchini, F. M. *et al.* (1997). Inflammatory cells in the bronchial glands of smokers with chronic bronchitis. *Am J Respir Crit Care Med* **156**, 1633–1639.

Santos, S., Peinado, V. I., Ramirez, J. *et al.* (2003). Enhanced expression of vascular endothelial growth factor in pulmonary arteries of smokers and patients with moderate chronic obstructive pulmonary disease. *Am J Respir Crit Care Med* **167**, 1250–1256.

Saunders, M. A., Mitchell, J. A., Seldon, P. M. *et al.* (1997). Release of granulocyte-macrophage colony stimulating factor by human cultured airway smooth muscle cells: suppression by dexamethasone. *Br J Pharmacol* **120**, 545–546.

Schmidt, D. T., Jorres, R. A., Ruhlmann, E. *et al.* (2001). Isolated airways from current smokers are hyper-responsive to histamine. *Clin Exp Allergy* **31**, 1041–1047.

Shan, X., Hu, A., Veler, H. *et al.* (2006). Regulation of Toll-like receptor 4-induced proasthmatic changes in airway smooth muscle function by opposing actions of ERK1/2 and p38 MAPK signaling. *Am J Physiol Lung Cell Mol Physiol* **291**, L324–L333.

Shapiro, S. D. (2002). Proteinases in chronic obstructive pulmonary disease. *Biochem Soc Trans* **30**, 98–102.

Shim, J. J., Dabbagh, K., Ueki, I. F. *et al.* (2001). IL-13 induces mucin production by stimulating epidermal growth factor receptors and by activating neutrophils. *Am J Physiol Lung Cell Mol Physiol* **280**, L134–L140.

Slats, A M, Janssen, K, van Schadewijk, A *et al.* (2007). Bronchial inflammation and airway responses to deep inspiration in asthma and chronic obstructive pulmonary disease. *Am J Respir Crit Care Med* **176**, 121–128.

Sommerhoff, C. P., Nadel, J. A., Basbaum, C. B. *et al.* (1990). Neutrophil elastase and cathepsin G stimulate secretion from cultured bovine airway gland serous cells. *J Clin Invest* **85**, 682–689.

Stocks, J., Bradbury, D., Corbett, L. *et al.* (2005). Cytokines upregulate vascular endothelial growth factor secretion by human airway smooth muscle cells: Role of endogenous prostanoids. *FEBS Lett* **579**, 2551–2556.

Sturrock, A., Huecksteadt, T. P., Norman, K. *et al.* (2007). Nox4 mediates TGF-beta1-induced retinoblastoma protein phosphorylation, proliferation, and hypertrophy in human airway smooth muscle cells. *Am J Physiol Lung Cell Mol Physiol* **292**, L1543–L1555.

Sukkar, M. B., Hughes, J. M., Armour, C. L. *et al.* (2001). Tumour necrosis factor-alpha potentiates contraction of human bronchus in vitro. *Respirology* **6**, 199–203.

Sukkar, M. B., Hughes, J. M., Johnson, P. R. *et al.* (2000). GM-CSF production from human airway smooth muscle cells is potentiated by human serum. *Mediators Inflamm* **9**(3–4), 161–168.

Sukkar, M. B., Issa, R., Xie, S. *et al.* (2004). Fractalkine/CX3CL1 production by human airway smooth muscle cells: induction by IFN-gamma and TNF-alpha and regulation by TGF-beta and corticosteroids. *Am J Physiol Lung Cell Mol.Physiol* **287**, L1230–L1240.

Sukkar, M. B., Xie, S., Khorasani, N. M. *et al.* (2006). Toll-like receptor 2, 3, and 4 expression and function in human airway smooth muscle. *J Allergy Clin Immunol* **118**, 641–648.

Takeyama, K., Dabbagh, K., Jeong, S. J. *et al.* (2000). Oxidative stress causes mucin synthesis via transactivation of epidermal growth factor receptor: role of neutrophils. *J Immunol* **164**, 1546–1552.

Takizawa, H., Tanaka, M., Takami, K. *et al.* (2001). Increased expression of transforming growth factor-beta1 in small airway epithelium from tobacco smokers and patients with chronic obstructive pulmonary disease (COPD). *Am J Respir Crit Care Med* **163**, 1476–1483.

Tashkin, D. P., Altose, M. D., Bleecker, E. R. *et al.* (1992). The lung health study: airway responsiveness to inhaled methacholine in smokers with mild to moderate airflow limitation. Lung Health Study Research Group. *Am Rev Respir Dis* **145**(2 pt 1), 301–310.

Tashkin, D. P., Altose, M. D., Connett, J. E. *et al.* (1996). Methacholine reactivity predicts changes in lung function over time in smokers with early chronic obstructive pulmonary disease. Lung Health Study Research Group. *Am.J Respir Crit Care Med* **153**(6 pt 1), 1802–1811.

Taube, C., Wei, X., Swasey, C. H. *et al.* (2004). Mast cells, Fc epsilon RI, and IL-13 are required for development of airway hyperresponsiveness after aerosolized allergen exposure in the absence of adjuvant. *J Immunol* **172**, 6398–6406.

Tiddens, H. A., Pare, P. D., Hogg, J. C. *et al.* (1995). Cartilaginous airway dimensions and airflow obstruction in human lungs. *Am J Respir Crit Care Med* **152**, 260–266.

Tliba, O., Cidlowski, J. A. and Amrani, Y. (2006). CD38 expression is insensitive to steroid action in cells treated with tumor necrosis factor-alpha and interferon-gamma by a mechanism involving the up-regulation of the glucocorticoid receptor beta isoform. *Mol Pharmacol* **69**, 588–596.

Tliba, O., Deshpande, D., Chen, H. *et al.* (2003). IL-13 enhances agonist-evoked calcium signals and contractile responses in airway smooth muscle. *Br J Pharmacol* **140**, 1159–1162.

Tran, T., McNeill, K. D., Gerthoffer, W. T. *et al.* (2006). Endogenous laminin is required for human airway smooth muscle cell maturation. *Respir Res* **7**, 117.

Vanaudenaerde, B. M., Wuyts, W. A., Dupont, L. J. *et al.* (2003). Interleukin-17 stimulates release of interleukin-8 by human airway smooth muscle cells in vitro: a potential role for interleukin-17 and airway smooth muscle cells in bronchiolitis obliterans syndrome. *J Heart Lung Transplant* **22**, 1280–1283.

van der Pouw Kraan, T. C., Kucukaycan, M., Bakker, A. M. *et al.* (2002). Chronic obstructive pulmonary disease is associated with the -1055 IL-13 promoter polymorphism. *Genes Immun* **3**, 436–439.

van Straaten, J. F., Coers, W., Noordhoek, J. A. *et al.* (1999). Proteoglycan changes in the extracellular matrix of lung tissue from patients with pulmonary emphysema. *Mod Pathol* **12**, 697–705.

Verhoeven, G. T., Verbraak, A. F., Boere-van der Straat, S. *et al.* (2000). Influence of lung parenchymal destruction on the different indexes of the methacholine dose-response curve in COPD patients. *Chest* **117**, 984–990.

Vignola, A. M, Chanez, P., Chiappara, G. *et al.* (1997). Transforming growth factor-beta expression in mucosal biopsies in asthma and chronic bronchitis. *Am J Respir Crit Care Med* **156**(2 pt 1), 591–599.

Vincenc, K. S., Black, J. L., Yan, K. *et al.* (1983). Comparison of in vivo and in vitro responses to histamine in human airways. *Am Rev Respir Dis* **128**, 875–879.

Voelkel, N. F., Vandivier, R. W. and Tuder, R. M. (2006). Vascular endothelial growth factor in the lung. *Am J Physiol Lung Cell Mol Physiol* **290**, L209–L221.

Wang, R. D., Tai, H., Xie, C. *et al.* (2003). Cigarette smoke produces airway wall remodeling in rat tracheal explants. *Am J Respir Crit Care Med* **168**, 1232–1236.

Wang, R. D., Wright, J. L. and Churg, A. (2005). Transforming growth factor-beta1 drives airway remodeling in cigarette smoke-exposed tracheal explants. *Am J Respir Cell Mol Biol* **33**, 387–393.

Wang, Z., Zheng, T., Zhu, Z. *et al.* (2000). Interferon gamma induction of pulmonary emphysema in the adult murine lung. *J Exp Med* **192**, 1587–1600.

Watson, M. L., Grix, S. P., Jordan, N. J. *et al.* (1998). Interleukin 8 and monocyte chemoattractant protein 1 production by cultured human airway smooth muscle cells. *Cytokine* **10**, 346–352.

Weaver, C. T., Hatton, R. D., Mangan, P. R. *et al.* (2007). IL-17 family cytokines and the expanding diversity of effector T cell lineages. *Annu Rev Immunol* **25**, 821–852.

Wen, F. Q., Liu, X., Manda, W. *et al.* (2003). Th2 Cytokine-enhanced and TGF-beta-enhanced vascular endothelial growth factor production by cultured human airway smooth muscle cells is attenuated by IFN-gamma and corticosteroids. *J Allergy Clin Immunol* **111**, 1307–1318.

White, T. A., Xue, A., Chini, E. N. *et al.* (2006). Role of transient receptor potential C3 in TNF-alpha-enhanced calcium influx in human airway myocytes. *Am J Respir Cell Mol Biol* **35**, 243–251.

Wills-Karp, M., Luyimbazi, J., Xu, X. *et al.* (1998). Interleukin-13: central mediator of allergic asthma. *Science* **282**, 2258–2261.

Wuyts, W. A., Vanaudenaerde, B. M., Dupont, L. J. *et al.* (2005). Interleukin-17-induced interleukin-8 release in human airway smooth muscle cells: role for mitogen-activated kinases and nuclear factor-kappaB. *J Heart Lung Transplant* **24**, 875–881.

Xie, S., Issa, R., Sukkar, M. B. *et al.* (2005). Induction and regulation of matrix metalloproteinase-12 in human airway smooth muscle cells. *Respir Res* **6**, 148.

Xie, S., Sukkar, M. B., Issa, R. *et al.* (2005). Regulation of TGF-β1-induced connective tissue growth factor expression in airway smooth muscle cells. *Am J Physiol Lung Cell Mol Physiol* **288**, L68–L76.

Xie, S., Sukkar, M. B., Issa, R. *et al.* (2007). Mechanisms of induction of airway smooth muscle hyperplasia by transforming growth factor-beta. *Am J Physiol Lung Cell Mol Physiol* **293**, L245–L253.

Xu, X., Rijcken, B., Schouten, J. P. *et al.* (1997). Airways responsiveness and development and remission of chronic respiratory symptoms in adults. *Lancet* **350**(9089), 1431–1434.

Yang, M., Rangasamy, D., Matthaei, K. I. *et al.* (2006). Inhibition of arginase I activity by RNA interference attenuates IL-13-induced airways hyperresponsiveness. *J Immunol* **177**, 5595–5603.

Zhang, K., Shan, L., Rahman, M. S. *et al.* (2007). Constitutive and inducible thymic stromal lymphopoietin expression in human airway smooth muscle cells: role in COPD. *Am J Physiol Lung Cell Mol Physiol* **293**, L375–L382.

Zheng, T., Zhu, Z., Wang, Z. *et al.* (2000). Inducible targeting of IL-13 to the adult lung causes matrix metalloproteinase- and cathepsin-dependent emphysema. *J Clin Invest* **106**, 1081–1093.

Zhu, Z., Homer, R. J., Wang, Z. *et al.* (1999). Pulmonary expression of interleukin-13 causes inflammation, mucus hypersecretion, subepithelial fibrosis, physiologic abnormalities, and eotaxin production. *J Clin Invest* **103**, 779–788.

Zimmermann, N., King, N. E., Laporte, J. *et al.* (2003). Dissection of experimental asthma with DNA microarray analysis identifies arginase in asthma pathogenesis. *J Clin Invest* **111**, 1863–1874.

Zimmermann, N. and Rothenberg, M. E. (2006). The arginine–arginase balance in asthma and lung inflammation. *Eur J Pharmacol* **533**(1–3), 253–262.

Zhou, D., Zheng, X., Wang, L. *et al.* (2003). Expression and effects of cardiotrophin-1 (CT-1) in human airway smooth muscle cells. *Br J Pharmacol* **140**, 1237–1244.

Zou, H., Xu, Y. J., Zhang, Z. X. (2005). Effect of vascular endothelial growth factor and its receptor KDR on human airway smooth muscle cells proliferation. *Chin Med J (Engl)* **118**, 591–594.

Zuyderduyn, S., Hiemstra, P. S., Rabe, K. F. (2004). TGF-beta differentially regulates Th2 cytokine-induced eotaxin and eotaxin-3 release by human airway smooth muscle cells. *J Allergy Clin Immunol* **114**, 791–798.

12

Glucocorticoid actions on airway smooth muscle

Ian M. Adcock, Loukia Tsaprouni and Pank Bhavsar

Cell and Molecular Biology, Airways Disease Section, National Heart and Lung Institute, Imperial College, London, UK

12.1 Introduction

Glucocorticoids are the most effective therapy for the treatment of many chronic inflammatory diseases such as asthma (Ito *et al.*, 2006). They have marked effects on airway smooth muscle (ASM) hypertrophy and hyperplasia and can modulate the expression of inflammatory mediators released from these cells (Chung *et al.*, 2004; Hirst *et al.*, 1998). Glucocorticoids act by binding to cytosolic glucocorticoid receptors (GR), which upon binding become activated and rapidly translocate to the nucleus. Within the nucleus, GR either induce transcription of genes, such as secretary leukocyte proteinase inhibitor (SLPI) (Abbinante-Nissen *et al.*, 1995) and mitogen-activated kinase phosphatase-1 (Lasa *et al.*, 2002), by binding to specific DNA elements (GRE) at the promoter/enhancer of responsive genes, or reduce inflammatory gene transcription induced by NF-κB or other pro-inflammatory transcription factors (Adcock *et al.*, 2000b). They also modulate the activity of many important signalling pathways to control other aspects of cell function (Ito *et al.*, 2006).

Airway Smooth Muscle   Edited by Kian Fan Chung

© 2008 John Wiley & Sons, Ltd

12.2 Airway smooth muscle (ASM) in airways disease

Henry Hyde Salter (1823-1871) first recognized the importance of ASM in asthma, particularly its ability to respond to noxious stimuli such as allergens by contraction and reducing the airway lumen (Hirst *et al.*, 2004; Munakata, 2006). Indeed, isolated ASM cells from asthmatic subjects contract much quicker and to a greater extent *in vitro* than cells from control subjects, perhaps as a result of increased myosin light-chain kinase (MLCK) expression (Ma *et al.*, 2002). More recently, ASM has been recognized as having a synthetic capacity allowing the expression of many key inflammatory mediators, adhesion molecules and receptors (Hirst *et al.*, 2004; Munakata, 2006). Whether the phenotypic changes in ASM cells which occur between contractile and synthetic cells in culture also occur in asthma *in vivo* remains to be determined (Munakata, 2006).

Increased ASM in asthma is well recognized as being associated with reduced airway patency in patients with both fatal and non-fatal asthma (Hirst *et al.*, 2004; Munakata, 2006). The increased ASM mass may result from a number of different factors, including hypertrophy, hyperplasia or migration (Johnson *et al.*, 2001), and the contribution of each process may vary with individual subjects (Ebina, *et al.*, 1993) and the duration of disease (Bai *et al.*, 2000). The presence of infiltrating inflammatory cells in close proximity to ASM has also been implicated in ASM remodelling and enhanced biosynthetic capacity in asthma (Brightling *et al.*, 2002) due to the ability of mast cell-released factors to induce ASM proliferation *in vitro* (Berger *et al.*, 1998). However, other studies have not confirmed an increase in ASM-associated mast cells in asthma (Niimi *et al.*, 2005). In addition, it is possible that the mitogens are derived from ASM cells themselves in an autocrine feed-forward mechanism (Johnson *et al.*, 2004). More recently, it has become apparent that changes in ASM remodelling occur in the small airways of patients with COPD and that this may be important in the obstruction observed in these patients (Chung, 2005; Ito *et al.*, 2005).

Mechanisms of hyperplasia

Changes in ASM cell number are certainly involved in the mechanisms of ASM remodelling in asthma (Hirst *et al.*, 2004; Munakata, 2006). This may involve cell proliferation, reduced apoptosis or potentially even mesenchymal cell migration (Holgate *et al.*, 2004). In cultured primary ASM cells, it is clear that growth factors, oxidative stress, muscle stretch, the matrix upon which cells are grown, and inflammatory stimuli can increase ASM proliferation (Bonacci *et al.*, 2003;

Hirst *et al.*, 2004; Munakata, 2006). However, it has proved difficult to observe markers of proliferation in ASM cells obtained from bronchial biopsies of asthmatic subjects (Hirst *et al.*, 2004; Munakata, 2006) or even in animal models of asthma (Hirst *et al.*, 2004; Munakata, 2006) despite clear evidence for ASM hyperplasia. In some animal models of asthma, increased ASM hyperplasia may result from reduced apoptosis (Hirst *et al.*, 2004; Munakata, 2006), although this again has not been reported in human disease (Druilhe *et al.*, 1998), perhaps because ASM cell apoptosis is inhibited by interaction with matrix proteins acting through β1-integrins (Freyer *et al.*, 2001).

This failure to observe an increase in the prevalence of proliferation markers or a reduction in apoptotic markers has led to the theory that the increase in cell ASM number is a result of mesenchymal cell migration and differentiation (Holgate *et al.*, 2000; 2007). The concept that the enhanced expression and release of growth factors, particularly EGF and TGFβ, from the epithelium of asthmatic subjects can have profound effects on mesenchymal cells, especially ASM cells, has been called the epithelial-mesenchymal trophic unit (EMTU) by Holgate and colleagues (Holgate, 2007, Holgate *et al.*, 2000).

Mechanisms of hypertrophy

Various studies have also reported an increase in the size of each ASM cell (hypertrophy) in the large airways of some, although not all, asthmatic subjects (Hirst *et al.*, 2004; Munakata, 2006). This may result from the actions of growth factors and/or inflammatory cytokines (Hirst *et al.*, 2004; Munakata, 2006).

Mediators and signalling pathways driving ASM growth

An important, recent study that looked at the structural alterations of airways in severe asthma (Benayoun *et al.*, 2003) found increases in ASM size and number, and the expression of MLCK was associated with reduced lung function in these patients. The authors were unable to show changes in the proliferation marker Ki67 in ASM despite this being observed in the epithelium and submucosa (Benayoun *et al.*, 2003). Importantly, allergen challenge causes the release of mitogen(s) into bronchoalveolar lavage (BAL) fluid from asthmatic patients, and this, in turn, caused marked increases in DNA synthesis, cell number, and cyclin D1 expression (Naureckas *et al.*, 1999; Xiong *et al.*, 1997). Numerous mitogens can provoke ASM proliferation, including growth factors, inflammatory

Table 12.1 Factors affecting ASM proliferation

Growth factors	Contractile agonists	Extracellular matrix (ECM) proteins	Other mediators
epidermal growth factor (EGF)	Endothelin-1	Collagen I, III, V	Lysosomal hydrolases
insulin-like growth factors (IGF)	Substance P	Fibronectin	A—Thrombin
platelet-derived growth factor (PDGF) isoforms	Phenylephrine	Tenascin	Tryptase
fibroblast growth factor 2 (FGF-2)	Serotonin	Hyaluronan	Sphingosine 1-phosphate
Basic fibroblast growth factor (bFGF)	Thromboxane A_2 Leukotriene D_4 Histamine	Verasican	Reactive oxygen species (ROS) Cytokines e.g. TNFα, IL-1β, IL-6

mediators, mechanical stress, ROS and contractile agents. These are described in Table 12.1. The effects of inflammatory cytokines, such as IL-1β, TNFα or IL-6, are variable, possibly due to actions of endogenous COX-2 products or other cytokines, such as IFNβ, which prevent DNA synthesis (Hirst *et al.*, 2004; Munakata, 2006).

The major proliferative stimuli for ASM cells (Table 12.1) can activate divergent intracellular signalling pathways, but they all converge on a few key steps or proteins. Thus, all mitogenic stimuli activate p21Ras, a 21-kDa guanosine triphosphatase (GTPase), either directly or subsequent to Src activation (Hirst *et al.*, 2004; Krymskaya *et al.*, 2005; Munakata, 2006) (Figure 12.1). Activation of p21Ras leads to Raf1 and phosphoinositide 3′-kinase (PI3K) stimulation and induction of the extracellular signal–regulated kinase (ERK) pathway. These kinases can both phosphorylate cyclin D1, either directly (ERK) or indirectly through activation of the S6 ribosomal kinase (p70^{S6K}), Rac1, and PKCζ (Hirst *et al.*, 2004; Munakata, 2006). G protein-coupled receptors (GPCRs) stimulate the production of diacylglycerol (DAG) and inositol triphosphate (IP$_3$), which, respectively, activate protein kinase C (PKC) and induce the release of stored Ca^{++} from the endoplasmic reticulum (ER). The combination of Ca^{++} mobilization and PKC activation is required for ASM proliferation (Hirst *et al.*, 2004; Munakata, 2006), although GPCR activation may also lead to p21Ras directly or indirectly through Src (Krymskaya *et al.*, 2005). PI3K and ERK appear to be the major pathways involved *in vitro*, but this remains to be confirmed *in vivo* in man.

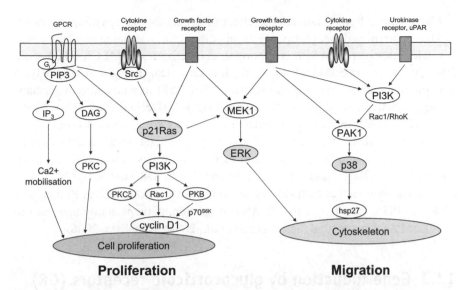

Figure 12.1 Schematic diagram of the signal pathways affecting ASM proliferation and migration in response to growth factors, urokinase (uPAR = urokinase receptor), cytokines or G-protein-coupled receptor (GPCR) activation. Key pathways involved in ASM proliferation include extracellular regulated kinase (ERK) and p21Ras. Both ERK and p21Ras affect nuclear cyclin D1, which is a key regulation of G1 cell-cycle progression. ERK affects cyclin D1 directly, whereas the effect of p21Ras is mediated through intermediary proteins such as phospho-inositide 3 kinase/protein kinase B (PI3K/Akt), Rac1 or PKCξ. ERK and the p38 MAPK/PAK pathway are the major regulators of cytoskeletal changes in ASM in response to chemotactic factors. This may be direct in the case of ERK or indirect via heat-shock protein (hsp)27 in the case of p38 MAPK

Mediators and signalling pathways driving ASM migration

The same factors are also involved in cellular migration (Goncharova *et al.*, 2006; Hirst *et al.*, 2004; Munakata, 2006). Thus, growth factors, including TGF, PDGF, FGF and EGF, inflammatory mediators such as plasminogen activators, urokinase, IL-1β and leukotriene E_4 drive chemotaxis and chemokinesis and are important for both migration and growth. However, this is not true in all cases, as thrombin increases proliferation but does not affect migration (Goncharova *et al.*, 2003). Vascular smooth muscle cells undergo migration in response to chemotactic signals by putting out filopodia and lamellipodia, a process regulated by profound changes in cytoskeletal remodelling (Madison, 2003), and it is hypothesized that similar mechanisms that drive vascular smooth muscle migration occur with ASM (Madison, 2003).

Many studies have examined the effect of urokinase on signalling pathways and have determined that phosphorylation of the actin-associated regulatory protein caldesmon by p38 MAPK is important, as is activation of MLCK (Hirst *et al.*, 2004; Madison, 2003; Munakata, 2006). It is now clear that p38 MAPK plays a key role in mediating the migratory response of ASM to many stimuli, perhaps acting downstream to p21-activated kinase 1 (PAK1) (Dechert *et al.*, 2001; Madison, 2003). The PAK family of molecules interacts with several Rho GTPases (RhoA, Rac, Cdc42) that regulate the actin polymerization-dependent appearance of filopodia and lamellipodia and membrane ruffles during cell migration (Dechert *et al.*, 2001; Madison, 2003). In addition to p38 MAPK and MLCK, it has become evident that pathways involved in ASM growth, such as Rho kinase, ERK and PI3K, also play a role in ASM migration, perhaps acting upstream to PAK1 or PAK1-associated factors (Hirst *et al.*, 2004; Munakata, 2006).

12.3 Gene induction by glucocorticoid receptors (GR)

Glucocorticoids have a major effect on ASM function due to their high expression of GR (Adcock *et al.*, 1996). Glucocorticoids act by freely diffusing across the cell membrane and binding to the cytoplasmic GR (Ito *et al.*, 2006). Two major GR isoforms exist (α and β), with the nuclear localized GRβ having a dominant negative effect on GRα via the formation of GRα/GRβ heterodimers. Evidence is accumulating that this isoform (GRβ) may be important in certain disease states in which GR nuclear translocation is deficient (Ito *et al.*, 2006; Zhou *et al.*, 2005).

Glucocorticoid binding to the cytoplasmic GR enables dissociation of chaperone proteins, including heat-shock protein (hsp) 90, and translocation to the nucleus (Ito *et al.*, 2006). Within the nucleus, GR dimerizes with another GR and binds to consensus DNA sites, termed glucocorticoid response elements (GREs), in the regulatory regions of glucocorticoid-responsive genes (Figure 12.2). This interaction allows GR to recruit activated transcriptional co-activator proteins, including steroid receptor coactivator-1 (SRC-1) and cAMP response element binding protein (CREB)-binding protein (CBP) through LxxLL motifs (Smith *et al.*, 2004).

GR transactivation and histone acetylation

Expression and repression of genes are associated with alterations in chromatin structure by enzymatic modification of core histones (Lee *et al.*, 2007; Li *et al.*,

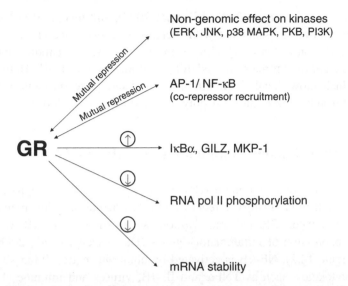

Figure 12.2 Upon activation by ligand binding, the glucocorticoid receptor (GR) can affect diverse signalling pathways involved in the synthetic and proliferative functions of ASM. Kinases such as extracellular regulated kinase (ERK), c-Jun N-terminal kinase (JNK), p38 mitogen-activated protein kinase (p38 MAPK), protein kinase B (PKB), and phosphoinositide-3 kinase (PI3K) are regulated by activated GR. Activation of these kinases can, in turn, modulate GR function. The regulation of inflammatory genes induced by transcription factors, such as activated protein (AP)-1 and nuclear factor (NF)-κB, is a major target for GR actions through effects on co-repressor recruitment. This repression is mutual, in that over-expression of AP-1 and NF-κB can suppress GR actions. These effects may, in part, be mediated by GR-mediated transcriptional induction of an NF-κB inhibitor (IκBα), an AP-1 DNA binding and activity inhibitor protein (glucocorticoid inducible leucine zipper (GILZ)), and a dual p38 MAPK and JNK inhibitor (MAPK phosphatase-1, MPK-1) in some cells. GR may also act by reducing the phosphorylation status of RNA polymerase II (RNA pol II) at some NF-κB-activated genes. Finally, GR may increase the levels of cell ribonucleases and mRNA destabilizing proteins, thereby altering mRNA stability

2007). Specific residues (lysines, arginines, and serines) within the N-terminal tails of core histones can be post-translationally modified by acetylation, methylation, ubiquitination, or phosphorylation, all of which have been implicated in the regulation of gene expression (Lee *et al.*, 2007; Li *et al.*, 2007). The 'histone code' refers to these modifications, which are set and maintained by histone-modifying enzymes and contribute to co-activator recruitment and subsequent increases in transcription (Jenuwein *et al.*, 2001; Rice *et al.*, 2001).

Transcriptional co-activators, such as CBP, SRC-1, TIF2, GRIP-1 and p300/CBP associated factor (PCAF), have intrinsic histone acetyltransferase

(HAT) activity (Lee *et al.*, 2007; Li *et al.*, 2007), and increased GR-mediated gene transcription is associated with increase in histone acetylation. Changes in the acetylation status of specific lysine residues forms a molecular tag for the recruitment of chromatin-remodelling enzymes such as Brg1 (Hebbar *et al.*, 2007), which allows local chromatin unwinding and the recruitment of the basal transcriptional complex and RNA polymerase II (Nie *et al.*, 2000).

Repression of inflammatory gene expression by GR

GR dimerization-deficient mice (Reichardt *et al.*, 1998; 2001) indicate that the major anti-inflammatory effects of glucocorticoids are due mainly to an interaction between GR and transcription factors such as NF-κB, which mediate the expression of inflammatory genes (De Bosccher *et al.*, 2006; Karin, 1998) (Figure 12.2). NF-κB is activated by numerous extracellular stimuli, including cytokines, such as TNFα and IL-1β, viruses and immune challenges (Baldwin, 2001). Activation of NF-κB involves stimulation of a phosphorylation cascade, resulting in phosphorylation and ubiquitination of a cytoplasmic inhibitor (IκBα), and release of NF-κB (generally a p65/p50 heterodimer) and its nuclear translocation (Ghosh *et al.*, 2002). NF-κB, as with GR, can induce histone acetylation in a temporal manner (Ito *et al.*, 2000; Lee *et al.*, 2006), leading to recruitment of distinct co-activator and remodelling complexes and the induction of inflammatory gene expression.

However, no all-activated NF-κB is the same. Thus, NF-κB activated by distinct cellular stimuli can control the expression of different patterns of genes (Covert *et al.*, 2005; Ogawa *et al.*, 2005; Werner *et al.*, 2005). LPS and TNFα induced distinct gene profiles as a result of differences in the amplitude and duration of NF-κB activation, rate of IκBα decay, and association with other factors such as IRF-3 (Covert *et al.*, 2005; Nelson *et al.*, 2004; Ogawa *et al.*, 2005; Werner *et al.*, 2005). Other signalling pathways, such as the MAPKs, may also affect the pattern and/or duration of NF-κB-mediated gene expression.

GR-NF-κB cross-talk

The precise mechanism for the ability of activated GR to repress NF-κB-induced gene transcription is still under debate, and it may alter depending upon GR expression levels (Simons, 2006) but includes binding to, or recruiting, nuclear receptor co-repressors (Ito *et al.*, 2000; Nie *et al.*, 2005b; Rosenfeld *et al.*, 2001), direct repression of co-activator complexes (Ito *et al.*, 2000; Pascual *et al.*, 2005), actions on histone phosphorylation status (Hasegawa *et al.*, 2005), or effects on RNA polymerase II phosphorylation (Luecke *et al.*, 2005; Nissen *et al.*, 2000) (Figure 12.2).

TNFα-stimulated CCL11 release from human ASM cells is partially blocked by both long-acting and short-acting β_2-agonists in a cAMP-dependent manner, and by glucocorticoids (Pang *et al.*, 2001). The combined use of β_2-agonists or rolipram and glucocorticoids completely abrogates TNFα-induced CCL11 release. The mechanism for this enhanced anti-inflammatory effect of combination treatment, at least in respect to TNFα-stimulated CCL11 release from human ASM, has been investigated by Knox and associates (Nie *et al.*, 2005b). Inhibition of TNFα-induced CCL11 expression by both salmeterol and fluticasone was transcriptional, and by using chromatin immunoprecipitation (ChIP) assays, Nie and colleagues were able to show that both drugs inhibited TNFα-induced histone H4 acetylation and p65 binding to the native CCL11 promoter in an additive manner (Nie *et al.*, 2005b).

Not surprisingly, these effects are context/gene dependent, and repression often depends upon factors complexed with NF-κB. Thus, activated GR represses a large set of functionally related inflammatory genes stimulated by p65/IRF-3 complexes in macrophages (Ogawa *et al.*, 2005). PPARγ and LXRs repress overlapping transcriptional targets in a p65/IRF-3-independent manner and cooperate with GR to suppress distinct subsets of pattern recognition receptor (PRR)-responsive genes (Ogawa *et al.*, 2005). Similar results can also be seen in ASM cells. Activation of PPARγ inhibits serum-induced human proliferation more effectively than dexamethasone and has a similar effect to dexamethasone in suppressing GM-CSF release (Patel *et al.*, 2003). However, PPARγ agonists, but not dexamethasone, can also inhibit G-CSF release, indicating that PPARγ activation has additional anti-inflammatory effects to those of glucocorticoids (Patel *et al.*, 2003). Combination treatment with salmeterol and/or fluticasone further enhanced the ability of the PPARγ agonists 15d-PGJ2 and troglitazone to suppress TNFα-induced CCL11 transcription (Nie *et al.*, 2005a). PPARγ and GR can form a complex, which was seen at the native CCL11 promoter and resulted in a reduction in NF-κB p65 DNA binding and a suppression of histone acetylation as determined by ChIP assays (Nie *et al.*, 2005a). Interestingly, a different mechanism was proposed for the suppression of TNFα-induced CCL2 expression, which was regulated at the post-transcriptional level under the same conditions (Nie *et al.* 2005a).

As described above, glucocorticoids do not completely suppress all NF-κB-stimulated genes, and targeting NF-κB itself may provide a greater anti-inflammatory effect than seen with glucocorticoids alone. Thus, IL-1β and TNFα both induce a range of pro-inflammatory mediators, including ICAM-1, COX-2, IL-6, IL-8, GM-CSF, CCL5, CCL2, CXCL1, CXCL7 and CXCL5, in primary ASM cells through an NF-κB-dependent process. Attenuation of NF-κB activity by biochemical and pharmacological means results in at least a comparable reduction in mediator release compared to that seen with dexamethasone, and in some cases (ICAM-1, MCP-1, GROα and NAP-2) a greater benefit is seen (Catley *et al.*, 2006). Furthermore, similar effects have been reported *in vivo* in an animal model of asthma (Birrell *et al.*, 2005).

Other mechanisms of GR action

Induction of the NF-κB inhibitor IκBα has been reported to be important for glucocorticoid actions in some cell types (Ito *et al.*, 2006). Indeed, dexamethasone-mediated suppression of TNFα-induced CD38 expression in ASM cells involves an enhancement of IκBα expression (Kang *et al.*, 2006). However, in the presence of IFNβ or IFNγ, TNFα-mediated CD38 induction over prolonged periods becomes insensitive to the repressive actions of glucocorticoids due to the induction of GRβ (Tliba *et al.*, 2006). In contrast, short term stimulation of ASM cells with IFNgamma/TNFalpha induces glucocorticoid insensitivity through an effect of IRF-1 independent of GRβ (Tliba *et al.* 2007).

In a similar manner to that described for NF-κB, the GR monomer can bind directly or indirectly with AP-1 (Ito *et al.*, 2006), which is also upregulated during inflammation (Demoly *et al.*, 1995). It is clear that the mechanism by which genes are induced affects how glucocorticoids alone and in combination with β₂-agonists will affect gene expression. Thus, TNFα-induced IL-6 expression is predominantly NF-κB dependent, whereas TNFα-induced CCL5 expression is AP-1- and NF-AT-dependent in human ASM cells (Ammit *et al.*, 2002). Dexamethasone completely attenuates TNFα-induced CCL5 expression but only partially inhibits IL-6 expression. In contrast, β₂-agonists enhances IL-6 expression via a CRE and represses CCL5 expression via an AP-1-independent mechanism.

In addition, glucocorticoids may play a role in repressing the action of mitogen-activated protein kinases (MAPKs), such as the extracellular signal-regulated kinase (ERK) and c-Jun N-terminal kinase (JNK) (Adcock, 2003). This may occur through induction of the dual specificity MAPK phosphatase-1 (MKP-1), which thereby attenuates p38 MAPK and JNK activation (Kassel *et al.*, 2001; Lasa *et al.*, 2001). Rogatsky and colleagues have in turn shown reciprocal inhibition of rat GR reporter gene activity by JNKs (Rogatsky *et al.*, 1998). Furthermore, we, and others, have shown that p38 MAPK-mediated GR phosphorylation can attenuate GR function (Irusen *et al.*, 2002; Szatmary *et al.*, 2004). In primary human ASM cells, the ability of dexamethasone to suppress IL-1β and TNFα-induced GROα expression is MKP-1 dependent (Issa *et al.*, 2007). In these studies siRNA directed against MKP-1 partially reversed dexamethasone suppression, whereas over-expression of MKP-1 resulted in repression of GROα release. In addition, the presence of IL-1β enhanced the ability of the activated GR to bind to the MKP-1 promoter and upregulate MKP-1 mRNA expression (Issa *et al.*, 2007).

Glucocorticoids also appear to exert anti-inflammatory actions that do not depend on the receptor's ability to regulate transcription in the nucleus but on their ability to destabilize the mRNA of inflammatory genes containing AU rich regions in their 3'-UTRs (Rhen *et al.*, 2005). This appears to occur in a p38 MAPK-mediated manner, particularly in cells stimulated by TNFα (Rhen *et al.*, 2005). Furthermore, dexamethasone and fluticasone, but not salmeterol, reduced the expression of α-smooth muscle actin and the short isoform of MLCK in TGF β-treated primary bronchial ASM cells (Goldsmith *et al.*, 2007). This effect occurred at the post-transcriptional level, as steady-state mRNA was unaltered. Fluticasone also significantly increased α-actin protein turnover and the incorporation of α-actin into filamentous actin, resulting in a loss of contractile function (Goldsmith *et al.*, 2007).

12.4 Actions of β_2-agonists and glucocorticoids on growth, proliferation, and migration of ASM

Regulation of cell-cycle progression by β_2-agonists

Both short-acting and long-acting β_2-agonists, acting through the β_2-receptor, block increases in ASM hyperplasia and hypertrophy and DNA synthesis *in vitro* in response to most mitogenic stimuli (Hirst *et al.*, 2004; Munakata, 2006). This has been proposed to be due to activation of CCAAT–enhancer-binding protein (C/EBP)α, which leads to induction of the cdk inhibitor p21$^{\mathrm{cip1/waf}}$ and reduction in Rb phosphorylation (Roth *et al.*, 2004; 2006).

Antiproliferative activity of glucocorticoids

In a recent study, a 6-week course of flunisolide was able to produce a significant decrease in α-smooth muscle actin expression in peripheral, but not central, airways of patients with mild/moderate asthma without affecting collagen deposition or TGFβ expression (Bergeron *et al.*, 2005).

The additional beneficial effect of combination therapy has been attributed to the ability of glucocorticoids also to stimulate C/EBPα expression in a distinctly temporal manner. Thus, the combination of low concentrations of both drugs results in a 'synchronized' activation of C/EBPα and GR, leading to an enhanced antiproliferative effect (Roth *et al.*, 2002).

Glucocorticoid-induced inhibition of mitogen-induced ASM proliferation can be studied *in vitro* (Bonacci *et al.*, 2006b; Vlahos *et al.*, 2003). Glucocorticoids decrease the transcription and the translation of cyclin D1 and thereby also reduce Rb phosphorylation, leading to G1 arrest possibly through actions on $p21^{cip1/waf1}$ (Fernandes *et al.*, 1999). However, this is likely to be stimulus specific, as $p21^{cip1/waf1}$ is not a target of glucocorticoids when cells are stimulated with thrombin or EGF (Vlahos *et al.*, 2003). Whether the addition of formoterol to budesonide or salmeterol to fluticasone enhances glucocorticoid responsiveness or the maximal response of ASM cells may depend upon the stimulus and/or the target gene function under investigation (Hirst *et al.*, 2004; Munakata, 2006), and a greater endeavour to examine these effects *in vivo* should be undertaken.

Influence of glucocorticoids on hypertrophy

Few, if any, studies have examined the effect of glucocorticoids on hypertrophy, but *in vitro* it appears that glucocorticoids, surprisingly, may increase ASM size (Hirst *et al.*, 2004). Extrapolation from these *in vitro* data to the *in vivo* situation may be difficult and may simply reflect differences in the concentrations of glucocorticoid that are seen *in vitro* and *in vivo* (Hirst *et al.*, 2004).

Factors limiting the antiproliferative actions of β₂-agonists and glucocorticoids

ASM proliferation induced by thrombin and other GPCR stimuli is more sensitive to repression by both β₂-agonists and glucocorticoids compared to ASM cells stimulated by other mitogens (Bonacci *et al.*, 2006a; 2006b; Tran *et al.*, 2005). In addition, the concurrent exposure to inflammatory stimuli can reduce the ability of glucocorticoids to repress ASM proliferation (Vlahos *et al.*, 1999), perhaps through changes in the transcriptome (Hakonarson *et al.*, 2001). Furthermore, IL-1β can affect β₂-receptor coupling both *in vitro* and *in vivo*, and this would reduce the antiproliferative effects of β_2-agonists (Koto *et al.*, 1996; Mak *et al.*, 2002; Shore, 2004).

Serum from asthmatics enhances production of the ECM proteins fibronectin, perlecan, laminin $\gamma 1$ and chondroitin sulfate compared to the effect seen with control serum (Johnson *et al.*, 2000). This process appears to be insensitive to beclomethasone in asthmatic subjects, suggesting that the interaction between the allergic process and ASM remodelling may affect glucocorticoid sensitivity. This interaction between ASM function and allergy is lent further support by the fact that TH_2 cytokines and $TGF\beta$, but not IL-10, can stimulate human ASM cells to transcribe VEGF (Wen *et al.*, 2003). This effect can be inhibited by both IFNγ and glucocorticoids. In contrast, TH_2 cytokines can reduce the ability of bradykinin to stimulate IL-6 transcription in human ASM cells (Huang *et al.*, 2003).

It is also becoming clear that the matrix upon which ASM cells grow affects not only proliferation to various mitogenic stimuli but also the response to glucocorticoids. In a series of important experiments, Stewart and colleagues (Bonacci *et al.*, 2003; 2006a; Hirst *et al.*, 2004) have shown that cells grown on collagen I grow more rapidly and are less responsive to the antiproliferative effects of glucocorticoids despite retaining glucocorticoid sensitivity to the production of inflammatory mediators (Bonacci *et al.*, 2003; 2006a; Hirst *et al.*, 2004). This effect is blocked by inhibitors of $\alpha 2\beta 1$ integrin function, enabling fluticasone to suppress cyclin D1 levels (Bonacci *et al.*, 2006a), and suggesting that activation of signalling pathways, such as focal adhesion kinase (FAK) or integrin-linked kinase, are selectively glucocorticoid non-responsive. Intriguingly, *in vitro* ASM cells from asthmatic subjects produce greater levels of collagen I and perlican than ASM cells from non-asthmatic subjects (Johnson *et al.*, 2004). In other studies, the same group were able to determine that the resistance of EGF-induced proliferation to inhibition by glucocorticoids is not associated with a failure to regulate cyclin D1 induction, nor does it appear to be explained by differential regulation of the levels of the cdk inhibitors $p21^{Cip1/waf1}$ and $p27^{Kip1}$ (Vlahos *et al.*, 2003).

Importantly, the differences in human ASM proliferation and glucocorticoid sensitivity seen with cells grown on thrombin compared either to other ECMs or to cells stimulated with cytokines are also seen with respect to the synthetic capacity of these cells (Tran *et al.*, 2005).

Effects of asthma drugs on ASM migration

Salmeterol and other cAMP elevating agents, such as PGE_2, reduce spontaneous ASM migration (Goncharova *et al.*, 2003). In addition, both glucocorticoids and

PGE_2 and cilomilast can suppress PDGF-induced ASM migration, and the presence of salmeterol, which has no effect alone, enhances the ability of fluticasone to reduce PDGF-induced ASM migration (Goncharova et al., 2003).

12.5 Summary and conclusions

Factors that promote the proliferation and migration of ASM cells are similar and activate a number of key intracellular signaling pathways, including PI3K, ERK and p38 MAPK, which are potential targets for drugs currently used to treat asthma. In addition, it is clear that ASM cells play a role in the inflammatory response in asthma by releasing numerous mediators through NF-κB- and p38 MAPK-mediated pathways. Glucocorticoids also have profound effects on the signalling pathways controlling gene expression in vitro. Whether the effects of β_2-agonists and glucocorticoids seen in vitro also occur in vivo in patients with asthma needs to be determined.

Acknowledgements

The literature in this area is extensive, and many important studies were omitted because of constraints on space, for which we apologize. We would like to thank other members of the Cell and Molecular Biology Group for their helpful discussions. Work in our group is supported by Asthma UK, the British Lung Foundation, the Clinical Research Committee (Brompton Hospital), the Medical Research Council (UK), the National Institutes of Health (USA), the Wellcome Trust, AstraZeneca, Boehringer Ingelheim, GlaxoSmithKline (UK), Mitsubishi Pharma (Japan), Novartis and Pfizer.

References

Abbinante-Nissen, J. M., Simpson, L. G., Leikauf, G. D. et al. (1995). Corticosteroids increase secretory leukocyte protease inhibitor transcript levels in airway epithelial cells. Am J Physiol 268, L601–L606.

Adcock, I. M. (2003). Glucocorticoids: new mechanisms and future agents. Curr Allergy Asthma Rep 3, 249–257.

Adcock, I. M., Gilbey, T., Gelder, C. M. et al. (1996). Glucocorticoid receptor localization in normal and asthmatic lung. Am J Respir Crit Care Med 154, 771–782.

Adcock, I. M. and Ito, K. (2000b). Molecular mechanisms of corticosteroid actions. *Monaldi Arch Chest Dis* **55**, 256–266.

Adcock, I. M. and Ito, K. (2000a). Molecular mechanisms of corticosteroid actions. *Monaldi Arch Chest Dis* **55**, 256–266.

Ammit, A. J., Lazaar, A.L., Irani, C. *et al.* (2002). Tumor necrosis factor-alpha-induced secretion of RANTES and interleukin-6 from human airway smooth muscle cells: modulation by glucocorticoids and beta-agonists. *Am J Respir Cell Mol Biol* **26**, 465–474.

Bai, T. R., Cooper, J., Koelmeyer, T. *et al.* (2000). The effect of age and duration of disease on airway structure in fatal asthma. *Am J Respir Crit Care Med* **162**, 663–669.

Baldwin, A. S., Jr. (2001). Series introduction: the transcription factor NF-kappaB and human disease. *J Clin Invest* **107**, 3–6.

Benayoun, L., Druilhe, A., Dombret, M. C. *et al.* (2003). Airway structural alterations selectively associated with severe asthma. *Am J Respir Crit Care Med* **167**, 1360–1368.

Berger, P., Walls, A. F., Marthan, R. *et al.* (1998). Immunoglobulin E-induced passive sensitization of human airways: an immunohistochemical study. *Am J Respir Crit Care Med* **157**, 610–616.

Bergeron, C., Hauber, H. P., Gotfried, M. *et al.* (2005). Evidence of remodeling in peripheral airways of patients with mild to moderate asthma: effect of hydrofluoroalkane-flunisolide. *J Allergy Clin Immunol* **116**, 983–989.

Birrell, M. A., Hardaker, E., Wong, S. *et al.* (2005). Ikappa-B kinase-2 inhibitor blocks inflammation in human airway smooth muscle and a rat model of asthma. *Am J Respir Crit Care Med* **172**, 962–971.

Bonacci, J. V., Harris, T. and Stewart, A. G. (2003). Impact of extracellular matrix and strain on proliferation of bovine airway smooth muscle. *Clin Exp Pharmacol Physiol* **30**, 324–328.

Bonacci, J. V., Schuliga, M., Harris, T. *et al.* (2006a). Collagen impairs glucocorticoid actions in airway smooth muscle through integrin signalling. *Br J Pharmacol* **149**, 365–373.

Bonacci, J. V. and Stewart, A. G. (2006b). Regulation of human airway mesenchymal cell proliferation by glucocorticoids and beta2-adrenoceptor agonists. *Pulm Pharmacol Ther* **19**, 32–38.

Brightling, C. E., Bradding, P., Symon, F. A. *et al.* (2002). Mast-cell infiltration of airway smooth muscle in asthma. *N Engl J Med* **346**, 1699–1705.

Catley, M. C., Sukkar, M. B., Chung, K. F. *et al.* (2006). Validation of the anti-inflammatory properties of small-molecule IkappaB kinase (IKK)-2 inhibitors by comparison with adenoviral-mediated delivery of dominant-negative IKK1 and IKK2 in human airways smooth muscle. *Mol Pharmacol* **70**, 697–705.

Chung, K. F. (2005). The role of airway smooth muscle in the pathogenesis of airway wall remodeling in chronic obstructive pulmonary disease. *Proc Am Thorac Soc* **2**, 347–354.

Chung, K. F. and Adcock, I. M. (2004). Combination therapy of long-acting beta2-adrenoceptor agonists and corticosteroids for asthma. *Treat Respir Med* **3**, 279–289.

Covert, M. W., Leung, T. H., Gaston, J. E. *et al.* (2005). Achieving stability of lipopolysaccharide-induced NF-kappaB activation. *Science* **309**, 1854–1857.

De, B. K., Vanden, B. W. and Haegeman, G. (2006). Cross-talk between nuclear receptors and nuclear factor kappaB. *Oncogene* **25**, 6868–6886.

Dechert, M. A., Holder, J. M. and Gerthoffer, W. T. (2001). p21-activated kinase 1 participates in tracheal smooth muscle cell migration by signaling to p38 MAPK. *Am J Physiol Cell Physiol* **281**, C123–C132.

Demoly, P., Chanez, P., Pujol, J. L. *et al.* (1995). Fos immunoreactivity assessment on human normal and pathological bronchial biopsies. *Respir Med* **89**, 329–335.

Druilhe, A., Wallaert, B., Tsicopoulos, A. *et al.* (1998). Apoptosis, proliferation, and expression of Bcl-2, Fas, and Fas ligand in bronchial biopsies from asthmatics. *Am J Respir Cell Mol Biol* **19**, 747–757.

Ebina, M., Takahashi, T., Chiba, T. *et al.* (1993). Cellular hypertrophy and hyperplasia of airway smooth muscles underlying bronchial asthma. A 3-D morphometric study. *Am Rev Respir Dis* **148**, 720–726.

Fernandes, D., Guida, E., Koutsoubos, V. *et al.* (1999). Glucocorticoids inhibit proliferation, cyclin D1 expression, and retinoblastoma protein phosphorylation, but not activity of the extracellular-regulated kinases in human cultured airway smooth muscle. *Am J Respir Cell Mol Biol* **21**, 77–88.

Freyer, A. M., Johnson, S. R. and Hall, I. P. (2001). Effects of growth factors and extracellular matrix on survival of human airway smooth muscle cells. *Am J Respir Cell Mol Biol* **25**, 569–576.

Ghosh, S. and Karin, M. (2002). Missing pieces in the NF-kappaB puzzle. *Cell* **109** Suppl, S81–S96.

Goldsmith, A. M., Hershenson, M. B., Wolbert, M. P. *et al.* (2007). Regulation of airway smooth muscle alpha-actin expression by glucocorticoids. *Am J Physiol Lung Cell Mol Physiol* **292**, L99–L106.

Goncharova, E. A., Billington, C. K., Irani, C. *et al.* (2003). Cyclic AMP-mobilizing agents and glucocorticoids modulate human smooth muscle cell migration. *Am J Respir Cell Mol Biol* **29**, 19–27.

Goncharova, E. A., Goncharov, D. A. and Krymskaya, V. P. (2006). Assays for in vitro monitoring of human airway smooth muscle (ASM) and human pulmonary arterial vascular smooth muscle (VSM) cell migration. *Nat Protoc* **1**, 2933–2939.

Hakonarson, H., Halapi, E., Whelan, R. *et al.* (2001). Association between IL-1beta/TNF-alpha-induced glucocorticoid-sensitive changes in multiple gene expression and altered responsiveness in airway smooth muscle. *Am J Respir Cell Mol Biol* **25**, 761–771.

Hasegawa, Y., Tomita, K., Watanabe, M. *et al.* (2005). Dexamethasone inhibits phosphorylation of histone H3 at serine 10. *Biochem Biophys Res Commun* **336**, 1049–1055.

Hebbar, P. B. and Archer, T. K. (2007). Chromatin-dependent cooperativity between site-specific transcription factors in vivo. *J Biol Chem* **282**, 8284–8291.

Hirst, S. J. and Lee, T. H. (1998). Airway smooth muscle as a target of glucocorticoid action in the treatment of asthma. *Am J Respir Crit Care Med* **158**, S201–S206.

Hirst, S. J., Martin, J. G., Bonacci, J. V. et al. (2004). Proliferative aspects of airway smooth muscle. *J Allergy Clin Immunol* **114**, S2–17.

Holgate, S. T. (2007). The epithelium takes centre stage in asthma and atopic dermatitis. *Trends Immunol* **28**, 248–251.

Holgate, S. T., Holloway, J., Wilson, S. et al. (2004). Epithelial-mesenchymal communication in the pathogenesis of chronic asthma. *Proc Am Thorac Soc* **1**, 93–98.

Holgate, S. T., Lackie, P., Wilson, S. et al. (2000). Bronchial epithelium as a key regulator of airway allergen sensitization and remodeling in asthma. *Am J Respir Crit Care Med* **162**, S113–S117.

Huang, C. D., Tliba, O., Panettieri, R. A., Jr. et al. (2003). Bradykinin induces interleukin-6 production in human airway smooth muscle cells: modulation by Th2 cytokines and dexamethasone. *Am J Respir Cell Mol Biol* **28**, 330–338.

Irusen, E., Matthews, J. G., Takahashi, A. et al. (2002). p38 Mitogen-activated protein kinase-induced glucocorticoid receptor phosphorylation reduces its activity: role in steroid-insensitive asthma. *J Allergy Clin Immunol* **109**, 649–657.

Issa, R., Xie, S., Khorasani, N. et al. (2007). Corticosteroid inhibition of growth-related oncogene protein-alpha via mitogen-activated kinase phosphatase-1 in airway smooth muscle cells. *J Immunol* **178**, 7366–7375.

Ito, K., Barnes, P. J. and Adcock, I. M. (2000). Glucocorticoid receptor recruitment of histone deacetylase 2 inhibits interleukin-1beta-induced histone H4 acetylation on lysines 8 and 12. *Mol Cell Biol* **20**, 6891–6903.

Ito, K., Chung, K. F. and Adcock, I. M. (2006). Update on glucocorticoid action and resistance. *J Allergy Clin Immunol* **117**, 522–543.

Ito, K., Ito, M., Elliott, W. M. et al. (2005). Decreased histone deacetylase activity in chronic obstructive pulmonary disease. *N Engl J Med* **352**, 1967–1976.

Jenuwein, T. and Allis, C. D. (2001). Translating the histone code. *Science* **293**, 1074–1080.

Johnson, P. R., Black, J. L., Carlin, S. et al. (2000). The production of extracellular matrix proteins by human passively sensitized airway smooth-muscle cells in culture: the effect of beclomethasone. *Am J Respir Crit Care Med* **162**, 2145–2151.

Johnson, P. R., Burgess, J. K., Underwood, P. A. et al. (2004). Extracellular matrix proteins modulate asthmatic airway smooth muscle cell proliferation via an autocrine mechanism. *J Allergy Clin Immunol* **113**, 690–696.

Johnson, P. R., Roth, M., Tamm, M. et al. (2001). Airway smooth muscle cell proliferation is increased in asthma. *Am J Respir Crit Care Med* **164**, 474–477.

Kang, B. N., Tirumurugaan, K. G., Deshpande, D. A. et al. (2006). Transcriptional regulation of CD38 expression by tumor necrosis factor-alpha in human airway smooth muscle cells: role of NF-kappaB and sensitivity to glucocorticoids. *FASEB J* **20**, 1000–1002.

Karin, M. (1998). New twists in gene regulation by glucocorticoid receptor: is DNA binding dispensable? *Cell* **93**, 487–490.

Kassel, O., Sancono A., Kratzschmar J. et al. (2001). Glucocorticoids inhibit MAP kinase via increased expression and decreased degradation of MKP-1. *EMBO J* **20**, 7108–7116.

Koto, H., Mak, J. C., Haddad, E. B. *et al.* (1996). Mechanisms of impaired beta-adrenoceptor-induced airway relaxation by interleukin-1beta in vivo in the rat. *J Clin Invest* **98**, 1780–1787.

Krymskaya, V. P., Goncharova, E. A., Ammit, A. J. *et al.* (2005). Src is necessary and sufficient for human airway smooth muscle cell proliferation and migration. *FASEB J* **19**, 428–430.

Lasa, M., Abraham, S. M., Boucheron, C. *et al.* (2002). Dexamethasone causes sustained expression of mitogen-activated protein kinase (MAPK) phosphatase 1 and phosphatase-mediated inhibition of MAPK p38. *Mol Cell Biol* **22**, 7802–7811.

Lasa, M., Brook, M., Saklatvala, J. *et al.* (2001). Dexamethasone destabilizes cyclooxygenase 2 mRNA by inhibiting mitogen-activated protein kinase p38. *Mol Cell Biol* **21**, 771–780.

Lee, K. K. and Workman, J. L. (2007). Histone acetyltransferase complexes: one size doesn't fit all. *Nat Rev Mol Cell Biol* **8**, 284–295.

Lee, K. Y., Ito, K., Hayashi, R. *et al.* (2006). NF-κB and activator protein 1 response elements and the role of histone modifications in IL-1β-induced TGF-β1 gene transcription. *J Immunol* **176**, 603–615.

Li, B., Carey, M. and Workman, J. L. (2007). The role of chromatin during transcription. *Cell* **128**, 707–719.

Luecke, H. F. and Yamamoto, K. R. (2005). The glucocorticoid receptor blocks P-TEFb recruitment by NFkappaB to effect promoter-specific transcriptional repression. *Genes Dev* **19**, 1116–1127.

Ma, X., Cheng, Z., Kong, H. *et al.* (2002). Changes in biophysical and biochemical properties of single bronchial smooth muscle cells from asthmatic subjects. *Am J Physiol Lung Cell Mol Physiol* **283**, L1181–L1189.

Madison, J. M. (2003). Migration of airway smooth muscle cells. *Am J Respir Cell Mol Biol* **29**, 8–11.

Mak, J. C., Hisada, T., Salmon, M. *et al.* (2002). Glucocorticoids reverse IL-1beta-induced impairment of beta-adrenoceptor-mediated relaxation and up-regulation of G-protein-coupled receptor kinases. *Br J Pharmacol* **135**, 987–996.

Munakata, M. (2006). Airway remodeling and airway smooth muscle in asthma. *Allergol Int* **55**, 235–243.

Naureckas, E. T., Ndukwu, I. M., Halayko, A. J. *et al.* (1999). Bronchoalveolar lavage fluid from asthmatic subjects is mitogenic for human airway smooth muscle. *Am J Respir Crit Care Med* **160**, 2062–2066.

Nelson, D. E., Ihekwaba, A. E., Elliott, M. *et al.* (2004). Oscillations in NF-kappaB signaling control the dynamics of gene expression. *Science* **306**, 704–708.

Nie, M., Corbett, L., Knox, A. J. *et al.* (2005a). Differential regulation of chemokine expression by peroxisome proliferator-activated receptor gamma agonists: interactions with glucocorticoids and beta2-agonists. *J Biol Chem* **280**, 2550–2561.

Nie, M., Knox, A. J. and Pang, L. (2005b). beta2-Adrenoceptor agonists, like glucocorticoids, repress eotaxin gene transcription by selective inhibition of histone H4 acetylation. *J Immunol* **175**, 478–486.

Nie, Z., Xue, Y., Yang, D. *et al.* (2000). A specificity and targeting subunit of a human SWI/SNF family-related chromatin-remodeling complex. *Mol Cell Biol* **20**, 8879–8888.

Niimi, A., Torrego, A., Nicholson, A. G. *et al.* (2005). Nature of airway inflammation and remodeling in chronic cough. *J Allergy Clin Immunol* **116**, 565–570.

Nissen, R. M. and Yamamoto, K. R. (2000). The glucocorticoid receptor inhibits NFkappaB by interfering with serine-2 phosphorylation of the RNA polymerase II carboxy-terminal domain. *Genes Dev* **14**, 2314–2329.

Ogawa, S., Lozach., J., Benner, C. *et al.* (2005). Molecular determinants of crosstalk between nuclear receptors and Toll-like receptors. *Cell* **122**, 707–721.

Pang, L. and Knox, A. J. (2001). Regulation of TNF-alpha-induced eotaxin release from cultured human airway smooth muscle cells by beta2-agonists and corticosteroids. *FASEB J* **15**, 261–269.

Pascual, G., Fong, A. L., Ogawa, S. *et al.* (2005). A SUMOylation-dependent pathway mediates transrepression of inflammatory response genes by PPAR-gamma. *Nature* **437**, 759–763.

Patel, H. J., Belvisi, M. G., Bishop-Bailey, D. *et al.* (2003). Activation of peroxisome proliferator-activated receptors in human airway smooth muscle cells has a superior anti-inflammatory profile to corticosteroids: relevance for chronic obstructive pulmonary disease therapy. *J Immunol* **170**, 2663–2669.

Reichardt, H. M., Kaestner, K. H., Tuckermann, J. *et al.* (1998). DNA binding of the glucocorticoid receptor is not essential for survival. *Cell* **93**, 531–541.

Reichardt, H. M., Tuckermann, J. P., Gottlicher, M. *et al.* (2001). Repression of inflammatory responses in the absence of DNA binding by the glucocorticoid receptor. *EMBO J* **20**, 7168–7173.

Rhen, T. and Cidlowski, J. A. (2005). Antiinflammatory action of glucocorticoids – new mechanisms for old drugs. *N Engl J Med* **353**, 1711–1723.

Rice, J. C. and Allis, C. D. (2001). Code of silence. *Nature* **414**, 258–261.

Rogatsky, I., Logan, S. K. and Garabedian, M. J. (1998). Antagonism of glucocorticoid receptor transcriptional activation by the c-Jun N-terminal kinase. *Proc Natl Acad Sci U S A* **95**, (2050–2055.

Rosenfeld, M. G. and Glass, C. K. (2001). Coregulator codes of transcriptional regulation by nuclear receptors. *J Biol Chem* **276**, 36865–36868.

Roth, M. and Black, J. L. (2006). Transcription factors in asthma: are transcription factors a new target for asthma therapy? *Curr Drug Targets* **7**, 589–595.

Roth, M., Johnson, P. R., Borger, P. *et al.* (2004). Dysfunctional interaction of C/EBPalpha and the glucocorticoid receptor in asthmatic bronchial smooth-muscle cells. *N Engl J Med* **351**, 560–574.

Roth, M., Johnson, P. R., Rudiger, J. J. *et al.* (2002). Interaction between glucocorticoids and beta2 agonists on bronchial airway smooth muscle cells through synchronised cellular signalling. *Lancet* **360**, 1293–1299.

Shore, S. A. (2004). Airway smooth muscle in asthma – not just more of the same. *N Engl J Med* **351**, 531–532.

Simons, S. S., Jr. (2006). How much is enough? Modulation of dose-response curve for steroid receptor-regulated gene expression by changing concentrations of transcription factor. *Curr Top Med Chem* **6**, 271–285.

Smith, C. L. and O'Malley, B. W. (2004). Coregulator function: a key to understanding tissue specificity of selective receptor modulators. *Endocr Rev* **25**, 45–71.

Szatmary, Z., Garabedian, M. J. and Vilcek, J. (2004). Inhibition of glucocorticoid receptor-mediated transcriptional activation by p38 mitogen-activated protein (MAP) kinase. *J Biol Chem* **279**, 43708–43715.

Tliba, O., Damera, G., Banerjee, A. *et al.* (2007). Cytokines Induce an Early Steroid Resistance in Airway Smooth Muscle Cells: Novel Role of IRF-1. *Am. J. Respir Cell Mol.* Biol. doi: 10.1165/rcmb.2007-0226OC.

Tliba, O., Cidlowski, J. A. and Amrani, Y. (2006). CD38 expression is insensitive to steroid action in cells treated with tumor necrosis factor-alpha and interferon-gamma by a mechanism involving the up-regulation of the glucocorticoid receptor beta isoform. *Mol Pharmacol* **69**, 588–596.

Tran, T., Fernandes, D. J., Schuliga, M. *et al.* (2005). Stimulus-dependent glucocorticoid-resistance of GM-CSF production in human cultured airway smooth muscle. *Br J Pharmacol* **145**, 123–131.

Vlahos, R., Lee, K. S., Guida, E. *et al.* (2003). Differential inhibition of thrombin- and EGF-stimulated human cultured airway smooth muscle proliferation by glucocorticoids. *Pulm Pharmacol Ther* **16**, 171–180.

Vlahos, R. and Stewart, A. G. (1999). Interleukin-1alpha and tumour necrosis factor-alpha modulate airway smooth muscle DNA synthesis by induction of cyclo-oxygenase-2: inhibition by dexamethasone and fluticasone propionate. *Br J Pharmacol* **126**, 1315–1324.

Wen, F. Q., Liu, X., Manda, W. *et al.* (2003). TH2 Cytokine-enhanced and TGF-beta-enhanced vascular endothelial growth factor production by cultured human airway smooth muscle cells is attenuated by IFN-gamma and corticosteroids. *J Allergy Clin Immunol* **111**, 1307–1318.

Werner, S. L., Barken, D. and Hoffmann, A. (2005). Stimulus specificity of gene expression programs determined by temporal control of IKK activity. *Science* **309**, 1857–1861.

Xiong, W., Pestell, R. G., Watanabe, G. *et al.* (1997). Cyclin D1 is required for S phase traversal in bovine tracheal myocytes. *Am J Physiol* **272**, L1205–L1210.

Zhou, J. and Cidlowski, J. A. (2005). The human glucocorticoid receptor: one gene, multiple proteins and diverse responses. *Steroids* **70**, 407–417.

13

β₂-Adrenergic receptors: effects on airway smooth muscle

Malcolm Johnson

GlaxoSmithKline Research and Development, UXBridge, Middlesex, UK

13.1 The β₂-adrenoceptor

The human β-adrenoceptor gene is situated on the long arm of chromosome 5 and codes for an intronless gene product of approximately 1200 base pairs (Kobilka *et al.*, 1987). The β-adrenoceptor comprises 413 amino-acid residues of approximately 46,500 Da. β-Adrenoceptors have been subdivided into at least three distinct groups – β_1, β_2 and β_3 – classically identified in cardiac, airway smooth muscle (ASM) and adipose tissue, respectively. This chapter focuses on the β_2-adrenergic receptor with regard to its function and response to β_2-agonists in ASM.

Like all G-protein-coupled receptors (GPCR), the β_2-receptor has seven transmembrane-spanning α-helices (Liggett, 2002). There are three extracellular loops, one being the amino-terminus, and three intracellular loops, with a carboxy-terminus (Figure 13.1). The receptor is *N*-glycosylated at amino acids 6, 15 and 187; these are important for insertion into the cell membrane

Airway Smooth Muscle   Edited by Kian Fan Chung
© 2008 John Wiley & Sons, Ltd

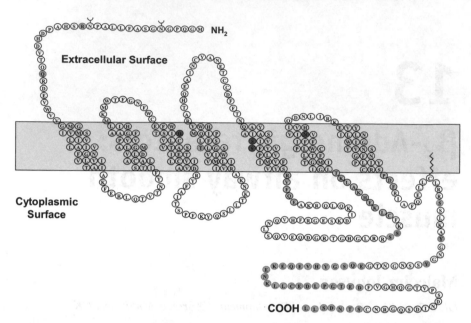

Figure 13.1 The structure of the human β₂-receptor. The sites of polymorphic mutations of the receptor are shown in green. The four amino-acid residues involved in β₂-agonist binding are shown in red. Regions or specific domains involved in Gs-protein coupling (blue), desensitization (pink) and downregulation (orange) are indicated (Johnson, 2006). (Reproduced from J Allergy Clin Immunol 2006;117: 18–24, courtesy of Elevier.) (For a colour reproduction of this figure, please see the colour section, located towards the centre of the book).

and for agonist-induced receptor trafficking (Mialet-Perez *et al.*, 2004). At amino-acid position 341, the cysteine of the human β₂-receptor is palmitoylated, acting to anchor the carboxy-terminus to the membrane (O'Dowd *et al.*, 1989). The region between the seventh transmembrane-spanning domain and the palmitoylated cysteine is also an α-helix, sometimes denoted as the fourth intracellular loop.

Autoradiographic studies have suggested that β₂-adrenoceptors are widely distributed, occurring in ASM at a density of 30,000-40,000/cell (Johnson, 2000). Radioligand-binding studies on lobectomy specimens have shown β₂-receptor density to increase through the respiratory tract, with high levels in the central airways and the alveolar region. Computed tomography (CT) scanning (Hoffman *et al.*, 1997) and positron emission tomography with a C11-labelled β-receptor ligand (Ueki *et al.*, 1993) have confirmed this β₂-receptor distribution.

13.2 β₂-Receptor activation

The regions of the β_2-adrenoceptor protein important for β_2-agonist binding and G-protein coupling in ASM have been identified by site-directed mutagenesis. The active site of the receptor with which β_2-agonists must interact in order to exert their biological effects is located approximately one-third of the way (15 angs Froms) into the receptor core (Figure 13.1). It is generally agreed that the residues of critical importance with respect to agonist binding to the active site are aspartate (Asp)-residue 113 (counted from the extracellular or N-terminus end) in the third domain; two serine (Ser) residues, 204 and 207, which are both on the fifth domain; and asparagine (Asn) 293, in the sixth domain (Liggett, 2002). Thus, a model has emerged for the agonist binding site of the β_2-adrenoceptor, in which the ligand is bound within the hydrophobic core of the protein in the transmembrane helices, and anchored by specific molecular interactions between amino-acid residues in the receptor and functional groups on the ligand. Asp forms an ion pair with the amino nitrogen, while the two Ser residues interact with the hydroxyl groups on the phenyl ring of the β_2-adrenoceptor agonist molecule. The β-hydroxy group in the β_2-agonist side chain binds to Asn 293 (Liggett, 2002).

There is now good evidence that β_2-adrenoceptors oscillate between two forms, activated and inactivated, and that under resting conditions these two forms are in equilibrium, but with the inactivated state predominant (Liggett, 2002). The β_2-receptor is in the activated form when it is associated with a molecule of guanosine triphosphate (GTP). The replacement of the GTP by guanosine diphosphate causes the receptor to return to its low-energy, inactivated form (Onaran *et al.*, 1993). β_2-Agonists have their effects, not through inducing a conformational change in the receptor, but rather by binding to and temporarily stabilizing receptors in their activated state, that is, bound to GTP. However, it is now clear that multiple signals are promoted by β_2-receptors and that the 'optimal' conformation for one effector signal may not be the same as for another.

13.3 β₂-Receptor signalling pathways

It has been the accepted dogma since the 1960s that β_2-receptor activation is mediated by increased intracellular cyclic adenosine monophosphate (cAMP) levels (Robinson *et al.*, 1967). This is the result of stimulation of adenylate cyclase, which catalyses the conversion of ATP into cAMP. The coupling of the

β_2-receptor to adenylate cyclase is affected through a trimeric Gs-protein (Johnson and Coleman, 1995), consisting of an α-subunit (which stimulates adenylate cyclase) and $\beta\gamma$-subunits (which transduce other signals). cAMP levels are then regulated through the activity of phosphodiesterase isozymes/isoforms, which degrade it to 5'-AMP.

The mechanism by which cAMP induces ASM cell relaxation is not fully understood, but it is believed that it catalyses the activation of protein kinase A (PKA), which in turn phosphorylates key regulatory proteins involved in the control of muscle tone (Johnson and Coleman, 1995). cAMP also results in inhibition of calcium ion (Ca^{2+}) release from intracellular stores, reduction of membrane Ca^{2+} entry, and sequestration of intracellular Ca^{2+}, leading to relaxation of the ASM. However, it has been suggested recently that some of the relaxant response to β_2-agonists may be mediated through cAMP-independent mechanisms, involving direct interaction of Gs with potassium channels, which are present in the ASM cell membrane. β_2-Receptors in ASM can also couple to Gi proteins (Daaka *et al.*, 1997), resulting in stimulation of the extracellular signal-regulation kinase (SRK) and p38 mitogen-activated protein kinase (MAPK) pathways. This mechanism may serve to uncouple the β_2-receptor from Gs and thus represent a means of terminating the β_2-agonist receptor signal and response. Recent evidence suggests that there may be cross-talk between β_2-receptors and other GPCR systems in smooth muscle cells. Under experimental conditions, where the β_2-receptor was either ablated or over-expressed, responses to Gq-coupled contractile agonists were inhibited and enhanced, respectively (McGraw *et al.*, 2003). Intracellular inositol-1,4,5-trisphosphate (IP_3) production and phospholipase C (PLC) levels were decreased in the absence of the β_2-receptor, and increased in the presence of high β_2-receptor density. There is, therefore, an interaction between β_2-receptors and PLC, termed antithetic regulation (McGraw *et al.*, 2003), which may cross-regulate stimulatory and inhibitory cell responses. This is in addition to the negative cross-talk between muscarinic M_2-receptors and β_2-receptors in ASM, where M_2-activation leads to attenuated cAMP accumulation (Sarria *et al.*, 2002).

13.4 Effects in airway smooth muscle (ASM)

Bronchodilation

The molecular structure of a β_2-agonist determines the manner in which it interacts with the β_2-adrenoceptor to induce bronchodilation. Short-acting β_2-agonists, such as salbutamol, which are hydrophilic in nature, access the active site of the

β_2-adrenoceptor directly from the extracellular aqueous compartment (Johnson, 2000). There is, therefore, a rapid onset of action. However, these drugs rapidly re-equilibrate, their residency time at the receptor active site is limited, and the resulting duration of action is short. The long-acting β_2-agonists have different mechanisms of action. Formoterol is moderately lipophilic, and it is taken up into the cell membrane in the form of a depot, from where it progressively leaches out to interact with the active site of the β_2-receptor (Anderson, 1993). The size of the depot is determined by the concentration or dose of formoterol applied. The onset of action of formoterol in ASM is similar to salbutamol, but the duration of activity is longer and concentration-dependent (Johnson, 2000). This profile has been confirmed clinically in asthmatic patients, where bronchodilation was observed for 8, 10 and 12 h following doses of 6, 12 and 24 μg, respectively (Ringdahl *et al.*, 1995).

Salmeterol, on the other hand, is over 10,000 times more lipophilic than salbutamol (Johnson, 2000). The molecule partitions rapidly (<1 min) into the cell membrane and then diffuses laterally to approach the active site of the β_2-adrenoceptor through the membrane. Therefore, the onset of action of salmeterol in ASM is slower than that of other β_2-agonists such as salbutamol and formoterol. The duration of action of salmeterol involves the interaction of the side chain of the molecule with an auxiliary binding site (exo-site), within the fourth domain of the β_2-adrenoceptor (Green *et al.*, 1996). Salmeterol appears to be inherently long-acting, in that its effects are independent of dose, as a result of exo-site binding (Johnson, 2000). This difference between salmeterol and formoterol has been observed in asthmatic patients. The onset of bronchodilation following salmeterol is slower than with formoterol. However, since salmeterol is still inducing bronchodilation 12 h after the previous dose, the asthmatic patient does not perceive the slower onset by the second dose of the drug.

The majority of β_2-adrenoceptor agonists used as bronchodilators have intermediate efficacy at the receptor, and if receptor numbers permit (as in ASM), they will behave as full agonists, but if receptor density is too low, or receptor coupling is inadequate, the β_2-agonist may behave in a partial manner; that is, it will be incapable of achieving the same maximum effect as an agonist of higher efficacy, and it may even behave as an antagonist. Examples of compounds of high efficacy (full agonists) are isoprenaline, fenoterol and formoterol, whereas salbutamol and terbutaline are of moderate efficacy (partial agonists). Salmeterol has an efficacy at β_2-adrenoceptors in ASM lower than that of salbutamol. Low efficacy in a β_2-adrenoceptor agonist does not, however, compromise clinical performance as a bronchodilator drug, but it may limit effectiveness in restoring airway calibre under conditions of acute severe bronchospasm.

Pharmacologically, a partial agonist must occupy a higher proportion of receptors than a full agonist to exert its maximum effect. *Ex vivo* studies in asthmatic patients have shown salmeterol at clinical doses to occupy only 4 per cent of the available pulmonary β_2-receptors (Johnson, 2006). This represents a receptor 'reserve' of 96 per cent and is reflected in a normal bronchodilator response to salbutamol, when used as rescue therapy, in patients chronically treated with salmeterol (Nelson *et al.*, 1999).

13.5 β2-Receptor desensitization

Associated with β_2-adrenoceptor activation in ASM is the autoregulatory process of receptor desensitization. This process operates as a safety device to prevent over-stimulation of receptors in the face of excessive β_2-agonist exposure. The principal mechanism of homologous short-term, β_2-agonist-promoted desensitization of the β_2-adrenoceptor is phosphorylation of the receptor by PKA, β-adrenoceptor kinase (βARK), or other closely related G protein-coupled receptor kinases (GRKs) (Johnson, 1998). Phosphorylation can occur at multiple serine and threonine residues in the third intracellular loop and in the proximal cytoplasmic tail (Figure 13.1). Such phosphorylation normally results in β-arrestin binding and partial uncoupling of the agonist-occupied form of the receptor from the stimulatory G_s protein, thereby limiting receptor function. A recent study (Moore *et al.*, 2007), however, has shown that despite phosphorylation of the β_2-receptor to an extent similar to formoterol, salmeterol did not induce significant β_2-receptor internalization (<5 per cent) or degradation as a result of lack of β-arrestin binding. In contrast, formoterol evoked β-arrestin binding and the loss of over 55 per cent of β_2-receptors from the cell surface (Moore *et al.*, 2007). Alternatively, the scaffolding action of β-arrestins may bring other proteins such as phosphodiesterase IV into the receptor microenvironment. This then acts to metabolize locally generated cAMP and terminate the effects of receptor activation. This simple form of desensitization is a transient process, and may be reversed within minutes of removal of the agonist.

After more prolonged agonist exposure, internalization of receptors occurs, resulting in some loss from the cell surface. This process, termed 'sequestration', may play a major role in short-term regulation of the receptor, since, while it is sequestered, dephosphorylation of the receptor occurs (Shenoy *et al.*, 2001).Internalization takes longer to reverse than uncoupling, but full reversal normally occurs within hours. After hours of agonist exposure, a net loss of cellular receptors occurs (denoted downregulation) via several mechanisms that are independent of receptor phosphorylation. Ubiquitination of the β_2-receptor

(Figure 13.1), via an E3-ligase, activates the process of receptor degradation (Shenoy *et al.*, 2001). Transcription at the β_2-receptor gene level and post-translational conversion of mRNA to protein is now required to restore the membrane complement of β_2-receptors. It is now well appreciated that in addition to desensitization processes that negatively regulate the function of the β_2-receptor protein itself, β_2-receptor agonists, acting through the cAMP pathway, also modulate β_2-receptor gene expression.

The level of βARK and GRK activity in ASM cells was only 10–20 per cent of that in mast cells (McGraw and Liggett, 1997). This predicts that ASM β_2-receptors will undergo minimal short-term, agonist-promoted desensitization. This may explain the clinical observation that repetitive administration of most β_2-agonists to asthmatics appears not to result in tolerance to the bronchodilatory response of β_2-receptors, expressed on bronchial smooth muscle. Tolerance to the bronchodilator effects of formoterol, but not salmeterol, has however, been reported in several studies (Bensch *et al.*, 2001; Lundback *et al.*, 1993; Shapiro *et al.*, 2000; Yates *et al.*, 1995). In a 12-week study, bronchodilation following 24 µg b.d. formoterol was reduced, whereas that to 12 µg b.d. was unchanged (Bensch *et al.*, 2001). Similarly, in the FACET study (Pauwels *et al.*, 1997), lung function (PEFR) was higher (30 per cent) during the first few days of formoterol (12 µg b.d.) treatment compared with values measured after week 1 (Figure 13.2).

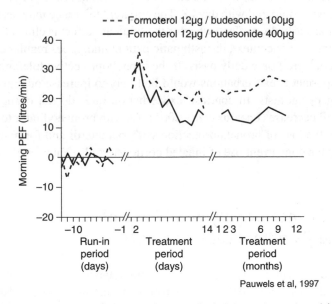

Figure 13.2 Bronchodilator tolerance to formoterol administered together with budesonide in mild-to-moderate asthma

Corticosteroids have facilitative effects on the β_2-receptor, both increasing gene transcription and regulating the numbers of the receptor on the cell surface (Mak *et al.*, 1995). It is of interest, therefore, that formoterol-induced tolerance was observed in the FACET study (Pauwels *et al.*, 1997) even in the presence of 400 μg b.d. budesonide. In contrast, two placebo-controlled studies (Lundback *et al.*, 1993; Shapiro *et al.*, 2000) found no difference in the FEV_1 response to salmeterol (50 μg b.d.) between day 1 and week 12. Following chronic treatment with salmeterol, baseline lung function (prior to the morning dose) was consistently increased (Shapiro *et al.*, 2000). This is another indication of the lack of tolerance at the β_2-receptor.

All β_2-agonists, including salbutamol (O'Connor *et al.*, 1992), salmeterol (Cheung *et al.*, 1992) and formoterol (Yates *et al.*, 1995), induce apparent tolerance to their bronchoprotective effects against provocational stimuli such as histamine, allergen, exercise and cold-air challenge. For example, the protective effects of salmeterol and formoterol are reduced by $\sim$50 per cent when the drugs are taken on a regular basis (Cheung *et al.*, 1992; Yates *et al.*, 1995). However, there is still a significant degree of bronchoprotection that does not wane on further treatment. Most of the studies have assessed the degree of bronchoprotection 1 h after dosing. In a study (Booth *et al.*, 1993), where the effects of salmeterol against methacholine challenge were measured at up to 1 h and at 1-12 h, there was significant loss of activity at the early, but not the later time points, when day 1 of dosing was compared with day 28. The apparent tolerance may be, therefore, a feature of an early dose effect, whereas the 'plateau' effect is clinically relevant to the patient. In this context, the asthmatic patient undergoes regular challenges (allergen, cold, etc.) on a daily basis. If the bronchoprotective tolerance were of clinical importance, exacerbations would be likely to increase on regular use of long-acting β_2-agonists. In contrast, exacerbation rates do not change or even show a small decrease (Lazarus *et al.*, 2001). As with bronchodilator tolerance to formoterol, the loss of bronchoprotection with salmeterol and formoterol is not affected by the concurrent use of inhaled corticosteroids (Yates *et al.*, 1996).

13.6 Influence of polymorphisms of the β₂-adrenoceptor in ASM

A number of polymorphisms of the β_2-receptor have recently been described which alter the behaviour of the receptor following agonist exposure (Hall, 1996). All of these differ from the accepted wild-type sequence by a single base change at different positions in the coding sequence of the gene. There are two genes for the β_2-adrenoceptor; therefore, an individual can be homozygous or heterozygous for a given polymorphism. The main clinical interest in these polymorphisms lies in the possibility that they may determine the extent to which the receptor downregulates in the airways and as such may modify bronchodilator responses.

The initial studies (Reishaus *et al.*, 1993) focused on amino-acid 16, which can be either arginine (Arg) or glycine (Gly), and the data suggest that the Gly 16 receptor downregulates following exposure to an agonist to a much greater extent than the Arg 16 form in primary cultured human ASM cells. Two clinical studies have supported the possibility that the Gly 16 form of the receptor is associated with markers of more severe asthma. Preliminary data from Dutch families with asthma suggest that Gly 16 may be associated with airway hyperresponsiveness (Holroyd *et al.*, 1995). In addition, patients with significant nocturnal worsening of their asthma were more likely to have the Gly 16 form of the receptor than asthmatic patients without nocturnal falls in peak flow rate (Hall, 1996). The allelic frequencies for Arg 16 and Gly 16 are 35 per cent and 65 per cent, respectively.

Some, but not all, studies have shown a relationship between Arg-Gly 16 polymorphisms and clinical responses to β_2-agonists. When compared with children homozygous for Gly 16, homozygotes for Arg 16 were 5.3 times, and heterozygotes for Arg/Gly 16 were 2.3 times, more likely to respond to salbutamol respectively (Martinez *et al.*, 1997). Similarly, homozygous Gly 16 was significantly more prone to bronchodilator tolerance than Arg 16 following administration of formoterol (24 μg b.d.) for 4 weeks (Tan *et al.*, 1997). However, other studies have shown the converse, in that patients with the resistant Arg-Arg (16) polymorphic form of the receptor exhibited a progressive reduction in bronchodilator response to regular administration of salbutamol and an improvement in lung function when the β_2-agonist was withdrawn (Israel *et al.*, 2000). Gly 16 homozygotes showed no such responses. Similarly, the presence of Arg–Arg, but not Gly–Gly, was associated with an increased frequency of asthma exacerbations in patients receiving salbutamol (Taylor *et al.*, 2000). This was not observed with salmeterol treatment. Of interest is a recent observation of a highly significant interaction between a coding non-synonymous polymorphism (lle 772 Met) of adenylate cyclase 9 (AC9) and bronchodilator response to salbutamol in paediatric asthma patients taking inhaled budesonide (Tantisira *et al.*, 2005). This suggests that genotype may influence β_2-receptor signalling when β_2-agonists and ICS are co-administered.

The second polymorphism is at position 27 and exists as either glutamine (Gln) or glutamate (Glu). The allelic frequency for Gln 27 and Glu 27 is 55 per cent and 45 per cent, respectively (Hall, 1996). In contrast to Gly 16, the Glu 27 form of the receptor appears to protect against downregulation. In primary cultured human ASM cells, following prolonged exposure to β_2-agonists, the Glu 27 form downregulated to a much lesser extent than the Gln 27 receptor, as assessed by changes in receptor number (Hall, 1996). The third polymorphism is at amino-acid 164, which can be either threonine (Thr) or isoleucine (Ile). This polymorphism is much rarer than that at amino-acid 16 or 27, with an allelic frequency of about 1 per cent (Green *et al.*, 1993).

Given that most individuals are heterozygous and that Arg–Gly 16 and Gln–Glu 27 polymorphisms may be in linkage disequilibrium, it is likely that large populations will have to be studied to determine the importance of β_2-adrenoceptor polymorphisms to the asthma phenotype.

13.7 Non-bronchodilator effects of β_2-agonists in ASM

β_2-Agonists have a spectrum of non-bronchodilator effects in ASM that may be relevant to their efficacy in asthma and COPD. These effects are mediated both directly through the β_2-adrenoceptor (Johnson and Rennard, 2001) and as a result of an interaction between β_2-adrenoceptors and glucocorticoid receptors (Johnson, 2004).

Long-acting β_2-agonists (LABA) prime the glucocorticoid receptor (GR) for subsequent corticosteroid binding by mitogen-activated protein kinase (MAPK)-dependent phosphorylation (Adcock *et al.*, 2002). Subsequently, the translocation of the GR from the cytosol of the cell to the nucleus, a fundamental step in the anti-inflammatory effect of corticosteroids, is increased (Usmani *et al.*, 2002). This effect is mediated by activation of the binding enhancer protein C/EBPα (Roth *et al.*, 2002). Salmeterol (Haque *et al.*, 2006) and formoterol (Jazrauri *et al.*, 2005) have been shown to enhance the extent and the duration of GR nuclear translocation. These *in vitro* findings have now been confirmed *in vivo* in patients with asthma and COPD treated with LABA and ICS. Usmani *et al.* (2005) showed that inhalation of salmeterol (50 μg) and fluticasone propionate (FP) (100 μg) increased the nuclear translocation of the GR in sputum macrophages taken from mild asthmatic patients, significantly more than the same dose of FP. The results were equivalent to a fivefold higher dose of the steroid alone. Haque *et al.* (2006) have confirmed these findings in COPD patients. These synergistic interactions between LABAs and CS result in enhanced anti-inflammatory effects of ICS in both asthma and COPD.

Effects on inflammatory mediator release

Human ASM cells produce high amounts of IL-8, a major neutrophil chemoattractant implicated in COPD and in asthma exacerbation. β_2-Agonists such as
salbutamol and salmeterol caused a time-dependent increase in basal IL-8 accumulation that was significant after 4 and 8 h, respectively (Pang and Knox, 2000).
The magnitude of IL-8 release by β_2-agonists was, however, smaller than that
following stimulation with TNFα, suggesting that an increase in cAMP is a weak
inducer of IL-8 generation by human ASM cells.

In contrast, steroids such as dexamethasone and FP resulted in a concentration-
dependent inhibition (Pang and Knox, 2000). However, when dexamethasone or
FP was used in combination with salmeterol, a further inhibition was observed that
was significantly greater than with the steroids alone (Figure 13.3). This represents
a synergistic interaction between LABA and CS in ASM, which may have implications in asthma and COPD, where the combined use of LABAs and ICS significantly reduces exacerbation, a condition associated with IL-8-driven neutrophilia.

The synergistic inhibition of IL-8 release was reversed by the β_2-selective antagonist, ICI 118551, confirming that the effect of salmeterol is mediated through
the β_2-receptor. KT 5720, a potent and selective inhibitor of cAMP-dependent
PKA, also significantly reduced the inhibitory effects of salmeterol and FP in

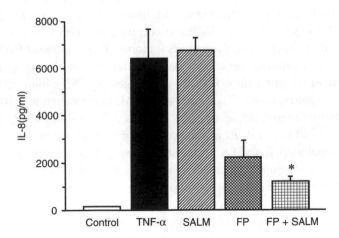

Figure 13.3   The salmeterol (SALM)/fluticasone propionate (FP) combination synergistically
(*p < 0.01) inhibited TNFα-induced IL-8 release from human airway smooth muscle cells
compared with SALM or FP alone (Pang and Knox, 2000). (Reproduced from Am J Respir Cell
Mol Biol 2000; 23:79–85, courtesy of the American Thoracic Society)

combination, whereas the phosphodiesterase IV inhibitor, rolipram, further increased the level of IL-8 inhibition (Pang and Knox, 2000). These observations indicate an involvement of the cAMP pathway in the synergy between salmeterol and CS in ASM cells.

Eosinophilia is a prominent feature of asthma and is also observed during exacerbations of COPD. Human ASM cells synthesize and release eotaxin, a potent eosinophil chemoattractant. β_2-Agonists had no effect on basal eotaxin, but significantly inhibited TNFα-induced release, with salmeterol being 10-fold more potent that salbutamol (Pang and Knox, 2001). Pre-treatment of cells with dexamethasone or FP inhibited responses to TNFα in a concentration-dependent manner, but did not abolish eotaxin release. When CS were used in combination with LABAs, a further additive inhibition of ASM cell eotaxin was observed, which was significantly greater than that of each agent alone (Pang and Knox, 2001). These effects were again reversed by ICI 118551 and further increased by rolipram, implicating both the β_2-receptor and the cAMP pathway.

Recent studies have demonstrated that TNFα stimulates the secretion of IL-6 and RANTES from ASM cells. The role of IL-6 may be that of an anti-inflammatory signal, modulating ASM in an autocrine manner. IL-6 release appears to be mediated by a NF-κB-dependent pathway, whereas RANTES release is mediated via activation of AP-1 (Panettieri, 2002). The LABA formoterol increased TNFα-induced IL-6 generation by approximately sixfold, with an EC_{50} of 0.01 μM (Ammit *et al.*, 2002). This was associated with binding to CRE and activation of the IL-6 promoter, but did not involve AP-1 or NF-κB. The lower-efficacy LABA, salmeterol, at higher concentrations, induced only a twofold increase in IL-6. In contrast, β_2-agonists inhibited TNFα-induced RANTES secretion in a concentration-dependent manner (Figure 13.4). Formoterol was the most effective inhibitor, with an IC_{50} of 1 μM, inhibiting TNFα-induced RANTES secretion by approximately 70 per cent at 10 μM. Isoprenaline was 10-fold less potent than formoterol, but at 100 μM abrogated TNFα-induced RANTES by approximately 80 per cent. At a concentration of 10 μM, salmeterol and salbutamol also inhibited RANTES secretion following TNFα stimulation (Ammit *et al.*, 2002). β_2-Agonists alone had no effect on basal RANTES secretion. The rank order of effectiveness of β_2-agonists in their ability to inhibit TNFα-induced RANTES secretion is similar to their ability to induce secretion of IL-6 and increase cytosolic levels of cAMP in ASM. However, RANTES appears to be more sensitive than IL-6 to small changes in cAMP concentration, and it did not appear to be mediated via AP-1.

CS have differential effects from β_2-agonists on the release of IL-6 and RANTES from ASM cells. Dexamethasone completely abrogated TNFα-induced

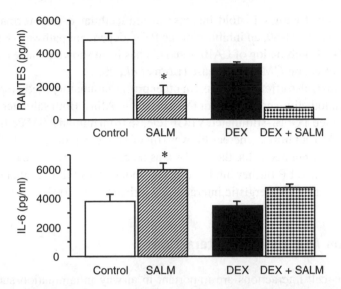

Figure 13.4 Salmeterol (SALM) significantly (*p < 0.05) inhibited TNFα-induced RANTES release and significantly (*p < 0.05) increased IL-6 generation by human airway smooth muscle cells. Dexamethasone (DEX) also inhibited RANTES, but had little effect on TNFα-induced IL-6 release. The SALM/DEX combination further enhanced RANTES inhibition over SALM or DEX alone, but increased IL-6 release compared with DEX alone (Ammit *et al.*, 2002). (Reproduced from Am. J. Respir. Cell Mol. Biol. 26, 465–474, (2002) courtesy of the American Thoracic Society)

RANTES secretion with an IC$_{50}$ of 1 nM, but, even at the highest concentration, it only partially inhibited IL-6. In combination with salmeterol, the same degree of inhibition of TNFα-induced RANTES release could be achieved with a 100-fold lower concentration of dexamethasone (Ammit *et al.*, 2002). However, TNFα-induced IL-6 secretion from ASM cells was significantly augmented by salmeterol/dexamethasone.

Similarly, studies of human ASM cells stimulated with IL-1β and TNFα have shown increased expression and release of granulocyte-macrophage colony-stimulating factor (GM-CSF), which stimulates maturation, surface activation and proliferation of several pro-inflammatory cells, and is particularly important for the survival of eosinophils. The β₂-agonist salbutamol did not significantly alter the basal release of GM-CSF, but attenuated cytokine-induced secretion with an IC$_{50}$ of 16 nM (Hallsworth *et al.*, 2001). Pretreatment with ICI 118551 prevented the inhibition due to salbutamol on IL-1β-stimulated GM-CSF release, confirming an effect mediated by membrane β₂-receptors. The effects of salbutamol were associated with a 15-fold increase in intracellular cAMP formation, and were inhibited by H-89, an inhibitor of the PKA-dependent pathway (Hallsworth *et al.*, 2001). Combination of LABAs and CS has been shown to have an additive inhibitory effect on GM-CSF release (Korn *et al.*, 2001).

β₂-Agonists also effect the release of other pro-inflammatory mediators, such as vascular endothelium growth factor (VEGF), from ASM. Both salmeterol and FP alone inhibited TGFβ₁-stimulated VEGF release from human ASM cells (Corbett *et al.*, 2002). In addition, whereas FP was still effective in reducing VEGF release from IL-1β-stimulated cells, the LABA was inactive. However, the combination of salmeterol and FP further inhibited IL-1β-induced VEGF over that of the CS alone, suggesting a synergistic interaction (Corbett *et al.*, 2002).

Effects on ASM cell–cell interactions

Direct cell–cell interactions are important in airway inflammation and remodelling and are mediated by adhesion molecules. ASM cells express ICAM-1 and VCAM-1 in response to inflammatory cytokines, such as TNFα. Isoprenaline, at a concentration (100 μM) previously shown to increase cAMP in human ASM cells, inhibited TNFα-induced expression of ICAM-1 and VCAM-1 by approximately 60 per cent. This was associated with a decreased activated T-cell adhesion to the ASM cells (Amrani *et al.*, 1999).

Migration of smooth muscle cells induced by inflammatory mediators may contribute to airway remodelling and also involves cell–cell interactions. Basal levels of human ASM cell migration were decreased by salmeterol by approximately 40 per cent, whereas the LABA had little effect on PDGF-induced migration of the cells (Krymskaya *et al.*, 2002). Treatment of cells with dexamethasone or FP also attenuated basal, but not PDGF-stimulated, ASM cell migration. However, the combination of salmeterol and FP significantly inhibited the PDGF-migratory response (Figure 13.5a), indicating a synergistic interaction between the LABA and CS (Krymskaya *et al.*, 2002).

Effect on ASM cell proliferation

Inhibition of mitogen-induced proliferation of ASM cells by β₂-agonists has been previously reported (Stewart *et al.*, 1999). β₂-agonists and other cAMP-elevating agents are thought to arrest the G1 phase of the cell cycle by post-transcriptionally inhibiting cyclin D1 protein levels by action on a proteosome-dependent

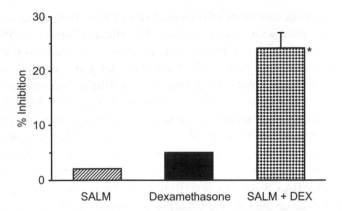

Figure 13.5 Effects of LABA/CS combination on human airway smooth muscle cell migration and proliferation. The salmeterol (SALM)/dexamethasone (DEX) combination synergistically ($^*p < 0.05$) inhibited PDGF-induced migration of human airways smooth muscle cells compared with SALM or DEX alone (Krymskaya *et al.*, 2002). (Reproduced from Am J Respir Crit Care Med 2002; 165:A346 courtesy of the American Thoracic Society)

degradation pathway. Alternatively, suppressed cyclin D1 gene expression may occur through phosphorylation and transactivation of CREB, suggesting that the effect of cAMP is exerted by way of *cis*-repression of cyclin D1 promoter (Panettieri, 2002). Salmeterol has been shown to inhibit thrombin-induced proliferation of human ASM cells. This was associated with a reduction in *c-myc* RNA expression (Stewart *et al.*, 1997). There was no effect, however, on other mitogens such as EGF. More recent studies have shown that β_2-agonists inhibit proliferation via activation of the CCAAT-enhancer-binding protein (C/EBP), which has a pivotal role in cell proliferation (Ramji and Foka, 2002) through stimulation of p21[WAF1/Cip1], a cyclin-dependent kinase inhibitor that exerts a brake on cell cycling. Salmeterol reduced cell proliferation in a dose-dependent manner with a maximum of 74 per cent of control at 10^{-6}M, and this was also associated with increased expression of p21[WAF1/Cip1] (Rudiger *et al.*, 2000). Incubation with antisense against C/EBPα mRNA reduced LABA-induced p21[WAF1/Cip1] expression. When LABA and CS were applied in combination, the effects on p21[WAF1/Cip1] expression and on cell proliferation were synergistic, even if the effects of either drug alone had reached a plateau (Rudiger *et al.*, 2000).

Budesonide also induced activation of p21[WAF1/Cip1] in human ASM cells stimulated with fetal bovine serum. By contrast, formoterol had no significant effect. However, the combination of formoterol and budesonide was shown to inhibit serum-induced proliferation in a concentration-dependent manner (Roth *et al.*,

2002). A synergistic inhibitory effect was observed when both drugs were used at $\leq 10^{-10}$M, compared with each drug alone. The effect of formoterol/budesonide was associated with enhanced nuclear translocation of GR, and this was apparent within 30 min and persisted for 6 h. The combination also resulted in stronger GR–GRE binding and greater DNA binding of C/EBPα than with either alone, and in enhanced activation of p21$^{WAF1/Cip1}$ (Roth *et al.*, 2002). The activation of p21$^{WAF1/Cip1}$ by the combination of formoterol and budesonide was reversed by the GR antagonist, RU-486, and by the β-blocker, propanolol, confirming an effect mediated by receptor-driven pathways (Figure 13.5b). Antisense oligonucleotides against either C/EBPα or p21$^{WAF1/Cip1}$ abrogated the enhanced antiproliferative effects of the LABA/CS combination (Roth *et al.*,2002). These findings show that LABAs both activate the GR and C/EBPα and inhibit the proliferation of ASM via the action of p21$^{WAF1/Cip1}$. Whereas formoterol and budesonide alone activate the two transcription factors with different kinetics, in combination, the drugs lead to a synchronous activation of the GR and C/EBPα, resulting in a synergistic stimulatory effect on p21$^{WAF1/Cip1}$ promoter activity and an inhibitory effect on serum-induced cell proliferation (Roth *et al.*, 2002).

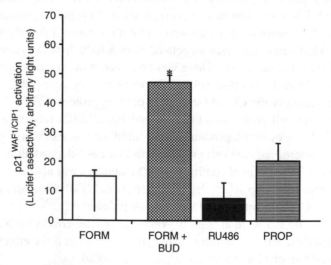

Figure 13.6 The formoterol (FORM)/budesonide (BUD) combination significantly (*p<0.01) increased p21$^{WAF1/Cip1}$ activation (expressed as luciferase activity in arbitary light units) in human airway smooth muscle cells compared with FORM alone. This effect was inhibited by the GR antagonist, RU486, and by the β-receptor antagonist, propanolol (PROP) (Roth *et al.*, 2002)

13.8 Conclusion

β_2-adrenergic agonists are used widely as bronchodilators and also in combination therapy with inhaled corticosteroids in the treatment of respiratory diseases, such as asthma and COPD. Knowledge of the function, response and regulation of the β_2-receptor is important to the clinician in interpreting patient response to both short- and long-acting β_2-agonists. Similarly, an understanding of the mechanisms of receptor desensitization, which may lead to tolerance to maintenance and rescue therapy, is useful. The presence of a particular genetic form of the β_2-receptor that may influence the clinical efficacy of regular β_2-agonists is of increasing interest in predicting good and poor responders. ASM cells contain both β_2-receptors and GR and are a potential target for β_2-agonist–corticosteroid interactions. With LABA/CS combinations, there is evidence of additive and/or synergistic inhibition of inflammatory mediator release, migration and proliferation of ASM cells. These effects may be relevant to the clinical efficacy of LABA/ICS combination therapy in asthma and COPD.

References

Adcock, I. M., Maneechotesuwan, K. and Usmani, O. (2002). Molecular interactions between glucocorticoids and long-acting β_2-agonists. *J Allergy Clin Immunol* **110**, S261–268.

Adcock, I., Maneechotesuwan, I. and Usmani, O. (2002). Effects of LABA on GR function: possible role of GR phosphorylation. *J Allergy Clin Immunol* **110**, S265–266.

Ammit, A. J., Lazaar, A. L., Irani, C. *et al.* (2002). Tumor necrosis factor-α-induced secretion of RANTES and interleukin-6 from human airway smooth muscle cells – modulation by glucocorticoids and β_2-agonists. *Am J Respir Cell Mol Biol* **26**, 465–474.

Amrani, Y., Lazaar, A. L. and Panettieri, R. A. (1999). Up-regulation of ICAM-1 by cytokines in human tracheal smooth muscle cells involves an NF-κB-dependent signalling pathway that is only partially sensitive to dexamethasone. *J Immunol* **163**, 2128–2134.

Anderson, G. P. (1993). Formoterol: pharmacology, molecular basis of agonism and mechanism of long duration of a highly potent and selective β_2-adrenoceptor agonist bronchodilator. *Life Sci* **52**, 2145–2160.

Bensch, G., Lapidus, R. J., Levine, B. E. *et al.* (2001). A randomised 12-week, double-blind, placebo-controlled study comparing formoterol dry powder inhaler with albuterol metered-dose inhaler. *Ann Allergy Asthma Immunol* **86**, 19–27.

Booth, H., Fishwick, K., Harkawat, R. *et al.* (1993). Changes in methacholine induced bronchoconstriction with the long-acting β_2-agonist salmeterol in mild to moderate asthmatic patients. *Thorax* **48**, 1121–1124.

Cheung, D., Timmens, M. K. and Zwinderman, A. H. (1992). Long term effects of long-acting β₂-adrenoceptor agonist, salmeterol on airway hyper-responsiveness in patients with mild asthma. *N Engl J Med* **327**, 1198–1203.

Corbett, L., Stocks, J., Pang, L. *et al.* (2002). TGFbeta increases VEGF release from cultured human airway smooth muscle cells: regulation by steroids and beta agonists. *Am J Respir Crit Care Med* **165**, A114.

Daaka, Y., Luttrell, L. M. and Lefkowitz, R. J. (1997). Switching of the coupling of the adrenergic receptor to different G-proteins by protein kinase A. *Nature* **390**, 88–91.

Green, S. A., Cole, G., Jacinto, M. *et al.* (1993). A polymorphism of the human β₂-adrenergic receptor within the fourth transmembrane domain alters ligand binding and functional properties of the receptor. *J Biol Chem* **268**, 23116–23121.

Green, S. A., Spasoff, A. P., Coleman, R. A. *et al.* (1996). Sustained activation of a G-protein coupled receptor via anchored agonist binding. *J Biol Chem* **271**, 24029–24035.

Hall, I. P. (1996). β₂-Adrenoceptor polymorphisms: are they clinically important? *Thorax* **51**, 351–353.

Hallsworth, M. P., Twort, C. H. C., Lee, T. H. *et al.* (2001). β₂-Adrenoceptor agonists inhibit release of eosinophil-activating cytokines from human airway smooth muscle cells. *Br J Pharmacol* **132**, 729–741.

Haque, R. A., Johnson, M., Adcock, I. M. *et al.* (2006). Addition of salmeterol to fluticasone propionate prolongs the retention of glucocorticoid receptors within the nucleus of BEAS-2B cells and enhances downstream glucocorticoid effects. *Proc Am Thorac Soc* **3**, A78.

Haque, R. A., Torego, A., Essilfie-Quaye, S. *et al.* (2006). Effect of salmeterol and fluticasone on glucocorticoid receptor translocation in sputum macrophages and peripheral blood mononuclear cells from patients with chronic obstructive pulmonary disease. *Proc Am Thorac Soc* **3**, A848.

Hoffman, E. A., Chiplunkar, R. and Casale, T. B. (1997). CT scanning confirms beta receptor distribution is greater for small versus large airways. *Am J Respir Crit Care Med* **155**, A855.

Holroyd, K. J., Levitt, R. C., Dragwa, C. *et al.* (1995). Evidence for β₂-adrenergic receptor polymorphism at amino acid 16 as a risk factor for bronchial hyper-responsiveness. *Am J Respir Crit Care Med* **151**, A673.

Israel, E., Drazen, J. M., Liggett, S. B. *et al.* (2000). The effect of polymorphisms of the β₂-adrenergic receptor on the response to regular use of salbutamol in asthma. *Am J Respir Crit Care Med* **162**, 75–80.

Jazrauri, E. P., Ito, K., Barnes, P. J. *et al.* (2005). Effect of budesonide and formoterol on glucocorticoid receptor DNA binding and transactivation in BEAS-2B cells. *Proc Am Thorac Soc* **2**, A108.

Johnson, M. The β₂-adrenoceptor. (1998). *Am J Respir Crit Care Med* **158**, S146–S153.

Johnson, M. (2000). Mechanisms of action of β₂-adrenoceptor agonists. In *Asthma and Rhinitis* (2nd edn), pp. 1541–1557, W. W. Busse and S. T. Holgate (eds). Blackwell: Cambridge.

Johnson, M. (2002). Combination therapy for asthma: complementary effects of long-acting β₂-agonists and corticosteroids. *Curr Opin Allergy Clin Immunol* **15**, 16–22.

Johnson, M. (2004). Interactions between corticosteroids and β₂-agonists in asthma and chronic obstructive pulmonary disease. *Proc Am Thorac Soc* **1**, 200–206.

Johnson, M. (2006). Molecular mechanisms of β₂-adrenergic receptor function, response and regulation. *J Allergy Clin Immunol* **117**, 18–24.

Johnson, M. and Coleman, R. A. (1994). Mechanisms of action of β₂-adrenoceptor agonists. In *Asthma and Rhinitis*, pp. 1278–1295, W. W. Busse and S. T. Holgate (eds). Blackwell: Cambridge.

Johnson, M. and Rennard, S. (2001). Alternative mechanisms for long-acting β₂-agonists in COPD. *Chest* **120**, 258–270.

Johnson, M., Butchers, P. R., Coleman, R. A. *et al.* (1993). The pharmacology of salmeterol. *Life Sci* **52**, 2131–2147.

Kobilka, B. K., Dixon, R. A., Frielle, H. G. *et al.* (1987). cDNA for the human β₂-adrenergic receptor: a protein with multiple spanning domains and encoded by a gene whose chromosomal location is shared with that of a receptor for platelet growth factor. *Proc Natl Acad Sci USA* **84**, 46–50.

Korn, S. H., Jerre, A. and Brattsand, R. (2001). Effects of formoterol and budesonide on GM-CSF and IL-8 secretion by triggered human bronchial epithelial cells. *Eur Respir J* **17**, 1070–1017.

Krymskaya, V. P., Goncharova, E. A. and Panettieri, R. A. (2002). cAMP-mobilizing and anti-inflammatory agents in migration of human airway smooth muscle (HASM) cells. *Am J Respir Crit Care Med* **165**, A346.

Lazarus, S. C., Boushey, H. A., Fahy, F. V. *et al.* (2001). Long-acting β₂-agonist monotherapy vs continued therapy with inhaled corticosteroids in patients with persistent asthma. *JAMA* **285**, 2583–2593.

Liggett, S. B. (2002). Update on current concepts of the molecular basis of β₂-adrenergic receptor signalling. *J Allergy Clin Immunol* **110**, S223–228.

Lundback, B., Rawlinson, D. W. and Palmer, J. B. D. (1993). Twelve month comparison of salmeterol and salbutamol as dry powder formulations in asthmatic patients. *Thorax* **48**, 148–153.

Mak, J. C. W., Nishikawa, M. and Barnes, P. J. (1995). Glucocorticosteroids increase β₂-adrenergic receptor transcription in human lung. *Am J Physiol* **268**, L41–L46.

Martinez, F. D., Graves, P. E., Baldini, M. *et al.* (1997). Association between genetic polymorphisms of the β₂-adrenoceptor and response to salbutamol in children with and without a history of wheezing. *J Clin Invest* **100**, 3184–3188.

McGraw, D. W., Almoosa, K. F., Paul, R. J. *et al.* (2003). Antithetic regulation by β-adrenergic receptors of Gq receptor signalling via phospholipase C underlies the airway β-agonist paradox. *J Clin Invest* **112**, 619–626.

McGraw, D. W. and Liggett, S. B. (1997). Heterogeneity of β-adrenergic receptor kinase expression in the lung accounts for cell-specific desensitisation of the β₂-adrenergic receptor. *J Biol Chem* **272**, 7338–7344.

Mialet-Perez, J., Green, S. A., Miller, W. E. *et al.* (2004). A primate-dominant third glycosylation site of the β₂-adrenergic receptor routes receptors to degradation during agonist regulation. *J Biol Chem* **279**, 38603–38607.

Moore, R. H., Millman, E. E., Godines, V. *et al.* (2007). Salmeterol stimulation dissociates β-receptor phosphorylation and internalisation. *Am J Respir Cell Mol Biol* **36**, 254–261.

Nelson, H. S., Berkowitz, R. B., Tinkelman, D. A. *et al.* (1999). Lack of subsensitivity to albuterol after treatment with salmeterol in patients with asthma. *Am J Respir Crit Care Med* **159**, 1556–1561.

O'Connor, B. J., Aikman, S. and Barnes, P. J. (1992). Tolerance to the non-bronchodilator effects of inhaled β₂-agonists. *N Engl J Med* **327**, 1204–1208.

O'Dowd, B. F., Hnatowich, M., Caron, M. G. *et al.* (1989). Palmitoylation of the human β₂-adrenergic receptor: mutation of Cys³⁴¹ in the carboxyl tail leads to an uncoupled nonpalmitoylated form of the receptor. *J Biol Chem* **164**, 7564–7569.

Onaran, H. O., Costa, T. and Rodbard, D. (1993). Subunits of guanine nucleotide-binding proteins and regulation of spontaneous receptor activity: thermodynamic model for the interaction between receptors and guanine nucleotide-binding protein subunits. *Mol Pharmacol* **43**, 245–256.

Panettieri, R. A. (2002). Airway smooth muscle: an immunomodulatory cell. *J Allergy Clin Immunol* **110**, S269–274.

Pang, L. and Knox, A. J. (2000). Synergistic inhibition by β₂-agonists and corticosteroids on tumour necrosis factor-α-induced interleukin-8 release from cultured human airway smooth muscle cells. *Am J Respir Cell Mol Biol* **23**, 79–85.

Pang, L. and Knox, A. J. (2001). Regulation of TNFα-induced eotaxin release from cultured human airway smooth muscle cells by β₂-agonists and corticosteroids. *FASEB J* **15**, 261–269.

Pauwels, R. A., Löfdahl, C. G., Postma, D. S. *et al.* (1997). Effects of inhaled formoterol and budesonide on exacerbations of asthma. *N Engl J Med* **337**, 1405–1411.

Ramji, D. P. and Foka, P. (2002). CCAAT/enhancer-binding proteins: structure, function and regulation. *Biochem J* **365**, 561–575.

Reishaus, E., Innis, M., MacIntyre, N. *et al.* (1993). Mutations in the gene encoding for the β₂-adrenergic receptor in normal and asthmatic subjects. *Am J Respir Cell Mol Biol* **8**, 334–339.

Ringdahl, N., Derom, E. and Pauwels, R. (1995). Onset and duration of action of single doses of formoterol inhaled via Turbuhaler in mild to moderate asthma. *Eur Respir J* **8**, 68S.

Robison, G. A., Butcher, R. W. and Sutherland, E. W. (1967). Adenyl cyclase as an adrenergic receptor. *Ann N Y Acad Sci* **319**, 703–723.

Roth, M., Johnson, P. R. A., Rudiger, J. J. *et al.* (2002). Interaction between glucocorticoids and β₂-agonists on bronchial airway smooth muscle cells through synchronised cellular signalling. *Lancet* **360**, 1293–1299.

Rudiger, J. J., Bihl, M. B., Cornelius, B. C. *et al.* (2000). Addition of β₂-agonists to glucocorticoid treatment augments glucocorticoid action via transcription factor C/EBPα. *Am J Respir Crit Care Med* **161**, A467.

Sarria, B., Naline, E., Zhang, Y. *et al.* (2002). Muscarinic M_2 receptors in acetylcholine–isoprenaline functional antagonism in human isolated bronchus. *Am J Physiol Lung Cell Mol Physiol* **283**, L1125–L1132.

Shapiro, G., Lumry, W., Wolfe, J. *et al.* (2000). Combined salmeterol 50 µg and fluticasone propionate 250 µg in the diskus device for the treatment of asthma. *Am J Respir Crit Care* **161**, 527–534.

Shenoy, S. K., McDonald, P. H., Kohout, T. A. *et al.* (2001). Regulation of receptor fate by ubiquitination of activated β_2-adrenergic receptor and β-arrestin. *Science* **294**, 1307–1313.

Stewart, A. G., Harris, T., Fernandes, D. J. *et al.* (1997). β_2-Adrenoceptor agonist-mediated inhibition of human airway smooth muscle cell proliferation: importance of the duration of β_2-adrenoceptor stimulation. *Br J Pharmacol* **121**, 361–368.

Stewart, A. G., Harris, T., Fernandes, D. J. *et al.* (1999). β_2-Adrenergic receptor agonists and cAMP arrest human cultured airway smooth muscle cells in the G_1 phase of the cell cycle: role of proteosome degradation of cyclin D1. *Mol Pharmacol* **56**, 1079–1086.

Tan, S., Hall, I. P., Dewar, J. *et al.* (1997). Association between β_2-adrenoceptor polymorphism and susceptibility to bronchodilator desensitisation in moderately severe stable asthmatics. *Lancet* **350**, 995–999.

Tantisira, K. G., Small, K. M., Litoajua, A. A. *et al.* (2005). Molecular properties and pharmacogenetics of a polymorphism of adenylyl cyclase 9 in asthma: interaction between beta-agonist and corticosteroid pathways. *Hum Mol Genet* **14**, 1671–1677.

Taylor, D. R., Drazen, J. M., Herbison, G. P. *et al.* (2000). Asthma exacerbations during long-term β-agonist use: influence of β_2-adrenoceptor polymorphism. *Thorax* **55**, 762–767.

Ueki, J., Rhodes, C. G., Hughes, J. M. B. *et al.* (1993). In vivo quantification of pulmonary beta-adrenoceptor density in humans with S-11C-CGP12177 and PET. *J Appl Physiol* **75**, 559–565.

Usmani, O., Maneechotesuwan, K., Adcock, I. *et al.* (2002). Glucocorticoid receptor activation following inhaled fluticasone and salmeterol. *Am J Respir Crit Care Med* **165**, A616.

Usmani, O., Maneechotesuwan, K., Adcock, I. M. *et al.* (2005). Glucocorticoid receptor activation following inhaled fluticasone and salmeterol. *Am J Respir Crit Care Med* **172**, 704–712.

Yates, D. H., Sussman, H. S., Shaw, M. J. *et al.* (1995). Regular formoterol treatment in mild asthma. *Am J Respir Crit Care Med* **152**, 1170–1174.

Yates, D. H., Kharitonov, S. A. and Barnes, P. J. (1996). An inhaled glucocorticoid does not prevent tolerance to the bronchoprotective effect of a long-acting inhaled β_2-agonist. *Am J Respir Crit Care Med* **154**, 1603–1607.

14

Asthma treatments: effects on the airway smooth muscle

Kian Fan Chung

Section of Airways Disease, National Heart and Lung Institute, Imperial College, London, UK

14.1 Introduction

The major role of the airway smooth muscle (ASM) in asthma has long been considered as contributing to airflow obstruction as a result of its contractile response, which may be enhanced. Treatment of asthma has mainly aimed to relieve airflow obstruction by bronchodilators that cause direct relaxation of ASM. The rapid beneficial effects of bronchodilators can be readily observed in the treatment of acute severe asthma, in which a rapid partial resolution of airflow obstruction can be achieved with nebulized β_2-adrenergic agonists or intravenous theophylline or magnesium sulphate. Corticosteroids are anti-inflammatory agents and are also an important component of the treatment of both chronic and acute asthma. In chronic asthma, corticosteroids, particularly by the inhaled route, lead to better control of asthma and improvement in airflow obstruction, with reduced risk of asthma exacerbation together with improvement in bronchial hyperreactivity (BHR). When used in the treatment of acute severe asthma, corticosteroids hasten the recovery from airflow obstruction, while they may also have a direct

Airway Smooth Muscle   Edited by Kian Fan Chung

bronchodilator effect. These therapeutic effects of corticosteroids in asthma have been attributed to their anti-inflammatory effects, particularly in a reduction in eosinophils, Th2 cell activation and mast cells, which in turn lead to a reduction in BHR and improved control of asthma.

The link between inflammation and BHR has been questioned by recent studies of specific anti-inflammatory agents, such as anti-IL-5 and anti-IgE antibodies, in the treatment of asthma that have not led to improvement in BHR (Djukanovic *et al.*, 2004; Leckie *et al.*, 2000). Research on the ASM cell in the last decade has established that it is abnormal in asthma in many aspects such as its pro-liferative nature and its potential to produce cytokines and chemokines (Chung, 2000). Thus, the ASM cell may contribute to airway pathophysiology, not only by narrowing the airways but also through its hyperproliferative state, and to the production of cytokines, chemokines, growth factors and extracellular matrix (ECM) proteins. Thus, ASM could be a direct target of anti-inflammatory agents, particularly corticosteroids, just as ASM is the direct target of β_2-adrenergic ago-nists to cause ASM relaxation. These effects may directly result in the therapeutic

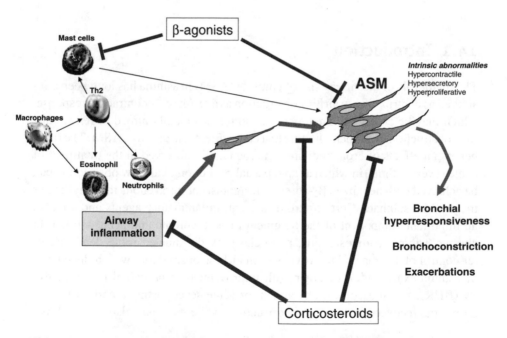

Figure 14.1   Schematic diagram to show site of action of β_2-adrenergic agonists on air-way smooth muscle (ASM) and on mast cells, and of corticosteroids on airway inflammatory cells and also on ASM cells in inhibiting directly the hypercontractile, hypersecretory and hyperproliferative state of the ASM in asthma. The direct effect on ASM cells is emphasized

benefits of anti-asthma treatment (Figure 14.1). This chapter will (i) examine the role of ASM in the pathogenesis of asthma and the reported abnormalities of ASM and (ii) discuss the effects of the major drugs used in the treatment of asthma with particular focus on the ASM cell.

14.2 Role and abnormalities of airway smooth muscle (ASM) in asthma and COPD

What is the role of ASM? The continuing debate among physiologists shows that this area remains contentious and unclear (Fredberg, 2007). Proposed functions of the ASM have included matching of ventilation/perfusion relationships, stabilizing airways, and assisting in mucus propulsion and clearance, but these propositions are not supported by any experimental evidence. This has led to the proposition that the ASM in healthy individuals represents the 'appendix' of the lungs, which can be disposed of without any harm (Mitzner, 2004). However, clearly in asthma, the ASM contributes to causing airflow obstruction and possibly other features of asthma.

Contractile response

The ASM in asthma may generate more force, given that ASM hyperplasia and hypertrophy are present (Benayoun et al., 2003; Ebina et al., 1993). An increase in maximal isometric contractile responses to constrictor agonists of asthmatic airways has been reported (Bai, 1990; DeJongste et al., 1987), while no changes have been observed (Cerrina et al., 1986). More recently, ASM cells grown from asthmatic airway biopsies in collagen gels were found to contract more to constrictor agonists than ASM cells from non-asthmatics (Matsumoto et al., 2007). Increased maximum shortening capacity and velocity of single human bronchial smooth muscle cells from asthmatics has been reported (Ma et al., 2002); this was associated with an increase in smooth muscle myosin light-chain kinase activity, which may lead to enhanced phosphorylation of the myosin light chain, allowing actinomysin ATP to be activated by actin, and leading to cross-bridge cycling (Kamm et al., 1985). The contractile response of ASM has to be examined in its natural environment. For instance, a reduced load on ASM due to uncoupling between ASM and surrounding parenchymal tissues, as a result of inflammation and oedema (Brown et al., 1995; James et al., 1989), may lead to greater force generation and airway luminal narrowing. Paradoxically, deep inspiration in asthmatics is ineffective in inducing bronchodilation following induced bronchoconstriction as occurs in non-asthmatics (Skloot and Togias, 2003). This effect in asthmatics has been attributed to the loss of interdependence caused by inflammation and oedema, which could uncouple the airways from the parenchyma such that the effect of deep inspiration in opening up the airways is no longer effective in asthma. This concept is supported by the improvement in bronchodilation following deep inspiration after a course of high-dose prednisone in asthma (Slats *et al.*, 2006). In summary, the increased contractile response and BHR of asthma are the result not only of the intrinsic contractile properties of ASM but also of the structural and mechanical properties of the airway wall (An *et al.*, 2007).

Increased ASM mass

The increase in the amount of ASM in the airways appear to be related to the degree of severity of asthma (Benayoun *et al.*, 2003; Pepe *et al.*, 2005; Woodruff *et al.*, 2004). Both hyperplasia and hypertrophy of ASM cells are present in asthma (Ebina *et al.*, 1993). The size of ASM cells and fibroblast accumulation below the true basement membrane were higher in severe asthmatics than mild-to-moderate asthmatics (Benayoun *et al.*, 2003), and the myosin light-chain kinase content of ASM was raised in severe asthmatics. The increase in proliferative activity of ASM cells from asthmatic patients observed under culture conditions (Johnson *et al.*, 2001) suggests that this phenotype is ingrained, since it is present with repeated passages.

Asthmatic ASM cells in culture lack the antiproliferative isoform of the CCAAT-enhancer-binding protein-α (cEBPα) (Roth *et al.*, 2004). cEBP-α is an important regulator of the cell cycle inhibitor p21$^{\text{waf/cip1}}$, and therefore its deficiency results in increased proliferation. Lack of cEBPα may also lead to enhanced expression of myosin light-chain kinase. This may also explain the lack of effect of corticosteroids in inhibiting proliferation of ASM cells from asthmatics, as corticosteroids are usually effective in non-asthmatic ASM cells (Roth *et al.*, 2004). In relation to the increased proliferation, increased expression of transforming growth factor (TGF)β and production of CTGF have been reported, and they could be responsible for the increased proliferation (Burgess *et al.*, 2003; Xie *et al.*, 2007).

Secretory ASM

ASM cells in culture have the capacity to release many cytokines and chemokines, and this response is dependent on the stimulatory signal (Jarai *et al.*, 2004).

CCL5/RANTES and CCL11/eotaxin expression in ASM is increased in asthma (Berkman *et al.*, 1996; Ghaffar *et al.*, 1999). The laser capture technique shows that the expression of the TGFβ gene is increased in ASM cells, as corroborated by the expression measured by immunohistochemistry. ASM cells can release CCL11/eotaxin (Chung *et al.*, 1999; Ghaffar *et al.*, 1999), CXCL10/IP-10 (Hardaker *et al.*, 2004), stem cell factor (Kassel *et al.*, 1999) and CX3CL1/fractalkine (Sukkar *et al.*, 2004). Increased release of CXCL10/IP-10 has been reported from ASM cells of asthmatics (Hardaker *et al.*, 2004); this may mediate lung mast cell migration to asthmatic ASM, underlying the increase in mast cell numbers reported in asthmatic ASM bundles (Brightling *et al.*, 2005). One important aspect of the release of chemokines from ASM cells is the potential autocrine effects (Figure 14.2). This is illustrated by the production of CXCL-8/IL-8 by ASM cells which can interact with its own receptors on ASM cells to induce ASM contraction and migration (Govindaraju *et al.*, 2006). The chemokines induced by ASM cells may also lead to their interaction with other inflammatory cells such as neutrophils, eosinophils, T cells and mast cells (Figure 14.2). An increased release of CCL11/eotaxin has been observed from asthmatic ASM cells, and this has been related to the altered secretion of ECM components from the

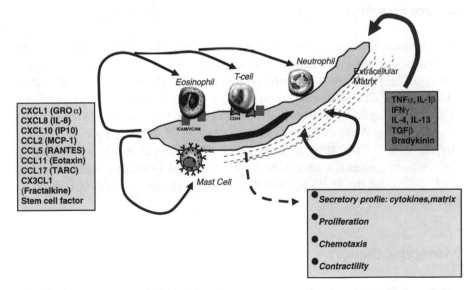

Figure 14.2   Schematic diagram showing the stimulants that activate ASM cells to produce a range of chemokines that, in turn, may have autocrine effects on the muscle cell or induce the interaction of inflammatory cells with the ASM cell. These responses may lead to phenotypic changes in the ASM, as may occur in asthma

asthmatic ASM cells (Chan *et al.*, 2006). Many growth factors are produced by ASM cells, including TGFβ, CTGF, and VEGF (Alagappan *et al.*, 2005; 2007; Burgess *et al.*, 2003; Faffe *et al.*, 2006; Jarai *et al.*, 2004; Stocks *et al.*, 2005; Xie *et al.*, 2005; 2007), and TGFβ may also stimulate the ASM to proliferation and hypertrophy (Goldsmith *et al.*, 2006; Xie *et al.*, 2007).

An important aspect of ASM cells is their capacity to regulate their production of ECM proteins and to interact with these (Hirst *et al.*, 2000). ASM cells from asthmatics produce more perlecan and collagen I and less laminin-β$_2$ and collagen IV (Johnson *et al.*, 2004). Enhanced ECM production may be the result of increased expression of CTGF (Burgess *et al.*, 2003). ECM proteins may increase the survival of ASM cells (Freyer *et al.*, 2001). Proteases which are responsible for degradation of ECM components are also secreted by ASM cells. Thus, ASM cells express pro-MMP-2, pro- and active MMP-3, MMP-9 and MT1-MMP, and MMP-12 (Elshaw *et al.*, 2004; Xie *et al.*, 2005). Increased expression of ADAM-33 in asthmatic ASM has been recently reported (Foley *et al.*, 2007). ADAM-15 is also expressed in ASM cells; the metalloprotease domain of ADAM15 also has an inhibitory effect on ASM migration induced by PDGF (Lu *et al.*, 2007).

Surface receptors

In addition to their secretory function, ASM cells also express many surface molecules that allow the cells to interact with immune or inflammatory cells. Adhesion of T cells to ASM cells may occur through the integrins, ICAM-1, VCAM-1 and CD44, leading to increased proliferation of ASM cells (Lazaar *et al.*, 1994). Ligation of OX40L, a member of the tumour necrosis factor (TNF) superfamily, on ASM cells leads to the release of IL-6 (Burgess *et al.*, 2004). Activation of the Toll-like receptors TLR2, TLR3 and TLR4 on ASM cells by their respective ligands leads to CXCL8/IL-8 and CXL11/eotaxin release (Sukkar *et al.*, 2006), and the TLR3 ligand dsRNA induces the release of CXCL10/IP-10 (Morris *et al.*, 2006).

Phenotypic changes

The major question in relation to the biology of ASM is whether the ASM cell can switch phenotype between hypercontractile, synthetic or proliferative states. Synthetic cells with a low density of contractile proteins and high content of cytoskeletal proteins are in a proliferative state and may lose their contractile ability (Hirst *et al.*, 2000). They may produce more inflammatory mediators and ECM proteins. On the other hand, hypercontractile cells have been obtained in canine trachealis smooth muscle cells cultured after prolonged withdrawal of serum (Ma *et al.*, 1998). The content of smooth muscle myosin light-chain kinase was increased 30-fold, and there was faster and greater shortening in response to contractile agents. This is reminiscent of the changes these investigators observed in asthmatic ASM cells (Ma *et al.*, 2002). However, this was not confirmed in a morphometric study that showed only hyperplasia of ASM, but not hypertrophy, in patients with mild-to-moderate asthma; in addition, gene analysis showed no differences in phenotypic markers (Woodruff *et al.*, 2004). In a chronic allergen-exposed rat model, an increase in maximal contractile response was observed in the presence of significant reduction of smooth muscle contractile proteins (Moir *et al.*, 2003), and alteration in the contractile apparatus is more likely to underlie this increased contractility (McVicker *et al.*, 2007). Therefore, hypercontractility may be associated with actin-myosin remodelling rather than with quantity of ASM contractile proteins.

14.3 Treatment of asthma

Treatment of mild asthma consists of short-acting β_2-adrenergic agonists used for the relief of symptoms of wheeze, together with the long-term use of inhaled corticosteroids. Mild asthma may also be treated with a leukotriene receptor antagonist. Asthma that is not controlled by inhaled corticosteroids requires the combination of inhaled corticosteroids and a long-acting β_2-agonist, which has been shown to be superior to high-dose inhaled corticosteroids in the control of asthma and lung function, and in reducing exacerbation of asthma. With more severe asthma, other asthma medications, such as leukotriene receptor antagonists and slow-release theophylline, may be added to combination inhalation therapy. This category of asthma patients may also need intermittent or chronic dosing with oral corticosteroids.

β_2-Adrenergic agonists

β_2-Adrenergic agonists cause the stimulation of adenylate cyclase, the production of cyclic adenosine monophosphate (AMP) and, consequently, the activation of protein kinase A, which in turn cause many changes that counteract the changes caused by bronchoconstrictor agents, such as reduction of cytosolic calcium levels through membrane K^+ and Ca^{++} channels (Kotlikoff and Kamm, 1996). Other non-voltage-dependent mechanisms, such as inhibition of Ca^{++} pumps and of the Rho A signalling pathway, and stimulation of the myosin light-chain phosphatase, also exist in the ASM cells of other species, but not in man (Janssen et al., 2004; Liu et al., 2006). The coupling of the β_2-receptor to adenylate cyclase is affected through the trimeric Gs-protein, and the cAMP levels are also regulated by phosphodiesterase enzymes that degrade it to inactive 5'-AMP.

It is quite readily accepted that β_2-adrenergic agonists act on ASM cells, since their effects in relieving airflow obstruction is rapid, occurring within minutes of inhalation, and such an effect could occur through relaxation of ASM cells under increased constrictor tone. The duration of bronchodilator effect distinguishes the short-acting (usually less than 4 h) from the long-acting β_2-adrenergic agonists (at least 12-h duration). β_2-Adrenergic agonists are the most effective bronchodilator available for the relief of airflow obstruction in asthma. The effects of β_2-adrenergic agonists may be reduced in asthmatic airways (Bai et al., 1992). One underlying mechanism could be through the uncoupling of the β_2-adrenergic receptor from adenylate cyclase through an increase in $G_{i\alpha}$, an effect that can be reproduced by IL-1β or by sensitization and allergen exposure (Koto et al., 1996).

β_2-Adrenergic agonists may also possess a number of anti-inflammatory properties on the ASM (Table 14.1). Salbutamol can inhibit TNFα-induced eotaxin and CCL5/RANTES and GM-CSF, while on the other hand, it increases TNFα-induced IL-6 release. Proliferation of ASM can be inhibited, and to a certain extent basal migration of ASM cells. The ECM may influence the response of ASM to β_2-adrenergic receptor stimulation. Thus, fibronectin increases the response to β_2-adrenergic stimulation, while collagen V or laminin reduces the accumulation of cyclic AMP through the modulation of $G_{i\alpha}$ (Freyer et al., 2004). β_2-Adrenergic agonists also have a stabilizing effect on mast cells (Peachell, 2006), but whether they can modulate mast cells in ASM bundles is not known. Whether these properties are relevant to the therapeutic actions of β_2-adrenergic agonists in asthma remains unclear, but this may be important in the use of β_2-adrenergic agonists in combination with corticosteroids.

Corticosteroids

The effect of corticosteroids on the ASM cell is tabulated in Table 14.2, and has been reviewed previously (Hirst and Lee, 1998). Inhaled corticosteroids ameliorate symptoms through better control of asthma, reduce the need for using the reliever β_2-adrenergic agonist, reverse airflow obstruction, improve BHR and

Table 14.1   Effect of β_2-adrenergic agonists on ASM cells

β_2-adrenergic agonist	Effect	Reference
Salbutamol Salmeterol	Inhibits TNFα-induced CCL11/eotaxin release through inhibition of histone acetylation	Nie *et al.*, 2005
Isoproterenol Salbutamol Formoterol Salmeterol	Increases basal and TNFα-induced IL-6 release, but inhibits TNFα-induced CCL5/RANTES	Ammit *et al.*, 2002
Fenoterol	Inhibits TNFα-induced CCL5/RANTES in the presence of indomethacin	Lazzeri *et al.*, 2001a; 2001b
Isoprenaline Salbutamol	Inhibits release of GM-CSF, RANTES and eotaxin but not CXCL8/IL-8 induced by IL-1β or TNFα	Hallsworth *et al.*, 2001
Fenoterol	Inhibits GM-CSF induced by IL-1β and TNFα, in the presence of indomethacin	Lazzeri *et al.*, 2001b[118]
Salbutamol Salmeterol	Inhibits TNFα- induced eotaxin	Pang and Knox, 2001
Salbutamol Salmeterol	Increases baseline CXCL8/IL-8 release, but has no effect on TNFα-induced CXCL8/IL-8	Pang and Knox, 2000
Proliferation		
Salbutamol Fenoterol	Inhibits proliferation induced by thrombin and epidermal growth factor	Stewart *et al.*, 1997; 1999; Tomlinson *et al.*, 1995
Salbutamol	Inhibits proliferation of ASM on collagen type I due to bFGF; does not inhibit GM-CSF release	Bonacci and Stewart, 2006
Levalbuterol	Inhibits proliferation to 5% fetal bovine serum; S-albuterol increases growth	Ibe *et al.*, 2006
Salmeterol	Inhibits basal migration by 30–60%, but not PDGF-stimulated migration	Goncharova *et al.*, 2003

reduce exacerbations of asthma. Corticosteroids do not ameliorate airflow obstruction through an effect on ASM tone. They may do so indirectly by inhibiting airway oedema. However, some *in vitro* studies report inhibitory effects on the signalling pathways of constrictor agents such as the inhibition of histamine-induced inositol phosphate formation (Hardy *et al.,* 1996) or the calcium influx induced

Table 14.2   Effect of corticosteroids on human airway smooth muscle cells *in vitro*

Cytokine/chemokine	Effect	Reference
CCL5/RANTES	Inhibition of CCL5/RANTES induced by TNFα and IFNγ by dexamethasone	John *et al.*, 1999
CCL11/eotaxin	Lack of inhibition of CCL11/eotaxin induced by IL-1β or TNFα by dexamethasone	Chung *et al.*, 1999
CCL11/eotaxin	Inhibition of TNFα-induced CCL11/eotaxin through histone acetylation	Pang *et al.*, 2001 Nie *et al.*, 2005
COX-2	Inhibition of IL-1β and TNFα-induced COX-2 expression	Vlahos *et al.*, 1999
COX-2	Inhibition of COX-2 induced by TNFα and IL-1β, and IFNγ by dexamethasone	Belvisi *et al.*, 1997
GM-CSF	Inhibition of TNFα- and IFNγ-induced GMCSF by dexamethasone	Saunders *et al.*, 1997
GM-CSF	Inhibition of IL-1β-induced GM-CSF through inhibition of p38 MAPK	Tran *et al.*, 2005
ICAM-1	Inhibition of IL-1β- and TNFα-induced ICAM-1 expression at 4 h by dexamethasone	Amrani, 1999
CXCL1/GRO-α	Inhibition of IL-1β- and TNFα-induced CXCL1/GRO-α release	Issa *et al.*, 2007
CD38	Inhibition of TNFα-induced increase in CD38 and ADP-ribosyl cyclase activity	Kang *et al.*, 2006
MCP-1, -2, -3	Inhibition of IL-1β, TNFα and IFNγ induction of MCP-1, -2, -3 and RANTES mRNA by dexamethasone	Pype *et al.*, 1999
CXCL8/IL-8	Inhibition by dexamethasone of TNFα- and IFNγ-induced IL-8	John *et al.*, 1998
CXCL8/IL-8	Inhibition of cigarette smoke- and TNFα-induced IL-8 release by dexamethasone and fluticasone	Oltmanns *et al.*, 2007
IL-6	Inhibition of TNFα-induced IL-6	Ammit *et al.*, 2002
IL-6	Inhibition of bradykinin-induced IL-6 by dexamethasone	Huang *et al.*, 2003
Fractalkine/CX3CL1	Increase of CX3CL1 induced by TNFα and IFNγ by dexamethasone	Sukkar *et al.*, 2004

CTGF	No effect of fluticasone or budesonide on TGFβ-induced CTGF, collagen I or fibronectin release	Burgess *et al.*, 2006
TLR-2,3,4	Inhibition of cytokine- and ligand-induced TLR-2, -3 and -4 expression and chemokine release by dexamethasone	Sukkar *et al.*, 2007
Proliferation/ migration		
TGFβ	Inhibition of TGFβ-induced ASM proliferation	Xie *et al.*, 2007
TGFβ	Inhibition of TGFβ-induced incorporation of α-actin and hypertrophy	Goldsmith *et al.*, 2007
PDGF	Inhibition of basal or PDGF-stimulated migration	Goncharova *et al.*, 2003
Serum	Inhibition of proliferation induced by 10% serum from asthmatic individual, but no effect on release of fibronectin and perlecan	Johnson *et al.*, 2000
Thrombin	Inhibition of proliferation, cyclin D1 expression, and retinoblastoma protein phosphorylation	Fernandes *et al.*, 1999
bFGF	No effect on proliferation of ASM cell culture on collagen 1 of fluticasone and dexamethasone	Bonacci and Stewart, 2006; Bonacci *et al.*, 2006
IL-1β	Steroids reversed IL-1β-induced β$_2$-adrenergic hyporesponsiveness	Moore *et al.*, 1999
Serum	Inhibition by corticosteroids of induced proliferation of ASM from non-asthmatics, but no effect on asthmatic ASM cells	Roth *et al.*, 2002

by bradykinin (Tanaka *et al.*, 1995). Corticosteroids inhibit the induced release of most cytokines and chemokines apart from CX3CL1 (Table 14.2) (Sukkar *et al.*, 2004). They also inhibit proliferation induced by TGFβ or PDGF, or that induced by serum through inhibition of cyclin D1 expression and phosphorylation of the retinoblastoma protein. Corticosteroids were less effective in inhibition of serum-induced proliferation of ASM cells cultured from asthmatic airways (Roth *et al.*, 2002), and when ASM cells were cultured in the presence of collagen 1, the inhibitory effect of steroids on serum-induced proliferation was abolished (Bonacci *et al.*, 2006). There have been no studies of the effect of corticosteroid therapy *in vivo* on ASM cells in patients with asthma. In a sensitized rat model of chronic allergen exposure, topical corticosteroids reversed the increased prolif-eration of ASM cells induced by chronic allergen exposure (Leung *et al.*, 2005), suggesting that corticosteroids may prevent ASM hyperplasia. TGFβ induces ASM hyperplasia *in vitro*, an effect that can be prevented by corticosteroids (Xie *et al.*, 2007). *In vivo*, the ASM hyperplasia observed in the chronic allergen model in the rat is also TGFβ-dependent (Leung *et al.*, 2006), and therefore the preventive effect of corticosteroids in this model may result from an inhibitory effect on the proliferative effects of TGFβ. TGFβ also induces hypertrophy of the ASM together with increase in α-smooth muscle actin and contractile capacity (Goldsmith *et al.*, 2006), but dexamethasone and fluticasone prevent the syn-thesis of α-smooth muscle actin, ASM size and increased contractile response (Goldsmith *et al.*, 2007). However, one must be aware that the presence of in-creased smooth muscle mass in patients with asthma is often observed despite constant treatment with corticosteroids. Whether corticosteroids have an influence on ASM mass in asthma needs to be studied.

Combination of β$_2$-adrenergic agonists and corticosteroids

The combination of inhaled corticosteroids and long-acting β$_2$-agonist caused a greater degree of bronchodilation than either pharmacological agent alone (Djukanovic *et al.*, 2004; Xie *et al.*, 2005), indicating potential interaction of these agents at the level of the ASM. Corticosteroids may increase the number of β$_2$-adrenoceptors on ASM cells (Mak *et al.*, 2002) and can reverse the impairment of β$_2$-adrenoceptor-mediated relaxation induced by IL-1β through upregulation of G-protein-coupled receptor kinases (Mak *et al.*, 2002). The interaction of cor-ticosteroids and β$_2$-agonists may also occur in terms of modulation of cytokine release from ASM cells, and this could be the basis for the beneficial effects of combination therapy in COPD. Long-acting β$_2$-agonists may prime the binding of corticosteroids to its receptor by mitogen-activated, protein kinase-dependent phosphorylation, which may result in increasing the extent and duration of the corticosteroid receptor nuclear translocation (Usmani *et al.*, 2005), but this is yet to be demonstrated in ASM cells. Corticosteroids usually inhibit the expres-sion of IL-8 from ASM, while the addition of β$_2$-adrenergic agonists enhances

this effect. Salmeterol potentiated the inhibition of TNFα-induced IL-8 by fluticasone, while on its own it had no effect (Pang and Knox, 2000). Salmeterol
also potentiated the inhibition of TNFα-induced CCL5/RANTES, a chemokine
expressed in asthma, by dexamethasone, while on its own it had some marked
inhibitory effect (Ammit *et al.*, 2000). However, this was specific to the expression of CCL5/RANTES, since for the release of IL-6, there was potentiation of
its release by salmeterol (Ammit *et al.*, 2000), indicating that effects on cytokine
release by combination treatment are cytokine- or stimulus-specific. ASM cells
exposed to cigarette smoke extracts induced CXCL8/IL-8 release and potentiated TNFα-tinduced CXCL8/IL-8 release, but inhibited TNFα-induced release
of CCL5/RANTES and eotaxin (Oltmanns *et al.*, 2007). The combination of salmeterol and dexamethasone synergistically inhibited PDGF-induced migration of
human ASM cells compared to salmeterol or dexamethasone only (Goncharova
et al., 2003). Similarly, the combination of formoterol and budesonide synergistically inhibited serum-induced proliferation and induced enhanced activation of
$p21^{WAF1/Cip1}$ stimulated with fetal bovine serum (FBS) in ASM cells (Roth *et al.*,
2002). Therefore, the evidence of synergy at the level of the ASM indicates that
part of the effectiveness of this combination therapy may result from actions at
the ASM level.

Theophylline and phosphodiesterase inhibitors

Theophylline has long been used as a bronchodilator in the treatment of asthma,
and therefore it has direct effects on ASM cells. The mechanisms of action of these
effects are unclear. It has also anti-inflammatory effects (Barnes, 2005), and may
reverse corticosteroid insensitivity through the recruitment of histone deacetylase
activity. Theophylline's place in the treatment of asthma is as an additive treatment
to inhaled corticosteroids, and also to the combination of inhaled corticosteroids
and long-acting β₂-agonists in moderate-to-severe asthma (Evans *et al.*, 1997).
Whether these are contributory effects at the level of ASM cells are unclear, since
there are relatively few studies of the effects of theophylline on ASM cells *in
vitro*.

Phosphodiesterase inhibition has been suggested as a potential mechanism
of theophylline, and more specific inhibitors of PDE, such as those of PDE_4,
are being investigated as new treatments for asthma (Chung, 2006). Roflumilast inhibits TGFβ-induced CTGF and collagen 1 and fibronectin release, an
effect not modulated by corticosteroids or long-acting β-agonists (Burgess *et al.*,

2006). Cilomilast can suppress PDGF-induced ASM migration (Goncharova *et al.*, 2003).

Leukotriene receptor antagonists

Leukotriene receptor antagonists have an antibronchoconstrictor effect and lead to bronchodilation (Hui and Barnes, 1991), indicating that asthmatic ASM is under a cysteinyl leukotriene tone. Apart from being potent ASM constrictors, cysteinyl leukotrienes have other effects on ASM cells, such as promoting ASM proliferation in concert with EGF (Espinosa *et al.*, 2003; Panettieri *et al.*, 1998) and augmenting ASM migration towards PDGF (Parameswaran *et al.*, 2002). Montelukast also prevented leukotriene (LT)D4-induced β_2-adrenoceptor desensitization. In *in vivo* sensitized rat and murine models, leukotriene receptor antagonists prevented proliferation and increase in ASM mass following allergen exposure (Henderson *et al.*, 2006; Salmon *et al.*, 1999; Wang *et al.*, 1993).

Anti-TNFα approaches

An open uncontrolled study using etanercept, a soluble fusion protein made of two p65 TNF receptors with an Fc fragment of human IgG1, showed an improvement in symptom scores, in FEV$_1$ of 12 per cent, and in BHR in corticosteroid-dependent asthma patients (Howarth *et al.*, 2005). These results were confirmed in a double-blind, controlled study of patients with refractory asthma using etanercept (Berry *et al.*, 2006). The TNFα antibody, infliximab, had no effect on lung function or symptoms in a group of moderate asthma patients but decreased the number of patients with exacerbation of asthma (Erin *et al.*, 2006). Thus, TNFα may be involved in the control of both moderate and severe asthma.

It is worth considering that blocking TNFα may have a direct effect on ASM cells. TNFα level in bronchoalveolar lavage fluid is increased in patients with severe asthma (Howarth *et al.*, 2005), and such patients also demonstrate increased expression of membrane-bound TNFα and TNFα receptor 1 in peripheral blood monocytes (Berry *et al.*, 2006). Sources of TNFα in the inflamed airway include mast cells, where it is stored in granules (Bradding *et al.*, 1994), epithelial cells, eosinophils, macrophages and neutrophils. TNFα increases transmembrane glycoprotein CD38 expression, of which ADP-ribosyl cyclase is an integral part; ADP-ribosyl cyclase is responsible for intracellular calcium mobilization during contractile responses to contractile agonists (Deshpande *et al.*, 2003). TNFα increases the maximal isotonic contractile response to methacholine in guinea pig tracheal preparations (Pennings *et al.*, 1998), while increasing the maximal isometric contractile response of human airways to acetylcholine (Sukkar *et al.*, 2001). Inhalation of TNFα by healthy volunteers or asthmatics leads to the development of BHR (Thomas and Heywood, 2002; Thomas *et al.*, 1995). In addition to effects on calcium signalling pathways, TNFα can induce the release of a whole array of cytokines and chemokines from ASM cells, including CCL11/eotaxin, CXCL1/GROα and CXCL8/IL-8 (Chung *et al.*, 1999; Issa *et al.*, 2006; John *et al.*, 1998).

Anti-IgE antibody

Omalizumab is a humanized murine anti-IgE antibody that binds to the FcεRI binding site of human IgE, such that IgE can no longer bind to cells such as basophils and mast cells. Free serum IgE levels fall to very low levels, and there is a dramatic reduction in the level of high-affinity IgE receptors on circulating basophils. Omalizumab may be considered an anti-inflammatory agent for asthma, since it causes inhibition of sputum eosinophils, submucosal eosinophils and T cells in patients with mild asthma (Djukanovic *et al.*, 2004). However, it had no effect on BHR.

Omalizumab reduced the early- and late-phase response to allergen in allergic, mild asthmatics, together with a reduction in allergen-induced sputum eosinophilia. In studies of patients with moderate-to-severe asthma, omalizumab reduced the number of exacerbations by half, while the patients were able to reduce the amount of inhaled corticosteroids used to control asthma (Busse *et al.*, 2001; Soler *et al.*, 2001).

Human ASM cells express FcεRI (Gounni, 2006) and upon FcεRI cross-linking, ASM cells release eosinophil mobilizing and chemoattractant factors such as IL-5 and eotaxin (Gounni *et al.*, 2005), as well as IL-4 and IL-13. These cytokines may therefore contribute to Th2-driven allergic inflammation. There is also mobilization of intracellular calcium, potentially promoting a hyperresponsive response. In addition, mast cells are present in increased numbers in the ASM bundle in asthma (Brightling *et al.*, 2002), and activation of mast cells by IgE cross-linking may lead to release of mast cell mediators such as histamine and leukotriene, but also other cytokines such as IL-4 and TNFα, that could interact with adjacent ASM cells. The effect of omalizumab treatment on the ASM cell has not been studied, but it is possible that the antibody may not get access to ASM cells.

Anti-IL5 antibody

The humanized anti-human IL-5 monoclonal antibodies, Sch-55700 and mepolizumab (SB-240563), have been studied in asthma. Mepolizumab inhibited allergen-induced sputum eosinophilia, but had no effect on late-phase response and BHR (Leckie *et al.*, 2000). In patients with severe persistent asthma, Sch-55700 reduced circulating blood eosinophils but had only a very small effect on FEV_1 (Kips *et al.*, 2003). Mepolizumab may have beneficial effects on the airway wall remodelling process, since it significantly reduced the expression of tenascin, lumican and procollagen III in the bronchial mucosal reticular basement membrane of mildly asthmatic patients (Flood-Page *et al.*, 2003). This could result from an inhibition of eosinophils expressing $TGF\beta1$ mRNA. However, the effect of anti-IL-5 antibodies on ASM cells is little known.

Bronchial thermoplasty

The most direct proof of the importance of ASM as a target of therapy for asthma comes from the effects of treatment aimed at 'removing' or nullifying the effect of ASM cells. The technique of bronchial thermoplasty involves the transient transfer of thermal energy to the airway wall, leading to the loss of ASM at the site of application (Miller *et al.*, 2005). This technique when applied to intrapulmonary airways of patients with asthma leads to a sustained improvement in BHR (Cox *et al.*, 2006), and in an open, controlled study, it led to an improvement in asthma control and reduction in the number of mild exacerbations (Cox *et al.*, 2007). The evaluation of airway inflammation as a result of this treatment has not been undertaken yet. Further double-blind, controlled studies are currently under way, but the results so far support the idea that ASM must be specifically targeted in asthma.

Corticosteroid resistance

Corticosteroid resistance can be demonstrated *in vitro* in ASM cells. In asthma, the enhanced proliferative phenotype of ASM derived from asthmatic subjects is not inhibited by corticosteroids *in vitro*, due to the specific absence of the transcription factor C/EBPα (Roth *et al.*, 2004). Interactions between integrins and ECM may lead to corticosteroid resistance. Fluticasone no longer inhibited the proliferation of ASM cells induced by bFGF when cells were cultured on type 1 collagen, but not when on laminin, an effect that involves $\alpha_1\beta_x$-integrins (Bonacci *et al.*, 2006). Upregulation of CD38 expression by TNFα is abrogated by fluticasone, dexamethasone or budesonide, but neither corticosteroid is active when CD38 is upregulated by TNFα and IFNγ (Tliba *et al.*, 2006). This response was shown to occur through the upregulation of the GRβ isoform. Further studies are needed to explain the phenomenon of relative CS resistance that occurs in severe asthma; while this has been demonstrated in macrophages/peripheral blood mononuclear cells by a relative reduction in suppression of cytokine release (Hew *et al.*, 2006), a similar response has yet to be shown in ASM cells. One possibility is that corticosteroids may not be effective in controlling the ASM hyperplasia or hypertrophy seen in asthmatic airways.

14.4 Conclusion

Understanding of the biology of ASM in airways disease such as asthma will continue, increasingly with study of ASM cells obtained from patients with asthma. The impact of existing treatments on ASM abnormalities in asthma of all severities can be studied by existing technologies. It will be particularly important to examine the abnormalities of ASM in severe asthma, and determine whether there is corticosteroid insensitivity. New therapies for asthma could be developed by studying the biology of asthmatic ASM, and the mechanisms by which bronchial thermoplasty improves asthma control and BHR. To prove that ASM cells are the most important target cells for therapeutic effects in the treatment of asthma, methods for delivering therapy to ASM cells specifically are needed, although this may prove to be difficult.

References

Alagappan, V. K., McKay, S., Widyastuti, A. *et al.* (2005). Proinflammatory cytokines upregulate mRNA expression and secretion of vascular endothelial growth factor in cultured human airway smooth muscle cells. *Cell Biochem Biophys* **43**, 119–129.

Alagappan, V. K., Willems-Widyastuti, A., Seynhaeve, A. L. *et al.* (2007). Vasoactive peptides upregulate mRNA expression and secretion of vascular endothelial growth factor in human airway smooth muscle cells. *Cell Biochem Biophys* **47**, 109–118.

Ammit, A. J., Hoffman, R. K., Amrani, Y. *et al.* (2000). Tumor necrosis factor-alpha-induced secretion of RANTES and interleukin-6 from human airway smooth-muscle cells. Modulation by cyclic adenosine monophosphate. *Am J Respir Cell Mol Biol* **23**, 794–802.

Ammit, A. J., Lazaar, A. L., Irani, C. *et al.* (2002). Tumor necrosis factor-alpha-induced secretion of RANTES and interleukin-6 from human airway smooth muscle cells: modulation by glucocorticoids and beta-agonists. *Am J Respir Cell Mol Biol* **26**, 465–474.

Amrani, Y., Lazaar, A. L., Panettieri, R. A. J. (1999). Up-regulation of ICAM-1 by cytokines in human tracheal smooth muscle cells involves an NF-kappaB-dependent signaling pathway that is only partially sensitive to dexamethasone. *J Immunol* **163**, 2128–2134.

An, S. S., Bai, T. R., Bates, J. H. *et al.* (2007). Airway smooth muscle dynamics: a common pathway of airway obstruction in asthma. *Eur Respir J* **29**, 834–860.

Bai, T. R. (1990). Abnormalities in airway smooth muscle in fatal asthma. *Am Rev Respir Dis* **141**, 552–557.

Bai, T. R., Mak, J. C. W. and Barnes, P. J. (1992). A comparison of β-adrenergic receptors and in vitro relaxant responses to isoproterenol in asthmatic airway smooth muscle. *Am J Respir Cell Mol Biol* **6**, 647–651.

Barnes, P. J. (2005). Theophylline in chronic obstructive pulmonary disease: new horizons. *Proc Am Thorac Soc* **2**, 334–339.

Belvisi, M. G., Saunders, M. A., Haddad, el-B. *et al.* (1997). Induction of cyclo-oxygenase-2 by cytokines in human cultured airway smooth muscle cells: novel inflammatory role of this cell type. *Br J Pharmacol* **120**, 910–916.

Benayoun, L., Druilhe, A., Dombret, M. C. *et al.* (2003). Airway structural alterations selectively associated with severe asthma. *Am J Respir Crit Care Med* **167**, 1360–1368.

Berkman, N., Krishnan, V. L., Gilbey, T. *et al.* (1996). Expression of RANTES mRNA and protein in airways of patients with mild asthma. *Am J Respir Crit Care Med* **154**, 1804–1811.

Berry, M. A., Hargadon, B., Shelley, M. *et al.* (2006). Evidence of a role of tumor necrosis factor alpha in refractory asthma. *N Engl J Med* **354**, 697–708.

Bonacci, J. V., Schuliga, M., Harris, T. *et al.* (2006). Collagen impairs glucocorticoid actions in airway smooth muscle through integrin signalling. *Br J Pharmacol* **149**, 365–373.

Bonacci, J. V. and Stewart, A. G. (2006). Regulation of human airway mesenchymal cell proliferation by glucocorticoids and beta2-adrenoceptor agonists. *Pulm Pharmacol Ther* **19**, 32–38.

Bradding, P., Roberts, J. A., Britten, K. M. *et al.* (1994). Interleukin-4, -5 and -6 and tumor necrosis factor-α in normal and asthmatic airways: evidence for the human mast cell as a source of these cytokines. *Am J Respir Cell Mol Biol* **10**, 471–480.

Brightling, C. E., Ammit, A. J., Kaur, D. *et al.* (2005). The CXCL10/CXCR3 axis mediates human lung mast cell migration to asthmatic airway smooth muscle. *Am J Respir Crit Care Med* **171**, 1103–1108.

Brightling, C. E., Bradding, P., Symon, F. A. *et al.* (2002). Mast-cell infiltration of airway smooth muscle in asthma. *N Engl J Med* **346**, 1699–1705.

Brown, R. H., Zerhouni, E. A. and Mitzner, W. (1995). Airway edema potentiates airway reactivity. *J Appl Physiol* **79**, 1242–1248.

Burgess, J. K., Carlin, S., Pack, R. A. *et al.* (2004). Detection and characterization of OX40 ligand expression in human airway smooth muscle cells: a possible role in asthma? *J Allergy Clin Immunol* **113**, 683–689.

Burgess, J. K., Johnson, P. R., Ge, Q. *et al.* (2003). Expression of connective tissue growth factor in asthmatic airway smooth muscle cells. *Am J Respir Crit Care Med* **167**, 71–77.

Burgess, J. K., Oliver, B. G., Poniris, M. H. *et al.* (2006). A phosphodiesterase 4 inhibitor inhibits matrix protein deposition in airways in vitro. *J Allergy Clin Immunol* **118**, 649–657.

Busse, W., Corren, J., Lanier, B. Q. *et al.* (2001). Omalizumab, anti-IgE recombinant humanized monoclonal antibody, for the treatment of severe allergic asthma. *J Allergy Clin Immunol* **108**, 184–190.

Cerrina, J., Ladurie, M. L., Lebat, G. *et al.* (1986). Comparison of human bronchial muscle response to histamine in vivo with histamine and isoproterenol agonists in vitro. *Am Rev Respir Dis* **134**, 57–61.

Chan, V., Burgess, J. K., Ratoff, J. C. *et al.* (2006). Extracellular matrix regulates enhanced eotaxin expression in asthmatic airway smooth muscle cells. *Am J Respir Crit Care Med* **174**, 379–385.

Chung, K. F. (2000). Airway smooth muscle cells: contributing to and regulating airway mucosal inflammation? *Eur Respir J* **15**, 961–968.

Chung, K. F. (2006). Phosphodiesterase inhibitors in airways disease. *Eur J Pharmacol* **533**, 110–117.

Chung, K. F., Patel, H. J., Fadlon, E. J. *et al.* (1999). Induction of eotaxin expression and release from human airway smooth muscle cells by IL-1beta and TNFalpha: effects of IL-10 and corticosteroids. *Br J Pharmacol* **127**, 1145–1150.

Cox, G., Miller, J. D., McWilliams, A. *et al.* (2006). Bronchial thermoplasty for asthma. *Am J Respir Crit Care Med* **173**, 965–969.

Cox, G., Thomson, N. C., Rubin, A. S. *et al.* (2007). Asthma control during the year after bronchial thermoplasty. *N Engl J Med* **356**, 1327–1337.

DeJongste, J. C., Mons, H., Bonta, I. L. *et al.* (1987). In vitro responses of airways from an asthmatic patient. *Eur J Respir Dis* **71**, 23–29.

Deshpande, D. A., Walseth, T. F., Panettieri, R. A. *et al.* (2003). CD38/cyclic ADP-ribose-mediated Ca^{2+} signaling contributes to airway smooth muscle hyper-responsiveness. *FASEB J* **17**, 452–454.

Djukanovic, R., Wilson, S. J., Kraft, M. *et al.* (2004).The effects of anti-IgE (omalizumab) on airways inflammation in allergic asthma. *Am J Respir Crit Care Med* **170**, 583–593.

Ebina, M., Takahashi, T., Chiba, T. *et al.* (1993). Cellular hypertrophy and hyperplasia of airway smooth muscle underlying bronchial asthma. *Am Rev Respir Dis* **148**, 720–726.

Elshaw, S. R., Henderson, N., Knox, A. J. *et al.* (2004). Matrix metalloproteinase expression and activity in human airway smooth muscle cells. *Br J Pharmacol* **142**, 1318–1324.

Erin, E. M., Leaker, B. R., Nicholson, G. C. *et al.* (2006). The effects of a monoclonal antibody directed against tumor necrosis factor-alpha in asthma. *Am J Respir Crit Care Med* **174**, 753–762.

Espinosa, K., Bosse, Y., Stankova, J. *et al.* (2003). CysLT1 receptor upregulation by TGF-beta and IL-13 is associated with bronchial smooth muscle cell proliferation in response to LTD4. *J Allergy Clin Immunol* **111**, 1032–1040.

Evans, D. J., Taylor, D. A., Zetterstrom, O. *et al.* (1997). A comparison of low-dose inhaled budesonide plus theophylline and high-dose inhaled budesonide for moderate asthma. *N Engl J Med* **337**, 1412–1418.

Faffe, D. S., Flynt, L., Bourgeois, K. *et al.* (2006). Interleukin-13 and interleukin-4 induce vascular endothelial growth factor release from airway smooth muscle cells: role of vascular endothelial growth factor genotype. *Am J Respir Cell Mol Biol* **34**, 213–218.

Fernandes, D., Guida, E., Koutsoubos, V. *et al.* (1999). Glucocorticoids inhibit proliferation, cyclin D1 expression, and retinoblastoma protein phosphorylation, but not activity of the extracellular-regulated kinases in human cultured airway smooth muscle. *Am J Respir Cell Mol Biol* **21**, 77–88.

Flood-Page, P., Menzies-Gow, A., Phipps, S. *et al.* (2003). Anti-IL-5 treatment reduces deposition of ECM proteins in the bronchial subepithelial basement membrane of mild atopic asthmatics. *J Clin Invest* **112**, 1029–1036.

Foley, S. C., Mogas, A. K., Olivenstein, R. *et al.* (2007). Increased expression of ADAM33 and ADAM8 with disease progression in asthma. *J Allergy Clin Immunol* **119**, 863–871.

Fredberg, J. J. (2007). Counterpoint: airway smooth muscle is not useful. *J Appl Physiol* **102**, 1709–1710.

Freyer, A. M., Billington, C. K., Penn, R. B. *et al.* (2004). Extracellular matrix modulates {beta}2-adrenergic receptor signaling in human airway smooth muscle cells. *Am J Respir Cell Mol Biol* **31**, 440–445.

Freyer, A. M., Johnson, S. R. and Hall, I. P. (2001). Effects of growth factors and extracellular matrix on survival of human airway smooth muscle cells. *Am J Respir Cell Mol Biol* **25**, 569–576.

Ghaffar, O., Hamid, Q., Renzi, P. M. *et al.* (1999). Constitutive and cytokine-stimulated expression of eotaxin by human airway smooth muscle cells. *Am J Respir Crit Care Med* **159**, 1933–1942.

Goldsmith, A. M., Bentley, J. K., Zhou, L. *et al.* (2006). Transforming growth factor-beta induces airway smooth muscle hypertrophy. *Am J Respir Cell Mol Biol* **34**, 247–254.

Goldsmith, A. M., Hershenson, M. B., Wolbert, M. P. *et al.* (2007). Regulation of airway smooth muscle alpha-actin expression by glucocorticoids. *Am J Physiol Lung Cell Mol Physiol* **292**, L99–L106.

Goncharova, E. A., Billington, C. K., Irani, C. *et al.* (2003). Cyclic AMP-mobilizing agents and glucocorticoids modulate human smooth muscle cell migration. *Am J Respir Cell Mol Biol* **29**, 19–27.

Gounni, A. S. (2006). The high-affinity IgE receptor (FcepsilonRI): a critical regulator of airway smooth muscle cells? *Am J Physiol Lung Cell Mol Physiol* **291**, L312–L321.

Gounni, A. S., Wellemans, V., Yang, J. *et al.* (2005). Human airway smooth muscle cells express the high affinity receptor for IgE (Fc epsilon RI): a critical role of Fc epsilon RI in human airway smooth muscle cell function. *J Immunol* **175**, 2613–2621.

Govindaraju, V., Michoud, M. C., Al Chalabi, M. *et al.* (2006). Interleukin-8: novel roles in human airway smooth muscle cell contraction and migration. *Am J Physiol Cell Physiol* **291**, C957–C965.

Hallsworth, M. P., Twort, C. H., Lee, T. H. *et al.* (2001). beta(2)-adrenoceptor agonists inhibit release of eosinophil-activating cytokines from human airway smooth muscle cells. *Br J Pharmacol* **132**, 729–741.

Hardy, E., Farahani, M. and Hall, I. P. (1996). Regulation of histamine H1 receptor coupling by dexamethasone in human cultured airway smooth muscle. *Br J Pharmacol* **118**, 1079–1084.

Hardaker, E. L., Bacon, A. M., Carlson, K. *et al.* (2004). Regulation of TNF-alpha- and IFN-gamma-induced CXCL10 expression: participation of the airway smooth muscle in the pulmonary inflammatory response in chronic obstructive pulmonary disease. *FASEB J* **18**, 191–193.

Henderson, W. R., Jr., Chiang, G. K., Tien, Y. T. *et al.* (2006). Reversal of allergen-induced airway remodeling by CysLT1 receptor blockade. *Am J Respir Crit Care Med* **173**, 718–728.

Hew, M., Bhavsar, P., Torrego, A. *et al.* (2006). Relative corticosteroid insensitivity of peripheral blood mononuclear cells in severe asthma. *Am J Respir Crit Care Med* **174**, 134–141.

Hirst, S. J. and Lee, T. H. (1998). Airway smooth muscle as a target of glucocorticoid action in the treatment of asthma. *Am J Respir Crit Care Med* **158**, S201–S206.

Hirst, S. J., Twort, C. H. and Lee, T. H. (2000). Differential effects of extracellular matrix proteins on human airway smooth muscle cell proliferation and phenotype. *Am J Respir Cell Mol Biol* **23**, 335–344.

Hirst, S. J., Walker, T. R. and Chilvers, E. R. (2000). Phenotypic diversity and molecular mechanisms of airway smooth muscle proliferation in asthma. *Eur Respir J* **16**, 159–177.

Howarth, P. H., Babu, K. S., Arshad, H. S. *et al.* (2005). Tumour necrosis factor (TNFalpha) as a novel therapeutic target in symptomatic corticosteroid dependent asthma. *Thorax* **60**, 1012–1018.

Huang, C. D., Tliba, O., Panettieri, R. A., Jr. *et al.* (2003). Bradykinin induces interleukin-6 production in human airway smooth muscle cells: modulation by Th2 cytokines and dexamethasone. *Am J Respir Cell Mol Biol* **28**, 330–338.

Hui, K. P. and Barnes, N. C. (1991). Lung function improvement in asthma with a cysteinyl-leukotriene receptor antagonist. *Lancet* **337**, 1062–1063.

Ibe, B. O., Portugal, A. M. and Raj, J. U. (2006). Levalbuterol inhibits human airway smooth muscle cell proliferation: therapeutic implications in the management of asthma. *Int Arch Allergy Immunol* **139**, 225–236.

Issa, R., Xie, S., Khorasani, N. *et al.* (2007). Corticosteroid inhibition of growth-related oncogene protein-{alpha} via mitogen-activated kinase phosphatase-1 in airway smooth muscle cells. *J Immunol* **178**, 7366–7375.

Issa, R., Xie, S., Lee, K. Y. *et al.* (2006). GRO-{alpha} regulation in airway smooth muscle by IL-1beta and TNF-{alpha}: role of NF-{kappa}B and MAP kinases. *Am J Physiol Lung Cell Mol Physiol* **291**, L66–L74.

James, A. L., Pare, P. D. and Hogg, J. C. (1989). The mechanics of airway narrowing in asthma. *Am Rev Respir Dis* **139**, 242–246. 26.

Janssen, L. J., Tazzeo, T. and Zuo, J. (2004). Enhanced myosin phosphatase and Ca(2+)-uptake mediate adrenergic relaxation of airway smooth muscle. *Am J Respir Cell Mol Biol* **30**, 548–554.

Jarai, G., Sukkar, M., Garrett, S. *et al.* (2004). Effects of interleukin-1beta, interleukin-13 and transforming growth factor-beta on gene expression in human airway smooth muscle using gene microarrays. *Eur J Pharmacol* **497**, 255–265.

John, M., Au, B. T., Jose, P. J. *et al.* (1998). Expression and release of interleukin-8 by human airway smooth muscle cells: inhibition by Th-2 cytokines and corticosteroids. *Am J Respir Cell Mol Biol* **18**, 84–90.

John, M., Hirst, S. J., Jose, P. J. *et al.* (1997). Human airway smooth muscle cells express and release RANTES in response to Th-1 cytokines: regulation by Th-2 cytokines and corticosteroids. *J Immunol* **158**, 1841–1847.

Johnson, P. R., Black, J. L., Carlin, S., *et al.* (2000). The production of extracellular matrix proteins by human passively sensitized airway smooth-muscle cells in culture: the effect of beclomethasone. *Am J Respir Crit Care Med* **162**, 2145–2151.

Johnson, P. R., Burgess, J. K, Underwood, P. A. *et al.* (2004). Extracellular matrix proteins modulate asthmatic airway smooth muscle cell proliferation via an autocrine mechanism. *J Allergy Clin Immunol* **113**, 690–696.

Johnson, P. R., Roth, M., Tamm, M. *et al.* (2001). Airway smooth muscle cell proliferation is increased in asthma. *Am J Respir Crit Care Med* **164**, 474–477.

Kamm, K. E. and Stull, J. T. (1985). The function of myosin and myosin light chain kinase phosphorylation in smooth muscle. *Annu Rev Pharmacol Toxicol* **25**, 593–620.

Kang, B. N., Tirumurugaan, K. G., Deshpande, D. A. *et al.* (2006). Transcriptional regulation of CD38 expression by tumor necrosis factor-alpha in human airway smooth muscle cells: role of NF-kappaB and sensitivity to glucocorticoids. *FASEB J* **20**, 1000–1002.

Kassel, O., Schmidlin, F., Duvernelle, C. *et al.* (1999). Human bronchial smooth muscle cells in culture produce stem cell factor. *Eur Respir J* **13**, 951–954.

Kips, J. C., O'Connor, B. J., Langley, S. J. *et al.* (2003). Effect of SCH55700, a humanized anti-human interleukin-5 antibody, in severe persistent asthma: a pilot study. *Am J Respir Crit Care Med* **167**, 1655–1659.

Kotlikoff, M. I. and Kamm, K. E. (1996). Molecular mechanisms of beta-adrenergic relaxation of airway smooth muscle. *Annu Rev Physiol* **58**, 115–141.

Koto, H., Mak, J. C., Haddad, E. B. *et al.* (1996). Mechanisms of impaired beta-adrenoceptor-induced airway relaxation by interleukin-1beta in vivo in the rat. *J Clin Invest* **98**, 1780–1787.

Lazaar, A. L., Albelda, S. M., Pilewski, J. M. *et al.* (1994). T lymphocytes adhere to airway smooth muscle cells via integrins and CD44 and induce smooth muscle cell DNA synthesis. *J Exp Med* **180**, 807–816.

Lazzeri, N., Belvisi, M. G., Patel, H. J. *et al.* (2001). RANTES release by human airway smooth muscle: effects of prostaglandin E(2) and fenoterol. *Eur J Pharmacol* **433**, 231–235.

Leckie, M. J., ten Brinke, A., Khan, J. *et al.* (2000). Effects of an interleukin-5 blocking monoclonal antibody on eosinophils, airway hyper-responsiveness, and the late asthmatic response. *Lancet* **356**(9248), 2114–2116.

Leung, S. Y., Eynott, P., Nath, P. *et al.* (2005). Effects of ciclesonide and fluticasone propionate on allergen-induced airway inflammation and remodeling features. *J Allergy Clin Immunol* **115**, 989–996.

Leung, S. Y., Niimi, A., Noble, A. *et al.* (2006). Effect of transforming growth factor-beta receptor I kinase inhibitor 2,4-disubstituted pteridine (SD-208) in chronic allergic airway inflammation and remodeling. *J Pharmacol Exp Ther* **319**, 586–594.

Liu, C., Zuo, J. and Janssen, L. J. (2006). Regulation of airway smooth muscle RhoA/ROCK activities by cholinergic and bronchodilator stimuli. *Eur Respir J* **28**, 703–711.

Lu, D., Xie, S., Sukkar, M. B. *et al.* (2007). Inhibition of airway smooth muscle adhesion and migration by the disintegrin domain of ADAM-15. *Am J Respir Cell Mol Biol* **37**, 494–500.

Ma, X., Cheng, Z., Kong, H. *et al.* (2002). Changes in biophysical and biochemical properties of single bronchial smooth muscle cells from asthmatic subjects. *Am J Physiol Lung Cell Mol Physiol* **283**, L1181–L1189.

Ma, X., Wang, Y. and Stephens, N. L. (1998). Serum deprivation induces a unique hypercontractile phenotype of cultured smooth muscle cells. *Am J Physiol* **274**, C1206–C1214.

Mak, J. C., Hisada, T., Salmon, M. *et al.* (2002). Glucocorticoids reverse IL-1beta-induced impairment of beta-adrenoceptor-mediated relaxation and up-regulation of G-protein-coupled receptor kinases. *Br J Pharmacol* **135**, 987–996.

Matsumoto, H., Moir, L. M., Oliver, B. G. *et al.* (2007). Comparison of gel contraction mediated by asthmatic and non-asthmatic airway smooth muscle cells. *Thorax* **62**, 848–854.

McVicker, C. G., Leung, S. Y., Kanabar, V. *et al.* (2007). Repeated allergen inhalation induces cytoskeletal remodeling in smooth muscle from rat bronchioles. *Am J Respir Cell Mol Biol* **36**, 721–727.

Miller, J. D., Cox, G., Vincic, L. *et al.* (2005). A prospective feasibility study of bronchial thermoplasty in the human airway. *Chest* **127**, 1999–2006.

Mitzner, W. (2004). Airway smooth muscle: the appendix of the lung. *Am J Respir Crit Care Med* **169**, 787–790.

Moir, L. M., Leung, S. Y., Eynott, P. R. *et al.* (2003). Repeated allergen inhalation induces phenotypic modulation of smooth muscle in bronchioles of sensitized rats. *Am J Physiol Lung Cell Mol Physiol* **284**, L148–L159.

Moore, P. E., Laporte, J. D., Gonzalez, S. *et al.* (1999). Glucocorticoids ablate IL-1beta-induced beta-adrenergic hyporesponsiveness in human airway smooth muscle cells. *Am J Physiol* **277**, L932–L942.

Morris, G. E., Parker, L. C., Ward, J. R. *et al.* (2006). Cooperative molecular and cellular networks regulate Toll-like receptor-dependent inflammatory responses. *FASEB J* **20**, 2153–2155.

Nie, M., Corbett, L., Knox, A. J. *et al.* (2005). Differential regulation of chemokine expression by peroxisome proliferator-activated receptor gamma agonists: interactions with glucocorticoids and beta2-agonists. *J Biol Chem* **280**, 2550–2561.

Nie, M., Knox, A. J. and Pang, L. (2005). beta2-Adrenoceptor agonists, like glucocorticoids, repress eotaxin gene transcription by selective inhibition of histone H4 acetylation. *J Immunol* **175**, 478–486.

Oltmanns, U., Walters, M., Sukkar, M. *et al.* (2007). Fluticasone, but not salmeterol, reduces cigarette smoke-induced production of interleukin-8 in human airway smooth muscle. *Pulm Pharmacol Ther* (in press).

Panettieri, R. A., Tan, E. M., Ciocca, V. *et al.* (1998). Effects of LTD4 on human airway smooth muscle cell proliferation, matrix expression, and contraction in vitro: differential sensitivity to cysteinyl leukotriene receptor antagonists. *Am J Respir Cell Mol Biol* **19**, 453–461.

Pang, L. and Knox, A. J. (2000). Synergistic inhibition by beta(2)-agonists and corticosteroids on tumor necrosis factor-alpha-induced interleukin-8 release from cultured human airway smooth-muscle cells. *Am J Respir Cell Mol Biol* **23**, 79–85.

Pang, L. and Knox, A. J. (2001). Regulation of TNF-alpha-induced eotaxin release from cultured human airway smooth muscle cells by beta2-agonists and corticosteroids. *FASEB J* **15**, 261–269.

Parameswaran, K., Cox, G., Radford, K. *et al.* (2002). Cysteinyl leukotrienes promote human airway smooth muscle migration. *Am J Respir Crit Care Med* **166**, 738–742.

Peachell, P. (2006). Regulation of mast cells by beta-agonists. *Clin Rev Allergy Immunol* **31**, 131–142.

Pennings, H. J., Kramer, K., Bast, A. *et al.* (1998). Tumour necrosis factor-alpha induces hyperreactivity in tracheal smooth muscle of the guinea-pig in vitro. *Eur Respir J* **12**, 45–49.

Pepe, C., Foley, S., Shannon, J. *et al.* (2005). Differences in airway remodeling between subjects with severe and moderate asthma. *J Allergy Clin Immunol* **116**, 544–549.

Pype, J. L., Dupont, L. J., Menten, P. *et al.* (1999). Expression of monocyte chemotactic protein (MCP)-1, MCP-2, and MCP-3 by human airway smooth-muscle cells. Modulation by corticosteroids and T-helper 2 cytokines. *Am J Respir Cell Mol Biol* **21**, 528–536.

Roth, M., Johnson, P. R., Borger, P. *et al.* (2004). Dysfunctional interaction of C/EBPalpha and the glucocorticoid receptor in asthmatic bronchial smooth-muscle cells. *N Engl J Med* **351**, 560–574.

Roth, M., Johnson, P. R., Rudiger, J. J. *et al.* (2002). Interaction between glucocorticoids and beta2 agonists on bronchial airway smooth muscle cells through synchronised cellular signalling. *Lancet* **360**, 1293–1299.

Salmon, M., Walsh, D. A., Huang, T. J. *et al.* (1999). Involvement of cysteinyl leukotrienes in airway smooth muscle cell DNA synthesis after repeated allergen exposure in sensitized brown Norway rats. *Br J Pharmacol* **127**, 1151–1158.

Saunders, M. A., Mitchell, J. A., Seldon, P. M. *et al.* (1997). Release of granulocyte-macrophage colony stimulating factor by human cultured airway smooth muscle cells: suppression by dexamethasone. *Br J Pharmacol* **120**, 545–546.

Skloot, G. and Togias, A. (2003). Bronchodilation and bronchoprotection by deep inspiration and their relationship to bronchial hyperresponsiveness. *Clin Rev Allergy Immunol* **24**, 55–72.

Slats, A. M., Sont, J. K., van Klink, R. H. *et al.* (2006). Improvement in bronchodilation following deep inspiration after a course of high-dose oral prednisone in asthma. *Chest* **130**, 58–65.

Soler, M., Matz, J., Townley, R. *et al.* (2001). The anti-IgE antibody omalizumab reduces exacerbations and steroid requirement in allergic asthmatics. *Eur Respir J* **18**, 254–261.

Stewart, A. G., Harris, T., Fernandes, D. J. *et al.* (1999). Beta2-adrenergic receptor agonists and cAMP arrest human cultured airway smooth muscle cells in the G(1) phase of the cell cycle: role of proteasome degradation of cyclin D1. *Mol Pharmacol* **56**, 1079–1086.

Stewart, A. G., Tomlinson, P. R. and Wilson, J. W. (1997). Beta 2-adrenoceptor agonist-mediated inhibition of human airway smooth muscle cell proliferation: importance of the duration of beta 2-adrenoceptor stimulation. *Br J Pharmacol* **121**, 361–368.

Stocks, J., Bradbury, D., Corbett, L. *et al.* (2005). Cytokines upregulate vascular endothelial growth factor secretion by human airway smooth muscle cells: role of endogenous prostanoids. *FEBS Lett* **579**, 2551–2556.

Sukkar, M. B., Hughes, J. M., Armour, C. L. *et al.* (2001). Tumour necrosis factor-alpha potentiates contraction of human bronchus in vitro. *Respirology* **6**, 199–203.

Sukkar, M. B., Issa, R., Xie, S. *et al.* (2004). Fractalkine/CX3CL1 production by human airway smooth muscle cells: induction by IFN-gamma and TNF-alpha and regulation by TGF-beta and corticosteroids. *Am J Physiol Lung Cell Mol Physiol* **287**, L1230–L1240.

Sukkar, M. B., Xie, S., Khorasani, N. M. *et al.* (2006). Toll-like receptor 2, 3, and 4 expression and function in human airway smooth muscle. *J Allergy Clin Immunol* **118**, 641-648.

Tanaka, H., Watanabe, K., Tamaru, N. *et al.* (1995). Arachidonic acid metabolites and glucocorticoid regulatory mechanism in cultured porcine tracheal smooth muscle cells. *Lung* **173**, 347–361.

Thomas, P. S. and Heywood, G. (2002). Effects of inhaled tumour necrosis factor alpha in subjects with mild asthma. *Thorax* **57**, 774–778.

Thomas, P. S., Yates, D. H. and Barnes, P. J. (1995). Tumor necrosis factor-alpha increases airway responsiveness and sputum neutrophilia in normal human subjects. *Am J Respir Crit Care Med* **152**, 76–80.

Tliba, O., Cidlowski, J. A. and Amrani, Y. (2006). CD38 expression is insensitive to steroid action in cells treated with tumor necrosis factor-alpha and interferon-gamma by a mechanism involving the up-regulation of the glucocorticoid receptor beta isoform. *Mol Pharmacol* **69**, 588–596.

Tomlinson, P. R., Wilson, J. W. and Stewart, A. G. (1995). Salbutamol inhibits the proliferation of human airway smooth muscle cells grown in culture: relationship to elevated cAMP levels. *Biochem Pharmacol* **49**, 1809–1819.

Usmani, O. S., Ito, K., Maneechotesuwan, K. *et al.* (2005). Glucocorticoid receptor nuclear translocation in airway cells after inhaled combination therapy. *Am J Respir Crit Care Med* **172**, 704–712.

Vlahos, R. and Stewart, A. G. (1999). Interleukin-1alpha and tumour necrosis factor-alpha modulate airway smooth muscle DNA synthesis by induction of cyclo-oxygenase-2: inhibition by dexamethasone and fluticasone propionate. *Br J Pharmacol* **126**, 1315–1324.

Wang, C. G., Du, T., Xu, L. J. *et al.* (1993). Role of leukotriene D4 in allergen-induced increases in airway smooth muscle in the rat. *Am Rev Respir Dis* **148**, 413–417.

Woodruff, P. G., Dolganov, G. M., Ferrando, R. E. *et al.* (2004). Hyperplasia of smooth muscle in mild to moderate asthma without changes in cell size or gene expression. *Am J Respir Crit Care Med* **169**, 1001–1006.

Xie, S., Issa, R., Sukkar, M. B. *et al.* (2005). Induction and regulation of matrix metalloproteinase-12 in human airway smooth muscle cells. *Respir Res* **6**, 148.

Xie, S., Sukkar, M. B., Issa, R. *et al.* (2005). Regulation of TGF-beta 1-induced connective tissue growth factor expression in airway smooth muscle cells. *Am J Physiol Lung Cell Mol Physiol* **288**, L68–L76.

Xie, S., Sukkar, M. B., Issa, R. *et al.* (2007). Mechanisms of induction of airway smooth muscle hyperplasia by transforming growth factor-beta. *Am J Physiol Lung Cell Mol Physiol* **293**, L245–L253.

Index

Notes: Page references in *italics* refer to Figures and Tables
ASM = airway smooth muscle

Airway Smooth Muscle   Edited by Kian Fan Chung
© 2008 John Wiley & Sons, Ltd

Index compiled by Annette Musker